If q is any function of several variables $x, \ldots, z$, then

$$\delta q = \sqrt{\left(\frac{\partial q}{\partial x}\,\delta x\right)^2 + \cdots + \left(\frac{\partial q}{\partial z}\,\delta z\right)^2}$$

(for independent random er

Statistical Definitions (Chapter 4)

If $x_1, \ldots, x_N$ denote N separate measurements of one quantity x, then we define:

$$\bar{x} = \frac{1}{N}\sum_{i=1}^{N} x_i = \text{mean};$$ (p. 98)

$$\sigma_x = \sqrt{\frac{1}{N-1}\sum(x_i - \bar{x})^2} = \text{standard deviation, or SD}$$ (p. 100)

$$\sigma_{\bar{x}} = \frac{\sigma_x}{\sqrt{N}} = \text{standard deviation of mean, or SDOM.}$$ (p. 102)

The Normal Distribution (Chapter 5)

For any limiting distribution $f(x)$ for measurement of a continuous variable x:

$f(x)\,dx$ = probability that any one measurement will give an answer between x and $x + dx$; (p. 128)

$\int_a^b f(x)\,dx$ = probability that any one measurement will give an answer between $x = a$ and $x = b$; (p. 128)

$\int_{-\infty}^{\infty} f(x)\,dx$ = 1 is the normalization condition. (p. 128)

The Gauss or normal distribution is

$$G_{X,\sigma}(x) = \frac{1}{\sigma\sqrt{2\pi}}\,e^{-(x-X)^2/2\sigma^2},$$ (p. 133)

where

X = center of distribution = true value of x

= mean after many measurements,

σ = width of distribution

= standard deviation after many measurements.

The probability of a measurement within t standard deviations of X is

$$\text{Prob(within } t\sigma) = \frac{1}{\sqrt{2\pi}}\int_{-t}^{t} e^{-z^2/2}\,dz = \text{normal error integral;}$$ (p. 136)

in particular

$$\text{Prob(within } 1\sigma) = 68\%.$$

AN INTRODUCTION TO
Error Analysis

THE STUDY OF UNCERTAINTIES
IN PHYSICAL MEASUREMENTS

SECOND EDITION

John R. Taylor

PROFESSOR OF PHYSICS
UNIVERSITY OF COLORADO

University Science Books
Sausalito, California

University Science Books
55D Gate Five Road
Sausalito, CA 94965
Fax: (415) 332-5393

Production manager: *Susanna Tadlock*
Manuscript editor: *Ann McGuire*
Designer: *Robert Ishi*
Illustrators: *John and Judy Waller*
Compositor: *Maple-Vail Book Manufacturing Group*
Printer and binder: *Maple-Vail Book Manufacturing Group*

This book is printed on acid-free paper.

Library of Congress Cataloging-in-Publication Data

Taylor, John R. (John Robert), 1939–
    An introduction to error analysis / John R. Taylor.—2nd ed.
      p. cm.
    Includes bibliographical references and index.
    ISBN 0-935702-42-3 (cloth).—ISBN 0-935702-75-X (pbk.)
    1. Physical measurements. 2. Error analysis (Mathematics)
  3. Mathematical physics. I. Title.
  QC39.T4   1997
  530.1′6—dc20
                                    96-953
                                       CIP

Printed in the United States of America
10 9 8 7 6 5 4 3 2 1

AN INTRODUCTION TO
Error Analysis

THE STUDY OF UNCERTAINTIES
IN PHYSICAL MEASUREMENTS

SECOND EDITION

John R. Taylor

PROFESSOR OF PHYSICS
UNIVERSITY OF COLORADO

University Science Books
Sausalito, California

University Science Books
55D Gate Five Road
Sausalito, CA 94965
Fax: (415) 332-5393

Production manager: *Susanna Tadlock*
Manuscript editor: *Ann McGuire*
Designer: *Robert Ishi*
Illustrators: *John and Judy Waller*
Compositor: *Maple-Vail Book Manufacturing Group*
Printer and binder: *Maple-Vail Book Manufacturing Group*

This book is printed on acid-free paper.

Library of Congress Cataloging-in-Publication Data

Taylor, John R. (John Robert), 1939–
    An introduction to error analysis / John R. Taylor.—2nd ed.
      p. cm.
    Includes bibliographical references and index.
    ISBN 0-935702-42-3 (cloth).—ISBN 0-935702-75-X (pbk.)
    1. Physical measurements.   2. Error analysis (Mathematics)
   3. Mathematical physics. I. Title.
    QC39.T4   1997
    530.1'6—dc20
                                        96-953
                                          CIP

Printed in the United States of America
10 9 8 7 6 5 4 3 2 1

To My Wife

Contents

Part I

Part II

Preface to the Second Edition

I first wrote *An Introduction to Error Analysis* because my experience teaching introductory laboratory classes for several years had convinced me of a serious need for a book that truly *introduced* the subject to the college science student. Several fine books on the topic were available, but none was really suitable for a student new to the subject. The favorable reception to the first edition confirmed the existence of that need and suggests the book met it.

The continuing success of the first edition suggests it still meets that need. Nevertheless, after more than a decade, every author of a college textbook must surely feel obliged to improve and update the original version. Ideas for modifications came from several sources: suggestions from readers, the need to adapt the book to the wide availability of calculators and personal computers, and my own experiences in teaching from the book and finding portions that could be improved.

Because of the overwhelmingly favorable reaction to the first edition, I have maintained its basic level and general approach. Hence, many revisions are simply changes in wording to improve clarity. A few changes are major, the most important of which are as follows:

(1) The number of problems at the end of each chapter is nearly doubled to give users a wider choice and teachers the ability to vary their assigned problems from year to year. Needless to say, any given reader does not need to solve anywhere near the 264 problems offered; on the contrary, half a dozen problems from each chapter is probably sufficient.

(2) Several readers recommended placing a few simple exercises regularly throughout the text to let readers check that they really understand the ideas just presented. Such exercises now appear as "Quick Checks," and I strongly urge students new to the subject to try them all. If any Quick Check takes much longer than a minute or two, you probably need to reread the preceding few paragraphs. The answers to all Quick Checks are given in the answer section at the back of the book. Those who find this kind of exercise distracting can easily skip them.

(3) Also new to this edition are complete summaries of all the important equations at the end of each chapter to supplement the first edition's brief summaries inside the front and back covers. These new summaries list all key equations from the chapter and from the problem sets as well.

(4) Many new figures appear in this edition, particularly in the earlier chapters. The figures help make the text seem less intimidating and reflect my conscious

effort to encourage students to think more visually about uncertainties. I have observed, for example, that many students grasp issues such as the consistency of measurements if they think visually in terms of error bars.

(5) I have reorganized the problem sets at the end of each chapter in three ways. First, the Answers section at the back of the book now gives answers to all of the odd-numbered problems. (The first edition contained answers only to selected problems.) The new arrangement is simpler and more traditional. Second, as a rough guide to the level of difficulty of each problem, I have labeled the problems with a system of stars: One star (★) indicates a simple exercise that should take no more than a couple of minutes if you understand the material. Two stars (★★) indicate a somewhat harder problem, and three stars (★★★) indicate a really searching problem that involves several different concepts and requires more time. I freely admit that the classification is extremely approximate, but students studying on their own should find these indications helpful, as may teachers choosing problems to assign to their students.

Third, I have arranged the problems by section number. As soon as you have read Section *N,* you should be ready to try any problem listed for that section. Although this system is convenient for the student and the teacher, it seems to be currently out of favor. I assume this disfavor stems from the argument that the system might exclude the deep problems that involve many ideas from different sections. I consider this argument specious; a problem listed for Section *N* can, of course, involve ideas from many earlier sections and can, therefore, be just as general and deep as any problem listed under a more general heading.

(6) I have added problems that call for the use of computer spreadsheet programs such as Lotus 123 or Excel. None of these problems is specific to a particular system; rather, they urge the student to learn how to do various tasks using whatever system is available. Similarly, several problems encourage students to learn to use the built-in functions on their calculators to calculate standard deviations and the like.

(7) I have added an appendix (Appendix E) to show two proofs that concern sample standard deviations: first, that, based on *N* measurements of a quantity, the best estimate of the true width of its distribution is the sample standard deviation with $(N-1)$ in the denominator, and second, that the uncertainty in this estimate is as given by Equation (5.46). These proofs are surprisingly difficult and not easily found in the literature.

It is a pleasure to thank the many people who have made suggestions for this second edition. Among my friends and colleagues at the University of Colorado, the people who gave most generously of their time and knowledge were David Alexander, Dana Anderson, David Bartlett, Barry Bruce, John Cumalat, Mike Dubson, Bill Ford, Mark Johnson, Jerry Leigh, Uriel Nauenberg, Bill O'Sullivan, Bob Ristinen, Rod Smythe, and Chris Zafiratos. At other institutions, I particularly want to thank R. G. Chambers of Leeds, England, Sharif Heger of the University of New Mexico, Steven Hoffmaster of Gonzaga University, Hilliard Macomber of the University of Northern Iowa, Mark Semon of Bates College, Peter Timbie of Brown University, and David Van Dyke of the University of Pennsylvania. I am deeply indebted to all of these people for their generous help. I am also most grateful to Bruce Armbruster of University Science Books for his generous encouragement and support. Above all, I want to thank my wife Debby; I don't know how she puts up with the stresses and strains of book writing, but I am so grateful she does.

J. R. Taylor

September 1996
Boulder, Colorado

Preface to the First Edition

All measurements, however careful and scientific, are subject to some uncertainties. Error analysis is the study and evaluation of these uncertainties, its two main functions being to allow the scientist to estimate how large his uncertainties are, and to help him to reduce them when necessary. The analysis of uncertainties, or "errors," is a vital part of any scientific experiment, and error analysis is therefore an important part of any college course in experimental science. It can also be one of the most interesting parts of the course. The challenges of estimating uncertainties and of reducing them to a level that allows a proper conclusion to be drawn can turn a dull and routine set of measurements into a truly interesting exercise.

This book is an introduction to error analysis for use with an introductory college course in experimental physics of the sort usually taken by freshmen or sophomores in the sciences or engineering. I certainly do not claim that error analysis is the most (let alone the only) important part of such a course, but I have found that it is often the most abused and neglected part. In many such courses, error analysis is "taught" by handing out a couple of pages of notes containing a few formulas, and the student is then expected to get on with the job solo. The result is that error analysis becomes a meaningless ritual, in which the student adds a few lines of calculation to the end of each laboratory report, not because he or she understands why, but simply because the instructor has said to do so.

I wrote this book with the conviction that any student, even one who has never heard of the subject, should be able to learn what error analysis is, why it is interesting and important, and how to use the basic tools of the subject in laboratory reports. Part I of the book (Chapters 1 to 5) tries to do all this, with many examples of the kind of experiment encountered in teaching laboratories. The student who masters this material should then know and understand almost all the error analysis he or she would be expected to learn in a freshman laboratory course: error propagation, the use of elementary statistics, and their justification in terms of the normal distribution.

Part II contains a selection of more advanced topics: least-squares fitting, the correlation coefficient, the χ^2 test, and others. These would almost certainly not be included officially in a freshman laboratory course, although a few students might become interested in some of them. However, several of these topics would be needed in a second laboratory course, and it is primarily for that reason that I have included them.

I am well aware that there is all too little time to devote to a subject like error analysis in most laboratory courses. At the University of Colorado we give a one-hour lecture in each of the first six weeks of our freshman laboratory course. These lectures, together with a few homework assignments using the problems at the ends of the chapters, have let us cover Chapters 1 through 4 in detail and Chapter 5 briefly. This gives the students a working knowledge of error propagation and the elements of statistics, plus a nodding acquaintance with the underlying theory of the normal distribution.

From several students' comments at Colorado, it was evident that the lectures were an unnecessary luxury for at least some of the students, who could probably have learned the necessary material from assigned reading and problem sets. I certainly believe the book could be studied without any help from lectures.

Part II could be taught in a few lectures at the start of a second-year laboratory course (again supplemented with some assigned problems). But, even more than Part I, it was intended to be read by the student at any time that his or her own needs and interests might dictate. Its seven chapters are almost completely independent of one another, in order to encourage this kind of use.

I have included a selection of problems at the end of each chapter; the reader does need to work several of these to master the techniques. Most calculations of errors are quite straightforward. A student who finds himself or herself doing many complicated calculations (either in the problems of this book or in laboratory reports) is almost certainly doing something in an unnecessarily difficult way. In order to give teachers and readers a good choice, I have included many more problems than the average reader need try. A reader who did one-third of the problems would be doing well.

Inside the front and back covers are summaries of all the principal formulas. I hope the reader will find these a useful reference, both while studying the book and afterward. The summaries are organized by chapters, and will also, I hope, serve as brief reviews to which the reader can turn after studying each chapter.

Within the text, a few statements—equations and rules of procedure—have been highlighted by a shaded background. This highlighting is reserved for statements that are important and are in their final form (that is, will not be modified by later work). You will definitely need to remember these statements, so they have been highlighted to bring them to your attention.

The level of mathematics expected of the reader rises slowly through the book. The first two chapters require only algebra; Chapter 3 requires differentiation (and partial differentiation in Section 3.11, which is optional); Chapter 5 needs a knowledge of integration and the exponential function. In Part II, I assume that the reader is entirely comfortable with all these ideas.

The book contains numerous examples of physics experiments, but an understanding of the underlying theory is not essential. Furthermore, the examples are mostly taken from elementary mechanics and optics to make it more likely that the student will already have studied the theory. The reader who needs it can find an account of the theory by looking at the index of any introductory physics text.

Error analysis is a subject about which people feel passionately, and no single treatment can hope to please everyone. My own prejudice is that, when a choice has to be made between ease of understanding and strict rigor, a physics text should choose the former. For example, on the controversial question of combining errors in quadrature versus direct addition, I have chosen to treat direct addition first, since the student can easily understand the arguments that lead to it.

In the last few years, a dramatic change has occurred in student laboratories with the advent of the pocket calculator. This has a few unfortunate consequences—most notably, the atrocious habit of quoting ridiculously *insignificant* figures just because the calculator produced them—but it is from almost every point of view a tremendous advantage, especially in error analysis. The pocket calculator allows one to compute, in a few seconds, means and standard deviations that previously would have taken hours. It renders unnecessary many tables, since one can now compute functions like the Gauss function more quickly than one could find them in a book of tables. I have tried to exploit this wonderful tool wherever possible.

It is my pleasure to thank several people for their helpful comments and suggestions. A preliminary edition of the book was used at several colleges, and I am grateful to many students and colleagues for their criticisms. Especially helpful were the comments of John Morrison and David Nesbitt at the University of Colorado, Professors Pratt and Schroeder at Michigan State, Professor Shugart at U.C. Berkeley, and Professor Semon at Bates College. Diane Casparian, Linda Frueh, and Connie Gurule typed successive drafts beautifully and at great speed. Without my mother-in-law, Frances Kretschmann, the proofreading would never have been done in time. I am grateful to all of these people for their help; but above all I thank my wife, whose painstaking and ruthless editing improved the whole book beyond measure.

J. R. Taylor

November 1, 1981
Boulder, Colorado

AN INTRODUCTION TO
Error Analysis

Part I

Part I introduces the basic ideas of error analysis as they are needed in a typical first-year, college physics laboratory. The first two chapters describe what error analysis is, why it is important, and how it can be used in a typical laboratory report. Chapter 3 describes error propagation, whereby uncertainties in the original measurements "propagate" through calculations to cause uncertainties in the calculated final answers. Chapters 4 and 5 introduce the statistical methods with which the so-called random uncertainties can be evaluated.

Chapter 1

Preliminary Description of Error Analysis

Error analysis is the study and evaluation of uncertainty in measurement. Experience has shown that no measurement, however carefully made, can be completely free of uncertainties. Because the whole structure and application of science depends on measurements, the ability to evaluate these uncertainties and keep them to a minimum is crucially important.

This first chapter describes some simple measurements that illustrate the inevitable occurrence of experimental uncertainties and show the importance of knowing how large these uncertainties are. The chapter then describes how (in some simple cases, at least) the magnitude of the experimental uncertainties can be estimated realistically, often by means of little more than plain common sense.

1.1 Errors as Uncertainties

In science, the word *error* does not carry the usual connotations of the terms *mistake* or *blunder*. Error in a scientific measurement means the inevitable uncertainty that attends all measurements. As such, errors are not mistakes; you cannot eliminate them by being very careful. The best you can hope to do is to ensure that errors are as small as reasonably possible and to have a reliable estimate of how large they are. Most textbooks introduce additional definitions of error, and these are discussed later. For now, error is used exclusively in the sense of uncertainty, and the two words are used interchangeably.

1.2 Inevitability of Uncertainty

To illustrate the inevitable occurrence of uncertainties, we have only to examine any everyday measurement carefully. Consider, for example, a carpenter who must measure the height of a doorway before installing a door. As a first rough measurement, he might simply look at the doorway and estimate its height as 210 cm. This crude "measurement" is certainly subject to uncertainty. If pressed, the carpenter might express this uncertainty by admitting that the height could be anywhere between 205 cm and 215 cm.

If he wanted a more accurate measurement, he would use a tape measure and might find the height is 211.3 cm. This measurement is certainly more precise than his original estimate, but it is obviously still subject to some uncertainty, because it is impossible for him to know the height to be exactly 211.3000 cm rather than 211.3001 cm, for example.

This remaining uncertainty has many sources, several of which are discussed in this book. Some causes could be removed if the carpenter took enough trouble. For example, one source of uncertainty might be that poor lighting hampers reading of the tape; this problem could be corrected by improving the lighting.

On the other hand, some sources of uncertainty are intrinsic to the process of measurement and can never be removed entirely. For example, let us suppose the carpenter's tape is graduated in half-centimeters. The top of the door probably will not coincide precisely with one of the half-centimeter marks, and if it does not, the carpenter must *estimate* just where the top lies between two marks. Even if the top happens to coincide with one of the marks, the mark itself is perhaps a millimeter wide; so he must estimate just where the top lies within the mark. In either case, the carpenter ultimately must estimate where the top of the door lies relative to the markings on the tape, and this necessity causes some uncertainty in the measurement.

By buying a better tape with closer and finer markings, the carpenter can reduce his uncertainty but cannot eliminate it entirely. If he becomes obsessively determined to find the height of the door with the greatest precision technically possible, he could buy an expensive laser interferometer. But even the precision of an interferometer is limited to distances of the order of the wavelength of light (about 0.5×10^{-6} meters). Although the carpenter would now be able to measure the height with fantastic precision, he still would not know the height of the doorway *exactly*.

Furthermore, as our carpenter strives for greater precision, he will encounter an important problem of principle. He will certainly find that the height is different in different places. Even in one place, he will find that the height varies if the temperature and humidity vary, or even if he accidentally rubs off a thin layer of dirt. In other words, he will find that there is no such thing as *the* height of the doorway. This kind of problem is called a *problem of definition* (the height of the door is not a well-defined quantity) and plays an important role in many scientific measurements.

Our carpenter's experiences illustrate a point generally found to be true, that is, that no physical quantity (a length, time, or temperature, for example) can be measured with complete certainty. With care, we may be able to reduce the uncertainties until they are extremely small, but to eliminate them entirely is impossible.

In everyday measurements, we do not usually bother to discuss uncertainties. Sometimes the uncertainties simply are not interesting. If we say that the distance between home and school is 3 miles, whether this means "somewhere between 2.5 and 3.5 miles" or "somewhere between 2.99 and 3.01 miles" is usually unimportant. Often the uncertainties are important but can be allowed for instinctively and without explicit consideration. When our carpenter fits his door, he must know its height with an uncertainty that is less than 1 mm or so. As long as the uncertainty is this small, the door will (for all practical purposes) be a perfect fit, and his concern with error analysis is at an end.

1.3   Importance of Knowing the Uncertainties

Our example of the carpenter measuring a doorway illustrates how uncertainties are always present in measurements. Let us now consider an example that illustrates more clearly the crucial importance of knowing how big these uncertainties are.

Suppose we are faced with a problem like the one said to have been solved by Archimedes. We are asked to find out whether a crown is made of 18-karat gold, as claimed, or a cheaper alloy. Following Archimedes, we decide to test the crown's density ρ knowing that the densities of 18-karat gold and the suspected alloy are

$$\rho_{\text{gold}} = 15.5 \text{ gram/cm}^3$$

and

$$\rho_{\text{alloy}} = 13.8 \text{ gram/cm}^3.$$

If we can measure the density of the crown, we should be able (as Archimedes suggested) to decide whether the crown is really gold by comparing ρ with the known densities ρ_{gold} and ρ_{alloy}.

Suppose we summon two experts in the measurement of density. The first expert, George, might make a quick measurement of ρ and report that his best estimate for ρ is 15 and that it almost certainly lies between 13.5 and 16.5 gram/cm^3. Our second expert, Martha, might take a little longer and then report a best estimate of 13.9 and a probable range from 13.7 to 14.1 gram/cm^3. The findings of our two experts are summarized in Figure 1.1.

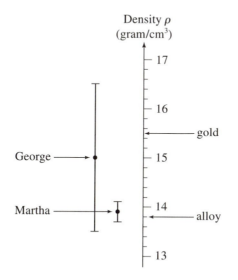

Figure 1.1. Two measurements of the density of a supposedly gold crown. The two black dots show George's and Martha's best estimates for the density; the two vertical error bars show their margins of error, the ranges within which they believe the density probably lies. George's uncertainty is so large that both gold and the suspected alloy fall within his margins of error; therefore, his measurement does not determine which metal was used. Martha's uncertainty is appreciably smaller, and her measurement shows clearly that the crown is not made of gold.

The first point to notice about these results is that although Martha's measurement is much more precise, George's measurement is probably also correct. Each expert states a range within which he or she is confident ρ lies, and these ranges overlap; so it is perfectly possible (and even probable) that both statements are correct.

Note next that the uncertainty in George's measurement is so large that his results are of no use. The densities of 18-karat gold and of the alloy both lie within his range, from 13.5 to 16.5 gram/cm^3; so no conclusion can be drawn from George's measurements. On the other hand, Martha's measurements indicate clearly that the crown is not genuine; the density of the suspected alloy, 13.8, lies comfortably inside Martha's estimated range of 13.7 to 14.1, but that of 18-karat gold, 15.5, is far outside it. Evidently, if the measurements are to allow a conclusion, the experimental uncertainties must not be too large. The uncertainties do not need to be extremely small, however. In this respect, our example is typical of many scientific measurements, for which uncertainties have to be reasonably small (perhaps a few percent of the measured value) but for which extreme precision is often unnecessary.

Because our decision hinges on Martha's claim that ρ lies between 13.7 and 14.1 gram/cm^3, she must give us sufficient reason to believe her claim. In other words, she must *justify* her stated range of values. This point is often overlooked by beginning students, who simply assert their uncertainties but omit any justification. Without a brief explanation of how the uncertainty was estimated, the assertion is almost useless.

The most important point about our two experts' measurements is this: Like most scientific measurements, they would both have been useless if they had not included reliable statements of their uncertainties. In fact, if we knew only the two best estimates (15 for George and 13.9 for Martha), not only would we have been unable to draw a valid conclusion, but we could actually have been misled, because George's result (15) seems to suggest the crown is genuine.

1.4   More Examples

The examples in the past two sections were chosen, not for their great importance, but to introduce some principal features of error analysis. Thus, you can be excused for thinking them a little contrived. It is easy, however, to think of examples of great importance in almost any branch of applied or basic science.

In the applied sciences, for example, the engineers designing a power plant must know the characteristics of the materials and fuels they plan to use. The manufacturer of a pocket calculator must know the properties of its various electronic components. In each case, somebody must measure the required parameters, and having measured them, must establish their reliability, which requires error analysis. Engineers concerned with the safety of airplanes, trains, or cars must understand the uncertainties in drivers' reaction times, in braking distances, and in a host of other variables; failure to carry out error analysis can lead to accidents such as that shown on the cover of this book. Even in a less scientific field, such as the manufacture of clothing, error analysis in the form of quality control plays a vital part.

In the basic sciences, error analysis has an even more fundamental role. When any new theory is proposed, it must be tested against older theories by means of one or more experiments for which the new and old theories predict different outcomes. In principle, a researcher simply performs the experiment and lets the outcome decide between the rival theories. In practice, however, the situation is complicated by the inevitable experimental uncertainties. These uncertainties must all be analyzed carefully and their effects reduced until the experiment singles out one acceptable theory. That is, the experimental results, with their uncertainties, must be *consistent* with the predictions of one theory and *inconsistent* with those of all known, reasonable alternatives. Obviously, the success of such a procedure depends critically on the scientist's understanding of error analysis and ability to convince others of this understanding.

A famous example of such a test of a scientific theory is the measurement of the bending of light as it passes near the sun. When Einstein published his general theory of relativity in 1916, he pointed out that the theory predicted that light from a star would be bent through an angle $\alpha = 1.8''$ as it passes near the sun. The simplest classical theory would predict no bending ($\alpha = 0$), and a more careful classical analysis would predict (as Einstein himself noted in 1911) bending through an angle $\alpha = 0.9''$. In principle, all that was necessary was to observe a star when it was aligned with the edge of the sun and to measure the angle of bending α. If the result were $\alpha = 1.8''$, general relativity would be vindicated (at least for this phenomenon); if α were found to be 0 or $0.9''$, general relativity would be wrong and one of the older theories right.

In practice, measuring the bending of light by the sun was extremely hard and was possible only during a solar eclipse. Nonetheless, in 1919 it was successfully measured by Dyson, Eddington, and Davidson, who reported their best estimate as $\alpha = 2''$, with 95% confidence that it lay between $1.7''$ and $2.3''$.[1] Obviously, this result was consistent with general relativity and inconsistent with either of the older predictions. Therefore, it gave strong support to Einstein's theory of general relativity.

At the time, this result was controversial. Many people suggested that the uncertainties had been badly underestimated and hence that the experiment was inconclusive. Subsequent experiments have tended to confirm Einstein's prediction and to vindicate the conclusion of Dyson, Eddington, and Davidson. The important point here is that the whole question hinged on the experimenters' ability to estimate reliably all their uncertainties and to convince everyone else they had done so.

Students in introductory physics laboratories are not usually able to conduct definitive tests of new theories. Often, however, they do perform experiments that test existing physical theories. For example, Newton's theory of gravity predicts that bodies fall with constant acceleration g (under the appropriate conditions), and students can conduct experiments to test whether this prediction is correct. At first, this kind of experiment may seem artificial and pointless because the theories have obvi-

---

[1]This simplified account is based on the original paper of F. W. Dyson, A. S. Eddington, and C. Davidson (*Philosophical Transactions of the Royal Society,* **220A,** 1920, 291). I have converted the *probable error* originally quoted into the 95% confidence limits. The precise significance of such confidence limits will be established in Chapter 5.

ously been tested many times with much more precision than possible in a teaching laboratory. Nonetheless, if you understand the crucial role of error analysis and accept the challenge to make the most precise test possible with the available equipment, such experiments can be interesting and instructive exercises.

1.5   Estimating Uncertainties When Reading Scales

Thus far, we have considered several examples that illustrate why every measurement suffers from uncertainties and why their magnitude is important to know. We have not yet discussed how we can actually evaluate the magnitude of an uncertainty. Such evaluation can be fairly complicated and is the main topic of this book. Fortunately, reasonable estimates of the uncertainty of some simple measurements are easy to make, often using no more than common sense. Here and in Section 1.6, I discuss examples of such measurements. An understanding of these examples will allow you to begin using error analysis in your experiments and will form the basis for later discussions.

The first example is a measurement using a marked scale, such as the ruler in Figure 1.2 or the voltmeter in Figure 1.3. To measure the length of the pencil in

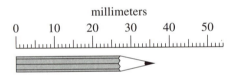

Figure 1.2. Measuring a length with a ruler.

Figure 1.2, we must first place the end of the pencil opposite the zero of the ruler and then decide where the tip comes to on the ruler's scale. To measure the voltage in Figure 1.3, we have to decide where the needle points on the voltmeter's scale. If we assume the ruler and voltmeter are reliable, then in each case the main prob-

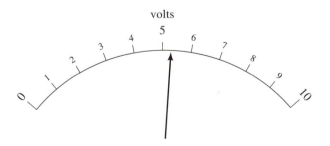

Figure 1.3. A reading on a voltmeter.

lem is to decide where a certain point lies in relation to the scale markings. (Of course, if there is any possibility the ruler and voltmeter are *not* reliable, we will have to take this uncertainty into account as well.)

The markings of the ruler in Figure 1.2 are fairly close together (1 mm apart). We might reasonably decide that the length shown is undoubtedly closer to 36 mm than it is to 35 or 37 mm but that no more precise reading is possible. In this case, we would state our conclusion as

$$\text{best estimate of length} = 36 \text{ mm,}$$
$$\text{probable range: 35.5 to 36.5 mm} \tag{1.1}$$

and would say that we have measured the length to the nearest millimeter.

This type of conclusion—that the quantity lies closer to a given mark than to either of its neighboring marks—is quite common. For this reason, many scientists introduce the convention that the statement "$l = 36$ mm" without any qualification is presumed to mean that l is closer to 36 than to 35 or 37; that is,

$$l = 36 \text{ mm}$$

means

$$35.5 \text{ mm} \leq l \leq 36.5 \text{ mm.}$$

In the same way, an answer such as $x = 1.27$ without any stated uncertainty would be presumed to mean that x lies between 1.265 and 1.275. In this book, I do not use this convention but instead always indicate uncertainties explicitly. Nevertheless, you need to understand the convention and know that it applies to any number stated without an uncertainty, especially in this age of pocket calculators, which display many digits. If you unthinkingly copy a number such as 123.456 from your calculator without any qualification, then your reader is entitled to assume the number is definitely correct to six significant figures, which is very unlikely.

The markings on the voltmeter shown in Figure 1.3 are more widely spaced than those on the ruler. Here, most observers would agree that you can do better than simply identify the mark to which the pointer is closest. Because the spacing is larger, you can realistically estimate where the pointer lies in the space between two marks. Thus, a reasonable conclusion for the voltage shown might be

$$\text{best estimate of voltage} = 5.3 \text{ volts,}$$
$$\text{probable range: 5.2 to 5.4 volts.} \tag{1.2}$$

The process of estimating positions between the scale markings is called *interpolation*. It is an important technique that can be improved with practice.

Different observers might not agree with the precise estimates given in Equations (1.1) and (1.2). You might well decide that you could interpolate for the length in Figure 1.2 and measure it with a smaller uncertainty than that given in Equation (1.1). Nevertheless, few people would deny that Equations (1.1) and (1.2) are *reasonable* estimates of the quantities concerned and of their probable uncertainties. Thus, we see that approximate estimation of uncertainties is fairly easy when the only problem is to locate a point on a marked scale.

1.6   Estimating Uncertainties in Repeatable Measurements

Many measurements involve uncertainties that are much harder to estimate than those connected with locating points on a scale. For example, when we measure a time interval using a stopwatch, the main source of uncertainty is not the difficulty of reading the dial but our own unknown reaction time in starting and stopping the watch. Sometimes these kinds of uncertainty can be estimated reliably, if we can repeat the measurement several times. Suppose, for example, we time the period of a pendulum once and get an answer of 2.3 seconds. From one measurement, we can't say much about the experimental uncertainty. But if we repeat the measurement and get 2.4 seconds, then we can immediately say that the uncertainty is probably of the order of 0.1 s. If a sequence of four timings gives the results (in seconds),

$$2.3, \ 2.4, \ 2.5, \ 2.4, \tag{1.3}$$

then we can begin to make some fairly realistic estimates.

First, a natural assumption is that the best estimate of the period is the *average*[2] *value,* 2.4 s.

Second, another reasonably safe assumption is that the correct period lies between the lowest value, 2.3, and the highest, 2.5. Thus, we might reasonably conclude that

$$\text{best estimate} \ = \ \text{average} \ = \ 2.4 \text{ s,}$$
$$\text{probable range: 2.3 to 2.5 s.} \tag{1.4}$$

Whenever you can repeat the same measurement several times, the spread in your measured values gives a valuable indication of the uncertainty in your measurements. In Chapters 4 and 5, I discuss statistical methods for treating such repeated measurements. Under the right conditions, these statistical methods give a more accurate estimate of uncertainty than we have found in Equation (1.4) using just common sense. A proper statistical treatment also has the advantage of giving an objective value for the uncertainty, independent of the observer's individual judgment.[3] Nevertheless, the estimate in statement (1.4) represents a simple, realistic conclusion to draw from the four measurements in (1.3).

Repeated measurements such as those in (1.3) cannot always be relied on to reveal the uncertainties. First, we must be sure that the quantity measured is really the *same* quantity each time. Suppose, for example, we measure the breaking strength of two supposedly identical wires by breaking them (something we can't do more than once with each wire). If we get two different answers, this difference *may* indicate that our measurements were uncertain *or* that the two wires were not really identical. By itself, the difference between the two answers sheds no light on the reliability of our measurements.

---

[2] I will prove in Chapter 5 that the best estimate based on several measurements of a quantity is almost always the average of the measurements.

[3] Also, a proper statistical treatment usually gives a *smaller* uncertainty than the full range from the lowest to the highest observed value. Thus, upon looking at the four timings in (1.3), we have judged that the period is "probably" somewhere between 2.3 and 2.5 s. The statistical methods of Chapters 4 and 5 let us state with 70% confidence that the period lies in the smaller range of 2.36 to 2.44 s.

Even when we can be sure we are measuring the same quantity each time, repeated measurements do not always reveal uncertainties. For example, suppose the clock used for the timings in (1.3) was running consistently 5% fast. Then, all timings made with it will be 5% too long, and no amount of repeating (with the same clock) will reveal this deficiency. Errors of this sort, which affect all measurements in the same way, are called *systematic* errors and can be hard to detect, as discussed in Chapter 4. In this example, the remedy is to check the clock against a more reliable one. More generally, if the reliability of any measuring device is in doubt, it should clearly be checked against a device known to be more reliable.

The examples discussed in this and the previous section show that experimental uncertainties sometimes can be estimated easily. On the other hand, many measurements have uncertainties that are *not* so easily evaluated. Also, we ultimately want more precise values for the uncertainties than the simple estimates just discussed. These topics will occupy us from Chapter 3 onward. In Chapter 2, I assume temporarily that you know how to estimate the uncertainties in all quantities of interest, so that we can discuss how the uncertainties are best reported and how they are used in drawing an experimental conclusion.

Chapter 2

How to Report and Use Uncertainties

Having read Chapter 1, you should now have some idea of the importance of experimental uncertainties and how they arise. You should also understand how uncertainties can be estimated in a few simple situations. In this chapter, you will learn some basic notations and rules of error analysis and study examples of their use in typical experiments in a physics laboratory. The aim is to familiarize you with the basic vocabulary of error analysis and its use in the introductory laboratory. Chapter 3 begins a systematic study of how uncertainties are actually evaluated.

Sections 2.1 to 2.3 define several basic concepts in error analysis and discuss general rules for stating uncertainties. Sections 2.4 to 2.6 discuss how these ideas could be used in typical experiments in an introductory physics laboratory. Finally, Sections 2.7 to 2.9 introduce fractional uncertainty and discuss its significance.

2.1 Best Estimate ± Uncertainty

We have seen that the correct way to state the result of measurement is to give a best estimate of the quantity and the range within which you are confident the quantity lies. For example, the result of the timings discussed in Section 1.6 was reported as

$$\text{best estimate of time } = 2.4 \text{ s,}$$
$$\text{probable range: } 2.3 \text{ to } 2.5 \text{ s.} \tag{2.1}$$

Here, the best estimate, 2.4 s, lies at the midpoint of the estimated range of probable values, 2.3 to 2.5 s, as it has in all the examples. This relationship is obviously natural and pertains in most measurements. It allows the results of the measurement to be expressed in compact form. For example, the measurement of the time recorded in (2.1) is usually stated as follows:

$$\text{measured value of time } = 2.4 \pm 0.1 \text{ s.} \tag{2.2}$$

This single equation is equivalent to the two statements in (2.1).

In general, the result of any measurement of a quantity x is stated as

$$\text{(measured value of } x) = x_{\text{best}} \pm \delta x. \tag{2.3}$$

This statement means, first, that the experimenter's best estimate for the quantity concerned is the number x_{best}, and second, that he or she is reasonably confident the quantity lies somewhere between $x_{best} - \delta x$ and $x_{best} + \delta x$. The number δx is called the *uncertainty,* or *error,* or *margin of error* in the measurement of x. For convenience, the uncertainty δx is always defined to be positive, so that $x_{best} + \delta x$ is always the *highest* probable value of the measured quantity and $x_{best} - \delta x$ the *lowest*.

I have intentionally left the meaning of the range $x_{best} - \delta x$ to $x_{best} + \delta x$ somewhat vague, but it can sometimes be made more precise. In a simple measurement such as that of the height of a doorway, we can easily state a range $x_{best} - \delta x$ to $x_{best} + \delta x$ within which we are *absolutely* certain the measured quantity lies. Unfortunately, in most scientific measurements, such a statement is hard to make. In particular, to be *completely* certain that the measured quantity lies between $x_{best} - \delta x$ and $x_{best} + \delta x$, we usually have to choose a value for δx that is too large to be useful. To avoid this situation, we can sometimes choose a value for δx that lets us state with a certain percent confidence that the actual quantity lies within the range $x_{best} \pm \delta x$. For instance, the public opinion polls conducted during elections are traditionally stated with margins of error that represent 95% confidence limits. The statement that 60% of the electorate favor Candidate A, with a margin of error of 3 percentage points (60 ± 3), means that the pollsters are 95% confident that the percent of voters favoring Candidate A is between 57 and 63; in other words, after many elections, we should expect the correct answer to have been *inside* the stated margins of error 95% of the times and *outside* these margins only 5% of the times.

Obviously, we cannot state a percent confidence in our margins of error until we understand the statistical laws that govern the process of measurement. I return to this point in Chapter 4. For now, let us be content with defining the uncertainty δx so that we are "reasonably certain" the measured quantity lies between $x_{best} - \delta x$ and $x_{best} + \delta x$.

Quick Check[1] 2.1. **(a)** A student measures the length of a simple pendulum and reports his best estimate as 110 mm and the range in which the length probably lies as 108 to 112 mm. Rewrite this result in the standard form (2.3). **(b)** If another student reports her measurement of a current as $I = 3.05 \pm 0.03$ amps, what is the range within which I probably lies?

2.2   Significant Figures

Several basic rules for stating uncertainties are worth emphasizing. First, because the quantity δx is an estimate of an uncertainty, obviously it should not be stated

---

[1] These "Quick Checks" appear at intervals through the text to give you a chance to check your understanding of the concept just introduced. They are straightforward exercises, and many can be done in your head. I urge you to take a moment to make sure you can do them; if you cannot, you should reread the preceding few paragraphs.

with too much precision. If we measure the acceleration of gravity g, it would be absurd to state a result like

$$\text{(measured } g) \; = \; 9.82 \pm 0.02385 \text{ m/s}^2. \tag{2.4}$$

The uncertainty in the measurement cannot conceivably be known to four significant figures. In high-precision work, uncertainties are sometimes stated with two significant figures, but for our purposes we can state the following rule:

Rule for Stating Uncertainties
Experimental uncertainties should almost always be rounded to one significant figure.   (2.5)

Thus, if some calculation yields the uncertainty $\delta g = 0.02385$ m/s^2, this answer should be rounded to $\delta g = 0.02$ m/s^2, and the conclusion (2.4) should be rewritten as

$$\text{(measured } g) \; = \; 9.82 \pm 0.02 \text{ m/s}^2. \tag{2.6}$$

An important practical consequence of this rule is that many error calculations can be carried out mentally without using a calculator or even pencil and paper.

The rule (2.5) has only one significant exception. If the leading digit in the uncertainty δx is a 1, then keeping two significant figures in δx may be better. For example, suppose that some calculation gave the uncertainty $\delta x = 0.14$. Rounding this number to $\delta x = 0.1$ would be a substantial proportionate reduction, so we could argue that retaining two figures might be less misleading, and quote $\delta x = 0.14$. The same argument could perhaps be applied if the leading digit is a 2 but certainly not if it is any larger.

Once the uncertainty in a measurement has been estimated, the significant figures in the measured value must be considered. A statement such as

$$\text{measured speed} \; = \; 6051.78 \pm 30 \text{ m/s} \tag{2.7}$$

is obviously ridiculous. The uncertainty of 30 means that the digit 5 might really be as small as 2 or as large as 8. Clearly the trailing digits 1, 7, and 8 have no significance at all and should be rounded. That is, the correct statement of (2.7) is

$$\text{measured speed} \; = \; 6050 \pm 30 \text{ m/s}. \tag{2.8}$$

The general rule is this:

Rule for Stating Answers
The last significant figure in any stated answer should usually be of the same order of magnitude (in the same decimal position) as the uncertainty.   (2.9)

For example, the answer 92.81 with an uncertainty of 0.3 should be rounded as

$$92.8 \pm 0.3.$$

If its uncertainty is 3, then the same answer should be rounded as

$$93 \pm 3,$$

and if the uncertainty is 30, then the answer should be

$$90 \pm 30.$$

An important qualification to rules (2.5) and (2.9) is as follows: To reduce inaccuracies caused by rounding, *any numbers to be used in subsequent calculations should normally retain at least one significant figure more than is finally justified.* At the end of the calculations, the final answer should be rounded to remove these extra, insignificant figures. An electronic calculator will happily carry numbers with far more digits than are likely to be significant in any calculation you make in a laboratory. Obviously, these numbers do not need to be rounded in the middle of a calculation but certainly must be rounded appropriately for the final answers.[2]

Note that the uncertainty in any measured quantity has the same dimensions as the measured quantity itself. Therefore, writing the units (m/s^2, cm^3, etc.) after both the answer *and* the uncertainty is clearer and more economical, as in Equations (2.6) and (2.8). By the same token, if a measured number is so large or small that it calls for scientific notation (the use of the form 3×10^3 instead of 3,000, for example), then it is simpler and clearer to put the answer and uncertainty in the same form. For example, the result

$$\text{measured charge} = (1.61 \pm 0.05) \times 10^{-19} \text{ coulombs}$$

is much easier to read and understand in this form than it would be in the form

$$\text{measured charge} = 1.61 \times 10^{-19} \pm 5 \times 10^{-21} \text{ coulombs.}$$

Quick Check 2.2. Rewrite each of the following measurements in its most appropriate form:
   (a) $v = 8.123456 \pm 0.0312$ m/s
   (b) $x = 3.1234 \times 10^4 \pm 2$ m
   (c) $m = 5.6789 \times 10^{-7} \pm 3 \times 10^{-9}$ kg.

2.3   Discrepancy

Before I address the question of how to use uncertainties in experimental reports, a few important terms should be introduced and defined. First, if two measurements

---

[2] Rule (2.9) has one more small exception. If the leading digit in the uncertainty is small (a 1 or, perhaps, a 2), retaining one extra digit in the final answer may be appropriate. For example, an answer such as 3.6 ± 1 is quite acceptable because one could argue that rounding it to 4 ± 1 would waste information.

of the same quantity disagree, we say there is a *discrepancy.* Numerically, we define the discrepancy between two measurements as their difference:

$$\text{discrepancy} = \text{difference between two measured} \atop \text{values of the same quantity.} \qquad (2.10)$$

More specifically, each of the two measurements consists of a best estimate and an uncertainty, and we define the discrepancy as the difference between the two best estimates. For example, if two students measure the same resistance as follows

>   Student A: 15 ± 1 ohms

and

>   Student B: 25 ± 2 ohms,

their discrepancy is

>   discrepancy = 25 − 15 = 10 ohms.

Recognize that a discrepancy may or may not be *significant.* The two measurements just discussed are illustrated in Figure 2.1(a), which shows clearly that the discrepancy of 10 ohms is *significant* because no single value of the resistance is compatible with both measurements. Obviously, at least one measurement is incorrect, and some careful checking is needed to find out what went wrong.

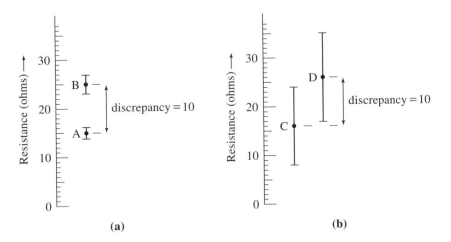

(a)   **(b)**

Figure 2.1. (a) Two measurements of the same resistance. Each measurement includes a best estimate, shown by a block dot, and a range of probable values, shown by a vertical error bar. The discrepancy (difference between the two best estimates) is 10 ohms and is *significant* because it is much larger than the combined uncertainty in the two measurements. Almost certainly, at least one of the experimenters made a mistake. (b) Two different measurements of the same resistance. The discrepancy is again 10 ohms, but in this case it is *insignificant* because the stated margins of error overlap. There is no reason to doubt either measurement (although they could be criticized for being rather imprecise).

Suppose, on the other hand, two other students had reported these results:

Student C: 16 ± 8 ohms

and

Student D: 26 ± 9 ohms.

Here again, the discrepancy is 10 ohms, but in this case the discrepancy is *insignificant* because, as shown in Figure 2.1(b), the two students' margins of error overlap comfortably and both measurements could well be correct. The discrepancy between two measurements of the same quantity should be assessed not just by its size but, more importantly, by how big it is *compared with the uncertainties in the measurements.*

In the teaching laboratory, you may be asked to measure a quantity that has been measured carefully many times before, and for which an accurate *accepted value* is known and published, for example, the electron's charge or the universal gas constant. This accepted value is not exact, of course; it is the result of measurements and, like all measurements, has some uncertainty. Nonetheless, in many cases the accepted value is much more accurate than you could possibly achieve yourself. For example, the currently accepted value of the universal gas constant R is

$$\text{(accepted } R) \ = \ 8.31451 \pm 0.00007 \text{ J/(mol·K).} \tag{2.11}$$

As expected, this value *is* uncertain, but the uncertainty is extremely small by the standards of most teaching laboratories. Thus, when you compare your measured value of such a constant with the accepted value, you can usually treat the accepted value as exact.[3]

Although many experiments call for measurement of a quantity whose accepted value is known, few require measurement of a quantity whose *true* value is known.[4] In fact, the true value of a measured quantity can almost *never* be known exactly and is, in fact, hard to define. Nevertheless, discussing the difference between a measured value and the corresponding true value is sometimes useful. Some authors call this difference the *true error.*

2.4   Comparison of Measured and Accepted Values

Performing an experiment without drawing some sort of conclusion has little merit. A few experiments may have mainly qualitative results—the appearance of an interference pattern on a ripple tank or the color of light transmitted by some optical system—but the vast majority of experiments lead to *quantitative* conclusions, that is, to a statement of numerical results. It is important to recognize that the statement of *a single measured number is completely uninteresting.* Statements that the density

---

[3] This is not always so. For example, if you look up the refractive index of glass, you find values ranging from 1.5 to 1.9, depending on the composition of the glass. In an experiment to measure the refractive index of a piece of glass whose composition is unknown, the accepted value is therefore no more than a rough guide to the expected answer.

[4] Here is an example: If you measure the ratio of a circle's circumference to its diameter, the true answer is exactly π. (Obviously such an experiment is rather contrived.)

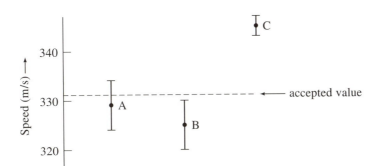

Figure 2.2. Three measurements of the speed of sound at standard temperature and pressure. Because the accepted value (331 m/s) is within Student A's margins of error, her result is satisfactory. The accepted value is just outside Student B's margin of error, but his measurement is nevertheless acceptable. The accepted value is *far outside* Student C's stated margins, and his measurement is definitely unsatisfactory.

of some metal was measured as 9.3 ± 0.2 gram/cm^3 or that the momentum of a cart was measured as 0.051 ± 0.004 kg·m/s are, by themselves, of no interest. An interesting conclusion must *compare two or more numbers*: a measurement with the accepted value, a measurement with a theoretically predicted value, or several measurements, to show that they are related to one another in accordance with some physical law. It is in such comparison of numbers that error analysis is so important. This and the next two sections discuss three typical experiments to illustrate how the estimated uncertainties are used to draw a conclusion.

Perhaps the simplest type of experiment is a measurement of a quantity whose accepted value is known. As discussed, this exercise is a somewhat artificial experiment peculiar to the teaching laboratory. The procedure is to measure the quantity, estimate the experimental uncertainty, and compare these values with the accepted value. Thus, in an experiment to measure the speed of sound in air (at standard temperature and pressure), Student A might arrive at the conclusion

$$\text{A's measured speed } = 329 \pm 5 \text{ m/s}, \tag{2.12}$$

compared with the

$$\text{accepted speed } = 331 \text{ m/s}. \tag{2.13}$$

Student A might choose to display this result graphically as in Figure 2.2. She should certainly include in her report both Equations (2.12) and (2.13) next to each other, so her readers can clearly appreciate her result. She should probably add an explicit statement that because the accepted value lies inside her margins of error, her measurement seems satisfactory.

The meaning of the uncertainty δx is that the correct value of x *probably* lies between $x_\text{best} - \delta x$ and $x_\text{best} + \delta x$; it is certainly *possible* that the correct value lies slightly outside this range. Therefore, a measurement can be regarded as satisfactory even if the accepted value lies slightly outside the estimated range of the measured value. For example, if Student B found the value

$$\text{B's measured speed} \; = \; 325 \pm 5 \; \text{m/s},$$

he could certainly claim that his measurement is consistent with the accepted value of 331 m/s.

On the other hand, if the accepted value is well outside the margins of error (the discrepancy is appreciably more than twice the uncertainty, say), there is reason to think something has gone wrong. For example, suppose the unlucky Student C finds

$$\text{C's measured speed} \; = \; 345 \pm 2 \; \text{m/s} \tag{2.14}$$

compared with the

$$\text{accepted speed} \; = \; 331 \; \text{m/s}. \tag{2.15}$$

Student C's discrepancy is 14 m/s, which is seven times bigger than his stated uncertainty (see Figure 2.2). He will need to check his measurements and calculations to find out what has gone wrong.

Unfortunately, the tracing of C's mistake may be a tedious business because of the numerous possibilities. He may have made a mistake in the measurements or calculations that led to the answer 345 m/s. He may have estimated his uncertainty incorrectly. (The answer 345 ± 15 m/s would have been acceptable.) He also might be comparing his measurement with the wrong accepted value. For example, the accepted value 331 m/s is the speed of sound at standard temperature and pressure. Because standard temperature is 0°C, there is a good chance the measured speed in (2.14) was *not* taken at standard temperature. In fact, if the measurement was made at 20°C (that is, normal room temperature), the correct accepted value for the speed of sound is 343 m/s, and the measurement would be entirely acceptable.

Finally, and perhaps most likely, a discrepancy such as that between (2.14) and (2.15) may indicate some undetected source of systematic error (such as a clock that runs consistently slow, as discussed in Chapter 1). Detection of such systematic errors (ones that consistently push the result in one direction) requires careful checking of the calibration of all instruments and detailed review of all procedures.

2.5   Comparison of Two Measured Numbers

Many experiments involve measuring two numbers that theory predicts should be equal. For example, the law of conservation of momentum states that the total momentum of an isolated system is constant. To test it, we might perform a series of experiments with two carts that collide as they move along a frictionless track. We could measure the total momentum of the two carts before (p) and after (q) they collide and check whether $p = q$ within experimental uncertainties. For a single pair of measurements, our results could be

$$\text{initial momentum } p \; = \; 1.49 \pm 0.03 \; \text{kg·m/s}$$

and

$$\text{final momentum } q \; = \; 1.56 \pm 0.06 \; \text{kg·m/s}.$$

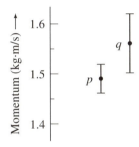

Figure 2.3. Measured values of the total momentum of two carts before (p) and after (q) a collision. Because the margins of error for p and q overlap, these measurements are certainly consistent with conservation of momentum (which implies that p and q should be equal).

Here, the range in which p probably lies (1.46 to 1.52) *overlaps* the range in which q probably lies (1.50 to 1.62). (See Figure 2.3.) Therefore, these measurements are consistent with conservation of momentum. If, on the other hand, the two probable ranges were not even close to overlapping, the measurements would be inconsistent with conservation of momentum, and we would have to check for mistakes in our measurements or calculations, for possible systematic errors, and for the possibility that some external forces (such as gravity or friction) are causing the momentum of the system to change.

If we repeat similar pairs of measurements several times, what is the best way to display our results? First, using a table to record a sequence of similar measurements is usually better than listing the results as several distinct statements. Second, the uncertainties often differ little from one measurement to the next. For example, we might convince ourselves that the uncertainties in all measurements of the initial momentum p are about $\delta p \approx 0.03$ kg·m/s and that the uncertainties in the final q are all about $\delta q \approx 0.06$ kg·m/s. If so, a good way to display our measurements would be as shown in Table 2.1.

Table 2.1. Measured momenta (kg·m/s).

Trial number	Initial momentum p (all ±0.03)	Final momentum q (all ±0.06)
1	1.49	1.56
2	3.10	3.12
3	2.16	2.05
etc.		

For each pair of measurements, the probable range of values for p overlaps (or nearly overlaps) the range of values for q. If this overlap continues for all measurements, our results can be pronounced consistent with conservation of momentum. Note that our experiment does not *prove* conservation of momentum; no experiment can. The best you can hope for is to conduct many more trials with progressively smaller uncertainties and that all the results are *consistent* with conservation of momentum.

In a real experiment, Table 2.1 might contain a dozen or more entries, and checking that each final momentum q is consistent with the corresponding initial momentum p could be tedious. A better way to display the results would be to add a fourth column that lists the differences $p - q$. If momentum is conserved, these values should be consistent with zero. The only difficulty with this method is that we must now compute the uncertainty in the difference $p - q$. This computation is performed as follows. Suppose we have made measurements

$$(\text{measured } p) \; = \; p_{\text{best}} \pm \delta p$$

and

$$(\text{measured } q) \; = \; q_{\text{best}} \pm \delta q.$$

The numbers p_{best} and q_{best} are our best estimates for p and q. Therefore, the best estimate for the difference $(p - q)$ is $(p_{\text{best}} - q_{\text{best}})$. To find the uncertainty in $(p - q)$, we must decide on the highest and lowest probable values of $(p - q)$. The highest value for $(p - q)$ would result if p had its *largest* probable value, $p_{\text{best}} + \delta p$, at the same time that q had its *smallest* value $q_{\text{best}} - \delta q$. Thus, the highest probable value for $p - q$ is

$$\text{highest probable value} \; = \; (p_{\text{best}} - q_{\text{best}}) + (\delta p + \delta q). \qquad (2.16)$$

Similarly, the lowest probable value arises when p is smallest $(p_{\text{best}} - \delta p)$, but q is largest $(q_{\text{best}} + \delta q)$. Thus,

$$\text{lowest probable value} \; = \; (p_{\text{best}} - q_{\text{best}}) - (\delta p + \delta q). \qquad (2.17)$$

Combining Equations (2.16) and (2.17), we see that the *uncertainty in the difference* $(p - q)$ is the **sum** $\delta p + \delta q$ *of the original uncertainties*. For example, if

$$p \; = \; 1.49 \pm 0.03 \text{ kg·m/s}$$

and

$$q \; = \; 1.56 \pm 0.06 \text{ kg·m/s},$$

then

$$p - q \; = \; -0.07 \pm 0.09 \text{ kg·m/s}.$$

We can now add an extra column for $p - q$ to Table 2.1 and arrive at Table 2.2.

Table 2.2. Measured momenta (kg·m/s).

Trial number	Initial p (all ± 0.03)	Final q (all ± 0.06)	Difference $p - q$ (all ± 0.09)
1	1.49	1.56	-0.07
2	3.10	3.12	-0.02
3	2.16	2.05	0.11
etc.			

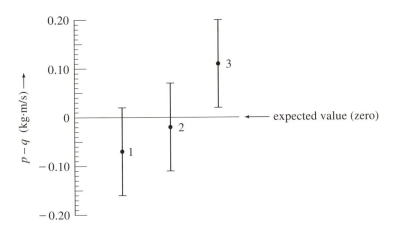

Figure 2.4. Three trials in a test of the conservation of momentum. The student has measured the total momentum of two carts before and after they collide (p and q, respectively). If momentum is conserved, the differences $p - q$ should all be zero. The plot shows the value of $p - q$ with its error bar for each trial. The expected value 0 is inside the margins of error in trials 1 and 2 and only slightly outside in trial 3. Therefore, these results are consistent with the conservation of momentum.

Whether our results are consistent with conservation of momentum can now be seen at a glance by checking whether the numbers in the final column are consistent with zero (that is, are less than, or comparable with, the uncertainty 0.09). Alternatively, and perhaps even better, we could plot the results as in Figure 2.4 and check visually. Yet another way to achieve the same effect would be to calculate the *ratios* q/p, which should all be consistent with the expected value $q/p = 1$. (Here, we would need to calculate the uncertainty in q/p, a problem discussed in Chapter 3.)

Our discussion of the uncertainty in $p - q$ applies to the difference of any two measured numbers. If we had measured any two numbers x and y and used our measured values to compute the difference $x - y$, by the argument just given, the resulting uncertainty in the difference would be the *sum* of the separate uncertainties in x and y. We have, therefore, established the following provisional rule:

Uncertainty in a Difference
(Provisional Rule)

If two quantities x and y are measured with uncertainties δx and δy, and if the measured values x and y are used to calculate the difference $q = x - y$, the *uncertainty in q is the sum of the uncertainties in x and y*:

$$\delta q \approx \delta x + \delta y. \tag{2.18}$$

I call this rule "provisional" because we will find in Chapter 3 that the uncertainty in the quantity $q = x - y$ is often somewhat smaller than that given by Equation

(2.18). Thus, we will be replacing the provisional rule (2.18) by an "improved" rule—in which the uncertainty in $q = x - y$ is given by the so-called quadratic sum of δx and δy, as defined in Equation (3.13). Because this improved rule gives a somewhat smaller uncertainty for q, you will want to use it when appropriate. For now, however, let us be content with the provisional rule (2.18) for three reasons: (1) The rule (2.18) is easy to understand—much more so than the improved rule of Chapter 3. (2) In most cases, the difference between the two rules is small. (3) The rule (2.18) always gives an upper bound on the uncertainty in $q = x - y$; thus, we know at least that the uncertainty in $x - y$ is never worse than the answer given in (2.18).

The result (2.18) is the first in a series of rules for the *propagation of errors*. To calculate a quantity q in terms of measured quantities x and y, we need to know how the uncertainties in x and y "propagate" to cause uncertainty in q. A complete discussion of error propagation appears in Chapter 3.

Quick Check 2.3.   In an experiment to measure the latent heat of ice, a student adds a chunk of ice to water in a styrofoam cup and observes the change in temperature as the ice melts. To determine the mass of ice added, she weighs the cup of water before and after she adds the ice and then takes the difference. If her two measurements were

$$(\text{mass of cup \& water}) = m_1 = 203 \pm 2 \text{ grams}$$

and

$$(\text{mass of cup, water, \& ice}) = m_2 = 246 \pm 3 \text{ grams},$$

find her answer for the mass of ice, $m_2 - m_1$, with its uncertainty, as given by the provisional rule (2.18).

2.6   Checking Relationships with a Graph

Many physical laws imply that one quantity should be proportional to another. For example, Hooke's law states that the extension of a spring is proportional to the force stretching it, and Newton's law says that the acceleration of a body is proportional to the total applied force. Many experiments in a teaching laboratory are designed to check this kind of proportionality.

If one quantity y is proportional to some other quantity x, a graph of y against x is a straight line through the origin. Thus, to test whether y is proportional to x, you can plot the measured values of y against those of x and note whether the resulting points do lie on a straight line through the origin. Because a straight line is so easily recognizable, this method is a simple, effective way to check for proportionality.

To illustrate this use of graphs, let us imagine an experiment to test Hooke's law. This law, usually written as $F = kx$, asserts that the extension x of a spring is proportional to the force F stretching it, so $x = F/k$, where k is the "force constant" of the spring. A simple way to test this law is to hang the spring vertically and suspend various masses m from it. Here, the force F is the weight mg of the load; so the extension should be

$$x = \frac{mg}{k} = \left(\frac{g}{k}\right) m. \tag{2.19}$$

The extension x should be proportional to the load m, and a graph of x against m should be a straight line through the origin.

  If we measure x for a variety of different loads m and plot our measured values of x and m, the resulting points almost certainly will not lie *exactly* on a straight line. Suppose, for example, we measure the extension x for eight different loads m and get the results shown in Table 2.3. These values are plotted in Figure 2.5(a),

Table 2.3. Load and extension.

Load m (grams) (δm negligible)	200	300	400	500	600	700	800	900
Extension x (cm) (all ± 0.3)	1.1	1.5	1.9	2.8	3.4	3.5	4.6	5.4

which also shows a possible straight line that passes through the origin and is reasonably close to all eight points. As we should have expected, the eight points do not lie exactly on any line. The question is whether this result stems from experimental uncertainties (as we would hope), from mistakes we have made, or even from the possibility the extension x is *not* proportional to m. To answer this question, we must consider our uncertainties.

  As usual, the measured quantities, extensions x and masses m, are subject to uncertainty. For simplicity, let us suppose that the masses used are known very accurately, so that the uncertainty in m is negligible. Suppose, on the other hand, that all measurements of x have an uncertainty of approximately 0.3 cm (as indicated in Table 2.3). For a load of 200 grams, for example, the extension would probably be in the range 1.1 ± 0.3 cm. Our first experimental point on the graph thus lies on the vertical line $m = 200$ grams, somewhere between $x = 0.8$ and $x = 1.4$ cm. This range is indicated in Figure 2.5(b), which shows an *error bar* through each point to indicate the range in which it probably lies. Obviously, we should expect to find a straight line that goes through the origin and *passes through or close to all the error bars*. Figure 2.5(b) has such a line, so we conclude that the data on which Figure 2.5(b) is based are consistent with x being proportional to m.

  We saw in Equation (2.19) that the slope of the graph of x against m is g/k. By measuring the slope of the line in Figure 2.5(b), we can therefore find the constant k of the spring. By drawing the steepest and least steep lines that fit the data reasonably well, we could also find the uncertainty in this value for k. (See Problem 2.18.)

  If the best straight line misses a high proportion of the error bars or if it misses any by a large distance (compared with the length of the error bars), our results

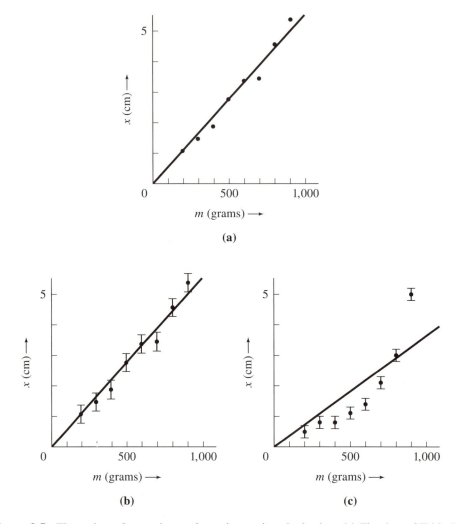

Figure 2.5. Three plots of extension x of a spring against the load m. **(a)** The data of Table 2.3 without error bars. **(b)** The same data with error bars to show the uncertainties in x. (The uncertainties in m are assumed to be negligible.) These data are consistent with the expected proportionality of x and m. **(c)** A different set of data, which are inconsistent with x being proportional to m.

would be *inconsistent* with x being proportional to m. This situation is illustrated in Figure 2.5(c). With the results shown there, we would have to recheck our measurements and calculations (including the calculation of the uncertainties) and consider whether x is *not* proportional to m for some reason. [In Figure 2.5(c), for instance, the first five points can be fitted to a straight line through the origin. This situation suggests that x may be proportional to m up to approximately 600 grams, but that Hooke's law breaks down at that point and the spring starts to stretch more rapidly.]

Thus far, we have supposed that the uncertainty in the mass (which is plotted along the horizontal axis) is negligible and that the only uncertainties are in x, as shown by the vertical error bars. If both x and m are subject to appreciable uncertainties the simplest way to display them is to draw vertical *and* horizontal error bars, whose lengths show the uncertainties in x and m respectively, as in Figure 2.6.

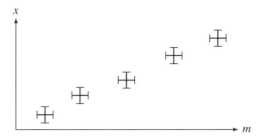

Figure 2.6. Measurements that have uncertainties in both variables can be shown by crosses made up of one error bar for each variable.

Each cross in this plot corresponds to one measurement of x and m, in which x probably lies in the interval defined by the vertical bar of the cross and m probably in that defined by the horizontal bar.

A slightly more complicated possibility is that some quantity may be expected to be proportional to a *power* of another. (For example, the distance traveled by a freely falling object in a time t is $d = \frac{1}{2}gt^2$ and is proportional to the square of t.) Let us suppose that y is expected to be proportional to x^2. Then

$$y = Ax^2, \tag{2.20}$$

where A is some constant, and a graph of y against x should be a parabola with the general shape of Figure 2.7(a). If we were to measure a series of values for y and x and plot y against x, we might get a graph something like that in Figure 2.7(b). Unfortunately, visually judging whether a set of points such as these fit a parabola (or any other curve, except a straight line) is very hard. A better way to check that $y \propto x^2$ is to plot y against x *squared*. From Equation (2.20), we see that such a plot should be a straight line, which we can check easily as in Figure 2.7(c).

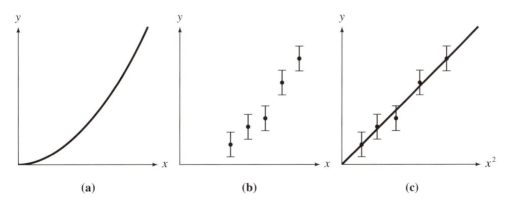

| (a) | (b) | (c) |

Figure 2.7. (a) If y is proportional to x^2, a graph of y against x should be a parabola with this general shape. (b) A plot of y against x for a set of measured values is hard to check visually for fit with a parabola. (c) On the other hand, a plot of y against x^2 should be a straight line through the origin, which is easy to check. (In the case shown, we see easily that the points *do* fit a straight line through the origin.)

In the same way, if $y = Ax^n$ (where n is any power), a graph of y against x^n should be a straight line, and by plotting the observed values of y against x^n, we can check easily for such a fit. There are various other situations in which a nonlinear relation (that is, one that gives a curved—nonlinear—graph) can be converted into a linear one by a clever choice of variables to plot. Section 8.6 discusses an important example of such "linearization," which is worth mentioning briefly here. Often one variable y depends *exponentially* on another variable x:

$$y = Ae^{Bx}.$$

(For example, the activity of a radioactive sample depends exponentially on time.) For such relations, the natural logarithm of y is easily shown to be linear in x; that is, a graph of $\ln(y)$ against x should be a straight line for an exponential relationship.

Many other, nongraphical ways are available to check the proportionality of two quantities. For example, if $y \propto x$, the ratio y/x should be constant. Thus, having tabulated the measured values of y and x, you could simply add a column to the table that shows the ratios y/x and check that these *ratios* are constant within their experimental uncertainties. Many calculators have a built-in function (called the correlation coefficient) to show how well a set of measurements fits a straight line. (This function is discussed in Section 9.3.) Even when another method is used to check that $y \propto x$, making the graphical check as well is an excellent practice. Graphs such as those in Figures 2.5(b) and (c) show clearly how well (or badly) the measurements verify the predictions; drawing such graphs helps you understand the experiment and the physical laws involved.

2.7   Fractional Uncertainties

The uncertainty δx in a measurement,

$$(\text{measured } x) = x_{\text{best}} \pm \delta x,$$

indicates the reliability or precision of the measurement. The uncertainty δx by itself does not tell the whole story, however. An uncertainty of one inch in a distance of one mile would indicate an unusually precise measurement, whereas an uncertainty of one inch in a distance of three inches would indicate a rather crude estimate. Obviously, the quality of a measurement is indicated not just by the uncertainty δx but also by the *ratio* of δx to x_{best}, which leads us to consider the *fractional uncertainty*,

$$\text{fractional uncertainty} = \frac{\delta x}{|x_{\text{best}}|}. \tag{2.21}$$

(The fractional uncertainty is also called the *relative uncertainty* or the *precision*.) In this definition, the symbol $|x_{\text{best}}|$ denotes the absolute value[5] of x_{best}. The uncer-

---

[5] The absolute value $|x|$ of a number x is equal to x when x is positive but is obtained by omitting the minus sign if x is negative. We use the absolute value in (2.21) to guarantee that the fractional uncertainty, like the uncertainty δx itself, is always positive, whether x_{best} is positive or negative. In practice, you can often arrange matters so that measured numbers are positive, and the absolute-value signs in (2.21) can then be omitted.

tainty δx is sometimes called the *absolute uncertainty* to avoid confusion with the fractional uncertainty.

In most serious measurements, the uncertainty δx is much smaller than the measured value x_{best}. Because the fractional uncertainty $\delta x/|x_{\text{best}}|$ is therefore usually a small number, multiplying it by 100 and quoting it as the *percentage uncertainty* is often convenient. For example, the measurement

$$\text{length } l \;=\; 50 \pm 1 \text{ cm} \tag{2.22}$$

has a fractional uncertainty

$$\frac{\delta l}{|l_{\text{best}}|} \;=\; \frac{1 \text{ cm}}{50 \text{ cm}} \;=\; 0.02$$

and a percentage uncertainty of 2%. Thus, the result (2.22) could be given as

$$\text{length } l \;=\; 50 \text{ cm} \pm 2\%.$$

Note that although the absolute uncertainty δl has the same units as l, the fractional uncertainty $\delta l/|l_{\text{best}}|$ is a *dimensionless* quantity, without units. Keeping this difference in mind can help you avoid the common mistake of confusing absolute uncertainty with fractional uncertainty.

The fractional uncertainty is an approximate indication of the quality of a measurement, whatever the size of the quantity measured. Fractional uncertainties of 10% or so are usually characteristic of fairly rough measurements. (A rough measurement of 10 inches might have an uncertainty of 1 inch; a rough measurement of 10 miles might have an uncertainty of 1 mile.) Fractional uncertainties of 1 or 2% are characteristic of reasonably careful measurements and are about the best to hope for in many experiments in the introductory physics laboratory. Fractional uncertainties much less than 1% are often hard to achieve and are rather rare in the introductory laboratory.

These divisions are, of course, extremely rough. A few simple measurements can have fractional uncertainties of 0.1% or less with little trouble. A good tape measure can easily measure a distance of 10 feet with an uncertainty of $\frac{1}{10}$ inch, or approximately 0.1%; a good timer can easily measure a period of an hour with an uncertainty of less than a second, or 0.03%. On the other hand, for many quantities that are very hard to measure, a 10% uncertainty would be regarded as an experimental triumph. Large percentage uncertainties, therefore, do not necessarily mean that a measurement is scientifically useless. In fact, many important measurements in the history of physics had experimental uncertainties of 10% or more. Certainly plenty can be learned in the introductory physics laboratory from equipment that has a minimum uncertainty of a few percent.

Quick Check 2.4.   Convert the errors in the following measurements of the velocities of two carts on a track into fractional errors and percent errors: **(a)** $v = 55 \pm 2$ cm/s; **(b)** $u = -20 \pm 2$ cm/s. **(c)** A cart's kinetic energy is measured as $K = 4.58$ J $\pm\ 2\%$; rewrite this finding in terms of its absolute uncertainty. (Because the uncertainties should be given to one significant figure, you ought to be able to do the calculations in your head.)

2.8   Significant Figures and Fractional Uncertainties

The concept of fractional uncertainty is closely related to the familiar notion of significant figures. In fact, the number of significant figures in a quantity is an approximate indicator of the fractional uncertainty in that quantity. To clarify this connection, let us review briefly the notion of significant figures and recognize that this concept is both approximate and somewhat ambiguous.

To a mathematician, the statement that $x = 21$ to two significant figures means unambiguously that x is closer to 21 than to either 20 or 22; thus, the number 21, with two significant figures, means 21 ± 0.5. To an experimental scientist, most numbers are numbers that have been read off a meter (or calculated from numbers read off a meter). In particular, if a digital meter displays two significant figures and reads 21, it *may* mean 21 ± 0.5, but it may also mean 21 ± 1 or even something like 21 ± 5. (Many meters come with a manual that explains the actual uncertainties.) Under these circumstances, the statement that a measured number has two significant figures is only a rough indicator of its uncertainty. Rather than debate exactly how the concept should be defined, I will adopt a middle-of-the-road definition that 21 with two significant figures means 21 ± 1, and more generally that a number with N significant figures has an uncertainty of about 1 in the Nth digit.

Let us now consider two numbers,

$$x = 21 \quad \text{and} \quad y = 0.21,$$

both of which have been certified accurate to two significant figures. According to the convention just agreed to, these values mean

$$x = 21 \pm 1 \quad \text{and} \quad y = 0.21 \pm 0.01.$$

Although the two numbers both have two significant figures, they obviously have very different uncertainties. On the other hand, they both have the same *fractional uncertainty,* which in this case is 5%:

$$\frac{\delta x}{x} = \frac{\delta y}{y} = \frac{1}{21} = \frac{0.01}{0.21} = 0.05 \text{ or } 5\%.$$

Evidently, the statement that the numbers 21 and 0.21 (or 210, or 2.1, or 0.0021, etc.) have two significant figures is equivalent to saying that they are 5% uncertain. In the same way, 21.0, with three significant figures, is 0.5% uncertain, and so on.

Unfortunately, this useful connection is only approximate. For example, the statement that $s = 10$, with two significant figures, means

$$s = 10 \pm 1 \quad \text{or} \quad 10 \pm 10\%.$$

At the opposite extreme, $t = 99$ (again with two significant figures) means

$$t = 99 \pm 1 \quad \text{or} \quad 99 \pm 1\%.$$

Evidently, the fractional uncertainty associated with two significant figures ranges from 1% to 10%, depending on the first digit of the number concerned.

The approximate correspondence between significant figures and fractional uncertainties can be summarized as in Table 2.4.

Table 2.4. Approximate correspondence between significant figures and fractional uncertainties.

Number of significant figures	Corresponding fractional uncertainty is	
	between	or roughly
1	10% and 100%	50%
2	1% and 10%	5%
3	0.1% and 1%	0.5%

2.9   Multiplying Two Measured Numbers

Perhaps the greatest importance of fractional errors emerges when we start multiplying measured numbers by each other. For example, to find the momentum of a body, we might measure its mass m and its velocity v and then multiply them to give the momentum $p = mv$. Both m and v are subject to uncertainties, which we will have to estimate. The problem, then, is to find the uncertainty in p that results from the known uncertainties in m and v.

First, for convenience, let us rewrite the standard form

$$(\text{measured value of } x) \ = \ x_{\text{best}} \pm \delta x$$

in terms of the fractional uncertainty, as

$$(\text{measured value of } x) \ = \ x_{\text{best}}\left(1 \pm \frac{\delta x}{|x_{\text{best}}|}\right). \qquad (2.23)$$

For example, if the fractional uncertainty is 3%, we see from (2.23) that

$$(\text{measured value of } x) \ = \ x_{\text{best}}\left(1 \pm \frac{3}{100}\right);$$

that is, 3% uncertainty means that x probably lies between x_{best} times 0.97 and x_{best} times 1.03,

$$(0.97) \times x_{\text{best}} \ \leq \ x \ \leq \ (1.03) \times x_{\text{best}}.$$

We will find this a useful way to think about a measured number that we will have to multiply.

Let us now return to our problem of calculating $p = mv$, when m and v have been measured, as

$$(\text{measured } m) \ = \ m_{\text{best}}\left(1 \pm \frac{\delta m}{|m_{\text{best}}|}\right) \qquad (2.24)$$

and

$$(\text{measured } v) \ = \ v_{\text{best}}\left(1 \pm \frac{\delta v}{|v_{\text{best}}|}\right) \qquad (2.25)$$

Because m_{best} and v_{best} are our best estimates for m and v, our best estimate for $p = mv$ is

$$\text{(best estimate for } p) = p_{best} = m_{best}v_{best}.$$

The largest probable values of m and v are given by (2.24) and (2.25) with the plus signs. Thus, the largest probable value for $p = mv$ is

$$\text{(largest value for } p) = m_{best}v_{best}\left(1 + \frac{\delta m}{|m_{best}|}\right)\left(1 + \frac{\delta v}{|v_{best}|}\right). \qquad (2.26)$$

The smallest probable value for p is given by a similar expression with two minus signs. Now, the product of the parentheses in (2.26) can be multiplied out as

$$\left(1 + \frac{\delta m}{|m_{best}|}\right)\left(1 + \frac{\delta v}{|v_{best}|}\right) = 1 + \frac{\delta m}{|m_{best}|} + \frac{\delta v}{|v_{best}|} + \frac{\delta m}{|m_{best}|}\frac{\delta v}{|v_{best}|}. \qquad (2.27)$$

Because the two fractional uncertainties $\delta m/|m_{best}|$ and $\delta v/|v_{best}|$ are small numbers (a few percent, perhaps), their product is extremely small. Therefore, the last term in (2.27) can be neglected. Returning to (2.26), we find

$$\text{(largest value of } p) = m_{best}v_{best}\left(1 + \frac{\delta m}{|m_{best}|} + \frac{\delta v}{|v_{best}|}\right).$$

The smallest probable value is given by a similar expression with two minus signs. Our measurements of m and v, therefore, lead to a value of $p = mv$ given by

$$\text{(value of } p) = m_{best}v_{best}\left(1 \pm \left[\frac{\delta m}{|m_{best}|} + \frac{\delta v}{|v_{best}|}\right]\right).$$

Comparing this equation with the general form

$$\text{(value of } p) = p_{best}\left(1 \pm \frac{\delta p}{|p_{best}|}\right),$$

we see that the best estimate for p is $p_{best} = m_{best}v_{best}$ (as we already knew) and that the *fractional uncertainty in p is the sum of the fractional uncertainties in m and v,*

$$\frac{\delta p}{|p_{best}|} = \frac{\delta m}{|m_{best}|} + \frac{\delta v}{|v_{best}|}.$$

If, for example, we had the following measurements for m and v,

$$m = 0.53 \pm 0.01 \text{ kg}$$

and

$$v = 9.1 \pm 0.3 \text{ m/s,}$$

the best estimate for $p = mv$ is

$$p_{best} = m_{best}v_{best} = (0.53) \times (9.1) = 4.82 \text{ kg·m/s.}$$

To compute the uncertainty in p, we would first compute the fractional errors

$$\frac{\delta m}{m_{\text{best}}} = \frac{0.01}{0.53} = 0.02 = 2\%$$

and

$$\frac{\delta v}{v_{\text{best}}} = \frac{0.3}{9.1} = 0.03 = 3\%.$$

The fractional uncertainty in p is then the sum:

$$\frac{\delta p}{p_{\text{best}}} = 2\% + 3\% = 5\%.$$

If we want to know the absolute uncertainty in p, we must multiply by p_{best}:

$$\delta p = \frac{\delta p}{p_{\text{best}}} \times p_{\text{best}} = 0.05 \times 4.82 = 0.241.$$

We then round δp and p_{best} to give us our final answer

$$(\text{value of } p) = 4.8 \pm 0.2 \text{ kg·m/s}.$$

The preceding considerations apply to any product of two measured quantities. We have therefore discovered our second general rule for the propagation of errors. If we measure any two quantities x and y and form their product, the uncertainties in the original two quantities "propagate" to cause an uncertainty in their product. This uncertainty is given by the following rule:

**Uncertainty in a Product
(Provisional Rule)**

If two quantities x and y have been measured with small fractional uncertainties $\delta x/|x_{\text{best}}|$ and $\delta y/|y_{\text{best}}|$, and if the measured values of x and y are used to calculate the product $q = xy$, then the *fractional uncertainty in q is the sum of the fractional uncertainties in x and y*,

$$\frac{\delta q}{|q_{\text{best}}|} \approx \frac{\delta x}{|x_{\text{best}}|} + \frac{\delta y}{|y_{\text{best}}|}. \qquad (2.28)$$

I call this rule "provisional," because, just as with the rule for uncertainty in a difference, I will replace it with a more precise rule later on. Two other features of this rule also need to be emphasized. First, the derivation of (2.28) required that the fractional uncertainties in x and y both be small enough that we could neglect their product. This requirement is almost always true in practice, and I will always assume it. Nevertheless, remember that if the fractional uncertainties are *not* much smaller than 1, the rule (2.28) may not apply. Second, even when x and y have different dimensions, (2.28) balances dimensionally because all fractional uncertainties are dimensionless.

In physics, we frequently multiply numbers together, and the rule (2.28) for finding the uncertainty in a product will obviously be an important tool in error analysis. For the moment, our main purpose is to emphasize that the uncertainty in any product $q = xy$ is expressed most simply in terms of fractional uncertainties, as in (2.28).

Quick Check 2.5. To find the area of a rectangular plate, a student measures its sides as $l = 9.1 \pm 0.1$ cm and $b = 3.3 \pm 0.1$ cm. Express these uncertainties as percent uncertainties and then find the student's answer for the area $A = lb$ with its uncertainty. (Find the latter as a percent uncertainty first and then convert to an absolute uncertainty. Do all error calculations in your head.)

Principal Definitions and Equations of Chapter 2

STANDARD FORM FOR STATING UNCERTAINTIES

The standard form for reporting a measurement of a physical quantity x is

$$(\text{measured value of } x) = x_{\text{best}} \pm \delta x,$$

where

$$x_{\text{best}} = (\text{best estimate for } x)$$

and

$$\delta x = (\text{uncertainty or error in the measurement}). \qquad [\text{See (2.3)}]$$

This statement expresses our confidence that the correct value of x probably lies in (or close to) the range from $x_{\text{best}} - \delta x$ to $x_{\text{best}} + \delta x$.

DISCREPANCY

The *discrepancy* between two measured values of the same physical quantity is

$$\text{discrepancy} = \text{difference between two measured values of the same quantity.} \qquad [\text{See (2.10)}]$$

FRACTIONAL UNCERTAINTY

If x is measured in the standard form $x_{\text{best}} \pm \delta x$, the *fractional uncertainty* in x is

$$\text{fractional uncertainty} = \frac{\delta x}{|x_{\text{best}}|}. \qquad [\text{See (2.21)}]$$

The *percent uncertainty* is just the fractional uncertainty expressed as a percentage (that is, multiplied by 100%).

We have found two provisional rules, (2.18) and (2.28), for error propagation that show how the uncertainties in two quantities x and y propagate to cause uncertainties in calculations of the difference $x - y$ or the product xy. A complete discussion of error propagation appears in Chapter 3, where I show that the rules (2.18) and (2.28) can frequently be replaced with more refined rules (given in Section 3.6). For this reason, I have not reproduced (2.18) and (2.28) here.

Problems for Chapter 2

Notes: *The problems at the end of each chapter are arranged by section number. A problem listed for a specific section may, of course, involve ideas from previous sections but does not require knowledge of later sections. Therefore, you may try problems listed for a specific section as soon as you have read that section.*

The approximate difficulty of each problem is indicated by one, two, or three stars. A one-star problem should be straightforward and usually involves a single concept. Two-star problems are more difficult or require more work (drawing a graph, for instance). Three-star problems are the most difficult and may require considerably more labor.

Answers to the odd-numbered problems can be found in the Answers Section at the back of the book.

For Section 2.1: Best Estimate $\pm$ Uncertainty

2.1. ★   In Chapter 1, a carpenter reported his measurement of the height of a doorway by stating that his best estimate was 210 cm and that he was confident the height was between 205 and 215 cm. Rewrite this result in the standard form $x_{\text{best}} \pm \delta x$. Do the same for the measurements reported in Equations (1.1), (1.2), and (1.4).

2.2. ★   A student studying the motion of a cart on an air track measures its position, velocity, and acceleration at one instant, with the results shown in Table 2.5. Rewrite these results in the standard form $x_{\text{best}} \pm \delta x$.

Table 2.5. Measurements of position, velocity, and acceleration; for Problem 2.2.

Variable	Best estimate	Probable range
Position, x	53.3	53.1 to 53.5 (cm)
Velocity, v	-13.5	-14.0 to -13.0 (cm/s)
Acceleration, a	93	90 to 96 (cm/s^2)

For Section 2.2: Significant Figures

2.3. ★   Rewrite the following results in their clearest forms, with suitable numbers of significant figures:

   (a) measured height $= 5.03 \pm 0.04329$ m
   (b) measured time $= 1.5432 \pm 1$ s
   (c) measured charge $= -3.21 \times 10^{-19} \pm 2.67 \times 10^{-20}$ C
   (d) measured wavelength $= 0.000,000,563 \pm 0.000,000,07$ m
   (e) measured momentum $= 3.267 \times 10^{3} \pm 42$ g·cm/s.

2.4. ★   Rewrite the following equations in their clearest and most appropriate forms:

   (a) $x = 3.323 \pm 1.4$ mm
   (b) $t = 1,234,567 \pm 54,321$ s
   (c) $\lambda = 5.33 \times 10^{-7} \pm 3.21 \times 10^{-9}$ m
   (d) $r = 0.000,000,538 \pm 0.000,000,03$ mm

For Section 2.3: Discrepancy

2.5. ★   Two students measure the length of the same rod and report the results 135 ± 3 mm and 137 ± 3 mm. Draw an illustration like that in Figure 2.1 to represent these two measurements. What is the discrepancy between the two measurements, and is it significant?

2.6. ★   Each of two research groups discovers a new elementary particle. The two reported masses are

$$m_1 = (7.8 \pm 0.1) \times 10^{-27} \text{ kg}$$

and

$$m_2 = (7.0 \pm 0.2) \times 10^{-27} \text{ kg.}$$

Draw an illustration like that in Figure 2.1 to represent these two measurements. The question arises whether these two measurements could actually be of the same particle. Based on the reported masses, would you say they are likely to be the same particle? In particular, what is the discrepancy in the two measurements (assuming they really are measurements of the same mass)?

For Section 2.4: Comparison of Measured and Accepted Values

2.7. ★ (a)   A student measures the density of a liquid five times and gets the results (all in gram/cm^3) 1.8, 2.0, 2.0, 1.9, and 1.8. What would you suggest as the best estimate and uncertainty based on these measurements? **(b)** The student is told that the accepted value is 1.85 gram/cm^3. What is the discrepancy between the student's best estimate and the accepted value? Do you think it is significant?

2.8. ★   Two groups of students measure the charge of the electron and report their results as follows:

$$\text{Group A:} \quad e = (1.75 \pm 0.04) \times 10^{-19} \text{ C}$$

and

Group B: $e = (1.62 \pm 0.04) \times 10^{-19}$ C.

What should each group report for the discrepancy between its value and the accepted value,

$$e = 1.60 \times 10^{-19} \text{ C}$$

(with negligible uncertainty)? Draw an illustration similar to that in Figure 2.2 to show these results and the accepted value. Which of the results would you say is satisfactory?

For Section 2.5: Comparison of Two Measured Numbers

2.9. ★   In an experiment on the simple pendulum, a student uses a steel ball suspended from a light string, as shown in Figure 2.8. The effective length l of the

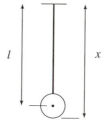

Figure 2.8. A simple pendulum; for Problem 2.9.

pendulum is the distance from the top of the string to the *center* of the ball, as shown. To find l, he first measures the distance x from the top of the string to the bottom of the ball and the radius r of the ball; he then subtracts to give $l = x - r$. If his two measurements are

$$x = 95.8 \pm 0.1 \text{ cm} \quad \text{and} \quad r = 2.30 \pm 0.02 \text{ cm},$$

what should be his answer for the length l and its uncertainty, as given by the provisional rule (2.18)?

2.10. ★   The time a carousel takes to make one revolution is measured by noting the starting and stopping times using the second hand of a wrist watch and subtracting. If the starting and stopping times are uncertain by ± 1 second each, what is the uncertainty in the time for one revolution, as given by the provisional rule (2.18)?

2.11. ★   In an experiment to check conservation of angular momentum, a student obtains the results shown in Table 2.6 for the initial and final angular momenta (L and L') of a rotating system. Add an extra column to the table to show the difference $L - L'$ and its uncertainty. Are the student's results consistent with conservation of angular momentum?

Table 2.6. Initial and final angular momenta (in kg·m²/s); for Problems 2.11 and 2.14.

Initial L	Final L'
3.0 ± 0.3	2.7 ± 0.6
7.4 ± 0.5	8.0 ± 1
14.3 ± 1	16.5 ± 1
25 ± 2	24 ± 2
32 ± 2	31 ± 2
37 ± 2	41 ± 2

2.12. ★   The acceleration a of a cart sliding down a frictionless incline with slope θ is expected to be $g \sin \theta$. To test this, a student measures the acceleration a of a cart on an incline for several different values of θ; she also calculates the corresponding expected accelerations $g \sin \theta$ for each θ and obtains the results shown in Table 2.7. Add a column to the table to show the discrepancies $a - g \sin \theta$ and their uncertainties. Do the results confirm that a is given by $g \sin \theta$? If not, can you suggest a reason they do not?

Table 2.7. Measured and expected accelerations; for Problem 2.12.

Trial number	Acceleration a (m/s²)	Expected acceleration $g \sin \theta$ (m/s²)
1	2.04 ± 0.04	2.36 ± 0.1
2	3.58 ± 0.06	3.88 ± 0.08
3	4.32 ± 0.08	4.57 ± 0.05
4	4.85 ± 0.09	5.05 ± 0.04
5	5.53 ± 0.1	5.72 ± 0.03

2.13. ★★   An experimenter measures the separate masses M and m of a car and trailer. He gives his results in the standard form $M_{best} \pm \delta M$ and $m_{best} \pm \delta m$. What would be his best estimate for the total mass $M + m$? By considering the largest and smallest probable values of the total mass, show that his uncertainty in the total mass is just the sum of δM and δm. State your arguments clearly; don't just write down the answer. (This problem provides another example of error propagation: The uncertainties in the measured numbers, M and m, propagate to cause an uncertainty in the sum $M + m$.)

For Section 2.6: Checking Relationships with a Graph

2.14. ★★   Using the data of Problem 2.11, make a plot of final angular momentum L' against initial angular momentum L for the experiment described there. (Include vertical and horizontal error bars, and be sure to include the origin. As with all graphs, label your axes, including units, use squared paper, and choose the scales so that the graph fills a good proportion of the page.) On what curve would you expect the points to lie? Do they lie on this curve within experimental uncertainties?

2.15. ★★ According to the ideal gas law, if the volume of a gas is kept constant, the pressure P should be proportional to the absolute temperature T. To check this proportionality, a student measures the pressure of a gas at five different temperatures (always with the same volume) and gets the results shown in Table 2.8. Plot these results in a graph of P against T, and decide whether they confirm the expected proportionality of P and T.

Table 2.8. Temperature and pressure of a gas; for Problem 2.15.

Temperature (K) (negligible uncertainty)	Pressure (atm) (all ±0.04)
100	0.36
150	0.46
200	0.71
250	0.83
300	1.04

2.16. ★★ You have learned (or will learn) in optics that certain lenses (namely, thin spherical lenses) can be characterized by a parameter called the focal length f and that if an object is placed at a distance p from the lens, the lens forms an image at a distance q, satisfying the *lens equation*, $1/f = (1/p) + (1/q)$, where f always has the same value for a given lens. To check if these ideas apply to a certain lens, a student places a small light bulb at various distances p from the lens and measures the location q of the corresponding images. She then calculates the corresponding values of f from the lens equation and obtains the results shown in Table 2.9. Make a plot of f against p, with appropriate error bars, and decide if it is true that this particular lens has a unique focal length f.

Table 2.9. Object distances p (in cm) and corresponding focal lengths f (in cm); for Problem 2.16.

Object distance p (negligible uncertainty)	Focal length f (all ±2)
45	28
55	34
65	33
75	37
85	40

2.17. ★★   The power P delivered to a resistance R by a current I is supposed to be given by the relation $P = RI^2$. To check this relation, a student sends several different currents through an unknown resistance immersed in a cup of water and measures the power delivered (by measuring the water's rise in temperature). Use the results shown in Table 2.10 to make plots of P against I and P against I^2, including error bars. Use the second plot to decide if this experiment is consistent with the expected proportionality of P and I^2.

Table 2.10. Current I and power P; for Problem 2.17.

Current I (amps) (negligible uncertainty)	Power P (watts) (all ± 50)
1.5	270
2.0	380
2.5	620
3.0	830
3.5	1280
4.0	1600

2.18. ★★★   If a stone is thrown vertically upward with speed v, it should rise to a height h given by $v^2 = 2gh$. In particular, v^2 should be proportional to h. To test this proportionality, a student measures v^2 and h for seven different throws and gets the results shown in Table 2.11. **(a)** Make a plot of v^2 against h, including vertical and horizontal error bars. (As usual, use squared paper, label your axes, and choose your scale sensibly.) Is your plot consistent with the prediction that $v^2 \propto h$? **(b)** The slope of your graph should be $2g$. To find the slope, draw what seems to be the best straight line through the points and then measure its slope. To find the uncertainty in the slope, draw the steepest and least steep lines that seem to fit the data reasonably. The slopes of these lines give the largest and smallest probable values of the slope. Are your results consistent with the accepted value $2g = 19.6$ m/s^2?

Table 2.11. Heights and speeds of a stone thrown vertically upward; for Problem 2.18.

h (m) all ± 0.05	v^2 (m^2/s^2)
0.4	7 ± 3
0.8	17 ± 3
1.4	25 ± 3
2.0	38 ± 4
2.6	45 ± 5
3.4	62 ± 5
3.8	72 ± 6

2.19. ★★★   In an experiment with a simple pendulum, a student decides to check whether the period T is independent of the amplitude A (defined as the largest angle that the pendulum makes with the vertical during its oscillations). He obtains the

Table 2.12. Amplitude and period of a pendulum; for Problem 2.19.

Amplitude A (deg)	Period T (s)
5 ± 2	1.932 ± 0.005
17 ± 2	1.94 ± 0.01
25 ± 2	1.96 ± 0.01
40 ± 4	2.01 ± 0.01
53 ± 4	2.04 ± 0.01
67 ± 6	2.12 ± 0.02

results shown in Table 2.12. **(a)** Draw a graph of T against A. (Consider your choice of scales carefully. If you have any doubt about this choice, draw two graphs, one including the origin, $A = T = 0$, and one in which only values of T between 1.9 and 2.2 s are shown.) Should the student conclude that the period is independent of the amplitude? **(b)** Discuss how the conclusions of part (a) would be affected if all the measured values of T had been uncertain by ± 0.3 s.

For Section 2.7: Fractional Uncertainties

2.20. ★   Compute the percentage uncertainties for the five measurements reported in Problem 2.3. (Remember to round to a reasonable number of significant figures.)

2.21. ★   Compute the percentage uncertainties for the four measurements in Problem 2.4.

2.22. ★   Convert the percent errors given for the following measurements into absolute uncertainties and rewrite the results in the standard form $x_{best} \pm \delta x$ rounded appropriately.
  (a) $x = 543.2$ m $\pm$ 4%
  (b) $v = -65.9$ m/s $\pm$ 8%
  (c) $\lambda = 671 \times 10^{-9}$ m $\pm$ 4%

2.23. ★   A meter stick can be read to the nearest millimeter; a traveling microscope can be read to the nearest 0.1 mm. Suppose you want to measure a length of 2 cm with a precision of 1%. Can you do so with the meter stick? Is it possible to do so with the microscope?

2.24. ★ (a)   A digital voltmeter reads voltages to the nearest thousandth of a volt. What will be its percent uncertainty in measuring a voltage of approximately 3 volts? **(b)** A digital balance reads masses to the nearest hundredth of a gram. What will be its percent uncertainty in measuring a mass of approximately 6 grams?

2.25. ★★   To find the acceleration of a cart, a student measures its initial and final velocities, v_i and v_f, and computes the difference $(v_f - v_i)$. Her data in two separate

Table 2.13. Initial and final velocities (all in cm/s and all $\pm 1\%$); for Problem 2.25.

	v_i	v_f
First run	14.0	18.0
Second run	19.0	19.6

trials are shown in Table 2.13. All have an uncertainty of $\pm 1\%$. **(a)** Calculate the absolute uncertainties in all four measurements; find the change $(v_f - v_i)$ and its uncertainty in each run. **(b)** Compute the percent uncertainty for each of the two values of $(v_f - v_i)$. Your answers, especially for the second run, illustrate the disastrous results of finding a small number by taking the difference of two much larger numbers.

For Section 2.8: Significant Figures and Fractional Uncertainties

2.26. ★ **(a)** A student's calculator shows an answer 123.123. If the student decides that this number actually has only three significant figures, what are its absolute and fractional uncertainties? (To be definite, adopt the convention that a number with N significant figures is uncertain by ± 1 in the Nth digit.) **(b)** Do the same for the number 1231.23. **(c)** Do the same for the number 321.321. **(d)** Do the fractional uncertainties lie in the range expected for three significant figures?

2.27. ★★ **(a)** My calculator gives the answer $x = 6.1234$, but I know that x has a fractional uncertainty of 2%. Restate my answer in the standard form $x_{\text{best}} \pm \delta x$ properly rounded. How many significant figures does the answer really have? **(b)** Do the same for $y = 1.1234$ with a fractional uncertainty of 2%. **(c)** Likewise, for $z = 9.1234$.

For Section 2.9: Multiplying Two Measured Numbers

2.28. ★ **(a)** A student measures two quantities a and b and obtains the results $a = 11.5 \pm 0.2$ cm and $b = 25.4 \pm 0.2$ s. She now calculates the product $q = ab$. Find her answer, giving both its percent and absolute uncertainties, as found using the provisional rule (2.28). **(b)** Repeat part (a) using $a = 5.0$ m $\pm 7\%$ and $b = 3.0$ N $\pm 1\%$.

2.29. ★ **(a)** A student measures two quantities a and b and obtains the results $a = 10 \pm 1$ N and $b = 272 \pm 1$ s. He now calculates the product $q = ab$. Find his answer, giving both its percent and absolute uncertainties, as found using the provisional rule (2.28). **(b)** Repeat part (a) using $a = 3.0$ ft $\pm 8\%$ and $b = 4.0$ lb $\pm 2\%$.

2.30. ★★ A well-known rule states that when two numbers are multiplied together, the answer will be reliable if rounded to the number of significant figures in the less precise of the original two numbers. **(a)** Using our rule (2.28) and the fact that significant figures correspond roughly to fractional uncertainties, prove that this rule is *approximately* valid. (To be definite, treat the case that the less precise number has two significant figures.) **(b)** Show by example that the answer can actually be somewhat less precise than the "well-known" rule suggests. (This reduced precision is especially true if several numbers are multiplied together.)

2.31. ★★ **(a)**   A student measures two numbers x and y as

$$x = 10 \pm 1 \quad \text{and} \quad y = 20 \pm 1.$$

What is her best estimate for their product $q = xy$? Using the largest probable values for x and y (11 and 21), calculate the largest probable value of q. Similarly, find the smallest probable value of q, and hence the range in which q probably lies. Compare your result with that given by the rule (2.28). **(b)** Do the same for the measurements

$$x = 10 \pm 8 \quad \text{and} \quad y = 20 \pm 15.$$

[Remember that the rule (2.28) was derived by assuming that the fractional uncertainties are much less than 1.]

Chapter 3

Propagation of Uncertainties

Most physical quantities usually cannot be measured in a single direct measurement but are instead found in two distinct steps. First, we measure one or more quantities that *can* be measured directly and from which the quantity of interest can be calculated. Second, we use the measured values of these quantities to calculate the quantity of interest itself. For example, to find the area of a rectangle, you actually measure its length *l* and height *h* and then calculate its area *A* as $A = lh$. Similarly, the most obvious way to find the velocity *v* of an object is to measure the distance traveled, *d*, and the time taken, *t*, and then to calculate *v* as $v = d/t$. Any reader with experience in an introductory laboratory can easily think of more examples. In fact, a little thought will show that almost all interesting measurements involve these two distinct steps of direct measurement followed by calculation.

When a measurement involves these two steps, the estimation of uncertainties also involves two steps. We must first estimate the uncertainties in the quantities measured directly and then determine how these uncertainties "propagate" through the calculations to produce an uncertainty in the final answer.[1] This propagation of errors is the main subject of this chapter.

In fact, examples of propagation of errors were presented in Chapter 2. In Section 2.5, I discussed what happens when two numbers *x* and *y* are measured and the results are used to calculate the difference $q = x - y$. We found that the uncertainty in *q* is just the *sum* $\delta q \approx \delta x + \delta y$ of the uncertainties in *x* and *y*. Section 2.9 discussed the product $q = xy$, and Problem 2.13 discussed the sum $q = x + y$. I review these cases in Section 3.3; the rest of this chapter is devoted to more general cases of propagation of uncertainties and includes several examples.

Before I address error propagation in Section 3.3, I will briefly discuss the estimation of uncertainties in quantities measured directly in Sections 3.1 and 3.2. The methods presented in Chapter 1 are reviewed, and further examples are given of error estimation in direct measurements.

Starting in Section 3.3, I will take up the propagation of errors. You will learn that almost all problems in error propagation can be solved using three simple rules.

---

[1] In Chapter 4, I discuss another way in which the final uncertainty can sometimes be estimated. If all measurements can be repeated several times, and if all uncertainties are known to be random in character, then the uncertainty in the quantity of interest can be estimated by examining the spread in answers. Even when this method is possible, it is usually best used as a check on the two-step procedure discussed in this chapter.

A single, more complicated, rule will also be presented that covers all cases and from which the three simpler rules can be derived.

This chapter is long, but its length simply reflects its great importance. Error propagation is a technique you will use repeatedly in the laboratory, and you need to become familiar with the methods described here. The only exception is that the material of Section 3.11 is not used again until Section 5.6; thus, if the ideas of this chapter are all new to you, consider skipping Section 3.11 on your first reading.

3.1   Uncertainties in Direct Measurements

Almost all direct measurements involve reading a scale (on a ruler, clock, or voltmeter, for example) or a digital display (on a digital clock or voltmeter, for example). Some problems in scale reading were discussed in Section 1.5. Sometimes the main sources of uncertainty are the reading of the scale and the need to interpolate between the scale markings. In such situations, a reasonable estimate of the uncertainty is easily made. For example, if you have to measure a clearly defined length l with a ruler graduated in millimeters, you might reasonably decide that the length could be read to the nearest millimeter but no better. Here, the uncertainty δl would be $\delta l = 0.5$ mm. If the scale markings are farther apart (as with tenths of an inch), you might reasonably decide you could read to one-fifth of a division, for example. In any case, the uncertainties associated with the reading of a scale can obviously be estimated quite easily and realistically.

Unfortunately, other sources of uncertainty are frequently much more important than difficulties in scale reading. In measuring the distance between two points, your main problem may be to decide where those two points really are. For example, in an optics experiment, you may wish to measure the distance q from the center of a lens to a focused image, as in Figure 3.1. In practice, the lens is usually several millimeters thick, so locating its center is hard; if the lens comes in a bulky mounting, as it often does, locating the center is even harder. Furthermore, the image may appear to be well-focused throughout a range of many millimeters. Even though the apparatus is mounted on an optical bench that is clearly graduated in millimeters, the uncertainty in the distance from lens to image could easily be a centimeter or so. Since this uncertainty arises because the two points concerned are not clearly defined, this kind of problem is called a *problem of definition*.

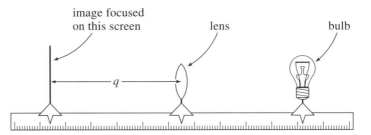

Figure 3.1. An image of the light bulb on the right is focused by the lens onto the screen at the left.

This example illustrates a serious danger in error estimation. If you look only at the scales and forget about other sources of uncertainty, you can badly underestimate the total uncertainty. In fact, the beginning student's most common mistake is to overlook some sources of uncertainty and hence *underestimate* uncertainties, often by a factor of 10 or more. Of course, you must also avoid *overestimating* errors. Experimenters who decide to play safe and to quote generous uncertainties on all measurements may avoid embarrassing inconsistencies, but their measurements may not be of much use. Clearly, the ideal is to find all possible causes of uncertainty and estimate their effects accurately, which is often not quite as hard as it sounds.

Superficially, at least, reading a digital meter is much easier than a conventional analog meter. Unless a digital meter is defective, it should display only significant figures. Thus, it is usually safe to say that the number of significant figures in a digital reading is precisely the number of figures displayed. Unfortunately, as discussed in Section 2.8, the exact meaning of significant figures is not always clear. Thus, a digital voltmeter that tells us that $V = 81$ microvolts could mean that the uncertainty is anything from $\delta V = 0.5$ to $\delta V = 1$ or more. Without a manual to tell you the uncertainty in a digital meter, a reasonable assumption is that the uncertainty in the final digit is ± 1 (so that the voltage just mentioned is $V = 81 \pm 1$).

The digital meter, even more than the analog scale, can give a misleading impression of accuracy. For example, a student might use a digital timer to time the fall of a weight in an Atwood machine or similar device. If the timer displays 8.01 seconds, the time of fall is apparently

$$t = 8.01 \pm 0.01 \text{ s.} \tag{3.1}$$

However, the careful student who repeats the experiment under nearly identical conditions might find a second measurement of 8.41 s; that is,

$$t = 8.41 \pm 0.01 \text{ s.}$$

One likely explanation of this large discrepancy is that uncertainties in the starting procedure vary the initial conditions and hence the time of fall; that is, the measured times really are different. In any case, the accuracy claimed in Equation (3.1) clearly is ridiculously too good. Based on the two measurements made, a more realistic answer would be

$$t = 8.2 \pm 0.2 \text{ s.}$$

In particular, the uncertainty is some 20 times larger than suggested in Equation (3.1) based on the original single reading.

This example brings us to another point mentioned in Chapter 1: Whenever a measurement can be repeated, it should usually be made several times. The resulting spread of values often provides a good indication of the uncertainties, and the average of the values is almost certainly more trustworthy than any one measurement. Chapters 4 and 5 discuss the statistical treatment of multiple measurements. Here, I emphasize only that if a measurement is repeatable, it should be repeated, both to obtain a more reliable answer (by averaging) and, more important, to get an estimate of the uncertainties. Unfortunately, as also mentioned in Chapter 1, repeating a measurement does not always reveal uncertainties. If the measurement is subject to a systematic error, which pushes all results in the same direction (such as a clock that runs slow), the spread in results will not reflect this systematic error. Eliminating such systematic errors requires careful checks of calibration and procedures.

3.2   The Square-Root Rule for a Counting Experiment

Another, different kind of direct measurement has an uncertainty that can be estimated easily. Some experiments require you to count events that occur at random but have a definite average rate. For example, the babies born in a hospital arrive in a fairly random way, but in the long run births in any one hospital probably occur at a definite average rate. Imagine that a demographer who wants to know this rate counts 14 births in a certain two-week period at a local hospital. Based on this result, he would naturally say that his best estimate for the expected number of births in two weeks is 14. Unless he has made a mistake, 14 is *exactly* the number of births in the two-week period he chose to observe. Because of the random way births occur, however, 14 obviously may *not* equal the actual average number of births in all two-week periods. Perhaps this number is 13, 15, or even a fractional number such as 13.5 or 14.7.

Evidently, the uncertainty in this kind of experiment is not in the observed number counted (14 in our example). Instead, the uncertainty is in how well this observed number approximates the true average number. The problem is to estimate how large this uncertainty is. Although I discuss the theory of these counting experiments in Chapter 11, the answer is remarkably simple and is easily stated here: The uncertainty in any counted number of random events, as an estimate of the true average number, is *the square root of the counted number*. In our example, the demographer counted 14 births in a certain two-week period. Therefore, his uncertainty is $\sqrt{14} \approx 4$, and his final conclusion would be

(average births in a two-week period) $= 14 \pm 4$.

To make this statement more general, suppose we count the occurrences of any event (such as the births of babies in a hospital) that occurs randomly but at a definite average rate. Suppose we count for a chosen time interval T (such as two weeks), and we denote the number of observed events by the Greek letter ν. (Pronounced "nu," this symbol is the Greek form of the letter n and stands for *number*.) Based on this experiment, our best estimate for the average number of events in time T is, of course, the observed number ν, and the uncertainty in this estimate is the square root of the number, that is, $\sqrt{\nu}$. Therefore, our answer for the average number of events in time T is

$$\text{(average number of events in time } T) = \nu \pm \sqrt{\nu}. \tag{3.2}$$

I refer to this important result as the *Square-Root Rule for Counting Experiments*.

Counting experiments of this type occur frequently in the physics laboratory. The most prominent example is in the study of radioactivity. In a radioactive material, each nucleus decays at a random time, but the decays in a large sample occur at a definite average rate. To find this rate, you can simply count the number ν of decays in some convenient time interval T; the expected number of decays in time T, with its uncertainty, is then given by the square-root rule, (3.2).

Quick Check 3.1. **(a)** To check the activity of a radioactive sample, an inspector places the sample in a liquid scintillation counter to count the number of decays in a two-minute interval and obtains 33 counts. What should he report as the number of decays produced by the sample in two minutes? **(b)** Suppose, instead, he had monitored the same sample for 50 minutes and obtained 907 counts. What would be his answer for the number of decays in 50 minutes? **(c)** Find the percent uncertainties in these two measurements, and comment on the usefulness of counting for a longer period as in part (b).

3.3   Sums and Differences; Products and Quotients

For the remainder of this chapter, I will suppose that we have measured one or more quantities $x, y, \ldots$, with corresponding uncertainties $\delta x, \delta y, \ldots$, and that we now wish to use the measured values of $x, y, \ldots$, to calculate the quantity of real interest, q. The calculation of q is usually straightforward; the problem is how the uncertainties, $\delta x, \delta y, \ldots$, propagate through the calculation and lead to an uncertainty δq in the final value of q.

SUMS AND DIFFERENCES

Chapter 2 discussed what happens when you measure two quantities x and y and calculate their sum, $x + y$, or their difference, $x - y$. To estimate the uncertainty in the sum or difference, we had only to decide on their highest and lowest probable values. The highest and lowest probable values of x are $x_{best} \pm \delta x$, and those of y are $y_{best} \pm \delta y$. Hence, the highest probable value of $x + y$ is

$$x_{best} + y_{best} + (\delta x + \delta y),$$

and the lowest probable value is

$$x_{best} + y_{best} - (\delta x + \delta y).$$

Thus, the best estimate for $q = x + y$ is

$$q_{best} = x_{best} + y_{best},$$

and its uncertainty is

$$\delta q \approx \delta x + \delta y. \tag{3.3}$$

A similar argument (be sure you can reconstruct it) shows that the uncertainty in the *difference* $x - y$ is given by the same formula (3.3). That is, the uncertainty in either the sum $x + y$ or the difference $x - y$ is the sum $\delta x + \delta y$ of the uncertainties in x and y.

If we have several numbers $x, \ldots, w$ to be added or subtracted, then repeated application of (3.3) gives the following provisional rule.

Uncertainty in Sums and Differences (Provisional Rule)

If several quantities $x, \ldots, w$ are measured with uncertainties $\delta x, \ldots, \delta w$, and the measured values used to compute

$$q = x + \cdots + z - (u + \cdots + w),$$

then the uncertainty in the computed value of q is the sum,

$$\delta q \approx \delta x + \cdots + \delta z + \delta u + \cdots + \delta w, \qquad (3.4)$$

of all the original uncertainties.

In other words, when you add or subtract any number of quantities, the uncertainties in those quantities always *add*. As before, I use the sign $\approx$ to emphasize that this rule is only provisional.

Example: Adding and Subtracting Masses

As a simple example of rule (3.4), suppose an experimenter mixes together the liquids in two flasks, having first measured their separate masses when full and empty, as follows:

M_1 = mass of first flask and contents   = 540 ± 10 grams

m_1 = mass of first flask empty   = 72 ± 1 grams

M_2 = mass of second flask and contents = 940 ± 20 grams

m_2 = mass of second flask empty   = 97 ± 1 grams

He now calculates the total mass of liquid as

$$\begin{aligned} M &= M_1 - m_1 + M_2 - m_2 \\ &= (540 - 72 + 940 - 97) \text{ grams } = 1{,}311 \text{ grams.} \end{aligned}$$

According to rule (3.4), the uncertainty in this answer is the sum of all four uncertainties,

$$\delta M \approx \delta M_1 + \delta m_1 + \delta M_2 + \delta m_2 = (10 + 1 + 20 + 1) \text{ grams} = 32 \text{ grams.}$$

Thus, his final answer (properly rounded) is

$$\text{total mass of liquid } = 1{,}310 \pm 30 \text{ grams.}$$

Notice how the much smaller uncertainties in the masses of the empty flasks made a negligible contribution to the final uncertainty. This effect is important, and we will discuss it later on. With experience, you can learn to identify in advance those uncertainties that are negligible and can be ignored from the outset. Often, this can greatly simplify the calculation of uncertainties.

PRODUCTS AND QUOTIENTS

Section 2.9 discussed the uncertainty in the product $q = xy$ of two measured quantities. We saw that, provided the fractional uncertainties concerned are small, the fractional uncertainty in $q = xy$ is the sum of the fractional uncertainties in x and y. Rather than review the derivation of this result, I discuss here the similar case of the quotient $q = x/y$. As you will see, the uncertainty in a quotient is given by the same rule as for a product; that is, the fractional uncertainty in $q = x/y$ is equal to the sum of the fractional uncertainties in x and y.

Because uncertainties in products and quotients are best expressed in terms of fractional uncertainties, a shorthand notation for the latter will be helpful. Recall that if we measure some quantity x as

$$\text{(measured value of } x) \ = \ x_{\text{best}} \pm \delta x$$

in the usual way, then the fractional uncertainty in x is defined to be

$$\text{(fractional uncertainty in } x) \ = \ \frac{\delta x}{|x_{\text{best}}|}.$$

(The absolute value in the denominator ensures that the fractional uncertainty is always positive, even when x_{best} is negative.) Because the symbol $\delta x / |x_{\text{best}}|$ is clumsy to write and read, from now on I will abbreviate it by omitting the subscript "best" and writing

$$\text{(fractional uncertainty in } x) \ = \ \frac{\delta x}{|x|}.$$

The result of measuring any quantity x can be expressed in terms of its fractional error $\delta x / |x|$ as

$$\text{(value of } x) \ = \ x_{\text{best}}(1 \pm \delta x / |x|).$$

Therefore, the value of $q = x/y$ can be written as

$$\text{(value of } q) \ = \ \frac{x_{\text{best}}}{y_{\text{best}}} \frac{1 \pm \delta x / |x|}{1 \pm \delta y / |y|}.$$

Our problem now is to find the extreme probable values of the second factor on the right. This factor is largest, for example, if the numerator has its largest value, $1 + \delta x / |x|$, and the denominator has its *smallest* value, $1 - \delta y / |y|$. Thus, the largest probable value for $q = x/y$ is

$$(\text{largest value of } q) \; = \; \frac{x_{\text{best}}}{y_{\text{best}}} \frac{1 + \delta x/|x|}{1 - \delta y/|y|}. \tag{3.5}$$

The last factor in expression (3.5) has the form $(1 + a)/(1 - b)$, where the numbers a and b are normally small (that is, much less than 1). It can be simplified by two approximations. First, because b is small, the binomial theorem[2] implies that

$$\frac{1}{(1 - b)} \; \approx \; 1 + b. \tag{3.6}$$

Therefore,

$$\frac{1 + a}{1 - b} \; \approx \; (1 + a)(1 + b) \; = \; 1 + a + b + ab$$
$$\approx \; 1 + a + b,$$

where, in the second line, we have neglected the product ab of two small quantities. Returning to (3.5) and using these approximations, we find for the largest probable value of $q = x/y$

$$(\text{largest value of } q) \; = \; \frac{x_{\text{best}}}{y_{\text{best}}} \left(1 + \frac{\delta x}{|x|} + \frac{\delta y}{|y|} \right).$$

A similar calculation shows that the smallest probable value is given by a similar expression with two minus signs. Combining these two, we find that

$$(\text{value of } q) \; = \; \frac{x_{\text{best}}}{y_{\text{best}}} \left(1 \pm \left[\frac{\delta x}{|x|} + \frac{\delta y}{|y|} \right] \right).$$

Comparing this equation with the standard form,

$$(\text{value of } q) \; = \; q_{\text{best}} \left(1 \pm \frac{\delta q}{|q|} \right),$$

we see that the best value for q is $q_{\text{best}} = x_{\text{best}}/y_{\text{best}}$, as we would expect, and that the fractional uncertainty is

$$\frac{\delta q}{|q|} \; \approx \; \frac{\delta x}{|x|} + \frac{\delta y}{|y|}. \tag{3.7}$$

We conclude that when we divide or multiply two measured quantities x and y, the fractional uncertainty in the answer is the sum of the fractional uncertainties in x and y, as in (3.7). If we now multiply or divide a series of numbers, repeated application of this result leads to the following provisional rule.

---

[2] The binomial theorem expresses $1/(1 - b)$ as the infinite series $1 + b + b^2 + \cdots$. If b is much less than 1, then $1/(1-b) \approx 1 + b$ as in (3.6). If you are unfamiliar with the binomial theorem, you can find more details in Problem 3.8.

**Uncertainty in Products and Quotients
(Provisional Rule)**

If several quantities $x, \ldots, w$ are measured with small uncertainties $\delta x, \ldots, \delta w$, and the measured values are used to compute

$$q = \frac{x \times \cdots \times z}{u \times \cdots \times w},$$

then the fractional uncertainty in the computed value of q is the sum,

$$\frac{\delta q}{|q|} \approx \frac{\delta x}{|x|} + \cdots + \frac{\delta z}{|z|} + \frac{\delta u}{|u|} + \cdots + \frac{\delta w}{|w|}, \qquad (3.8)$$

of the fractional uncertainties in $x, \ldots, w$.

Briefly, when quantities are multiplied or divided the *fractional uncertainties add.*

Example: A Problem in Surveying

In surveying, sometimes a value can be found for an inaccessible length l (such as the height of a tall tree) by measuring three other lengths l_1, l_2, l_3 in terms of which

$$l = \frac{l_1 l_2}{l_3}.$$

Suppose we perform such an experiment and obtain the following results (in feet):

$$l_1 = 200 \pm 2, \qquad l_2 = 5.5 \pm 0.1, \qquad l_3 = 10.0 \pm 0.4.$$

Our best estimate for l is

$$l_{\text{best}} = \frac{200 \times 5.5}{10.0} = 110 \text{ ft.}$$

According to (3.8), the fractional uncertainty in this answer is the sum of the fractional uncertainties in l_1, l_2, and l_3, which are 1%, 2%, and 4%, respectively. Thus

$$\frac{\delta l}{l} \approx \frac{\delta l_1}{l_1} + \frac{\delta l_2}{l_2} + \frac{\delta l_3}{l_3} = (1 + 2 + 4)\%$$
$$= 7\%,$$

and our final answer is

$$l = 110 \pm 8 \text{ ft.}$$

Quick Check 3.2. Suppose you measure the three quantities x, y, and z as follows:

$$x = 8.0 \pm 0.2, \qquad y = 5.0 \pm 0.1, \qquad z = 4.0 \pm 0.1.$$

Express the given uncertainties as percentages, and then calculate $q = xy/z$ with its uncertainty δq [as given by the provisional rule (3.8)].

3.4 Two Important Special Cases

Two important special cases of the rule (3.8) deserve mention. One concerns the product of two numbers, one of which has *no* uncertainty; the other involves a power (such as x^3) of a measured number.

MEASURED QUANTITY TIMES EXACT NUMBER

Suppose we measure a quantity x and then use the measured value to calculate the product $q = Bx$, where the number B has *no uncertainty*. For example, we might measure the diameter of a circle and then calculate its circumference, $c = \pi \times d$; or we might measure the thickness T of 200 identical sheets of paper and then calculate the thickness of a single sheet as $t = (1/200) \times T$. According to the rule (3.8), the fractional uncertainty in $q = Bx$ is the sum of the fractional uncertainties in B and x. Because $\delta B = 0$, this implies that

$$\frac{\delta q}{|q|} = \frac{\delta x}{|x|}.$$

That is, the fractional uncertainty in $q = Bx$ (with B known exactly) is the same as that in x. We can express this result differently if we multiply through by $|q| = |Bx|$ to give $\delta q = |B| \delta x$, and we have the following useful rule:[3]

Measured Quantity Times Exact Number

If the quantity x is measured with uncertainty δx and is used to compute the product

$$q = Bx,$$

where B has no uncertainty, then the uncertainty in q is just $|B|$ times that in x,

$$\delta q = |B| \delta x. \tag{3.9}$$

---

[3] This rule (3.9) was derived from the rule (3.8), which is provisional and will be replaced by the more complete rules (3.18) and (3.19). Fortunately, the same conclusion (3.9) follows from these improved rules. Thus (3.9) is already in its final form.

This rule is especially useful in measuring something inconveniently small but available many times over, such as the thickness of a sheet of paper or the time for a revolution of a rapidly spinning wheel. For example, if we measure the thickness T of 200 sheets of paper and get the answer

$$\text{(thickness of 200 sheets)} = T = 1.3 \pm 0.1 \text{ inches,}$$

it immediately follows that the thickness t of a single sheet is

$$\text{(thickness of one sheet)} = t = \frac{1}{200} \times T$$
$$= 0.0065 \pm 0.0005 \text{ inches.}$$

Notice how this technique (measuring the thickness of several identical sheets and dividing by their number) makes easily possible a measurement that would otherwise require quite sophisticated equipment and that this technique gives a remarkably small uncertainty. Of course, the sheets must be known to be equally thick.

Quick Check 3.3.   Suppose you measure the diameter of a circle as

$$d = 5.0 \pm 0.1 \text{ cm}$$

and use this value to calculate the circumference $c = \pi d$. What is your answer, with its uncertainty?

POWERS

The second special case of the rule (3.8) concerns the evaluation of a power of some measured quantity. For example, we might measure the speed v of some object and then, to find its kinetic energy $\frac{1}{2}mv^2$, calculate the square v^2. Because v^2 is just $v \times v$, it follows from (3.8) that the fractional uncertainty in v^2 is *twice* the fractional uncertainty in v. More generally, from (3.8) the general rule for any power is clearly as follows.

Uncertainty in a Power

If the quantity x is measured with uncertainty δx and the measured value is used to compute the power

$$q = x^n,$$

then the fractional uncertainty in q is n times that in x,

$$\frac{\delta q}{|q|} = n \frac{\delta x}{|x|}. \tag{3.10}$$

The derivation of this rule required that n be a positive integer. In fact, however, the rule generalizes to include *any* exponent n, as we will see later in Equation (3.26).

Quick Check 3.4.   To find the volume of a certain cube, you measure its side as 2.00 ± 0.02 cm. Convert this uncertainty to a percent and then find the volume with its uncertainty.

Example: Measurement of g

Suppose a student measures g, the acceleration of gravity, by measuring the time t for a stone to fall from a height h above the ground. After making several timings, she concludes that

$$t = 1.6 \pm 0.1 \text{ s},$$

and she measures the height h as

$$h = 46.2 \pm 0.3 \text{ ft}.$$

Because h is given by the well-known formula $h = \frac{1}{2}gt^2$, she now calculates g as

$$g = \frac{2h}{t^2}$$

$$= \frac{2 \times 46.2 \text{ ft}}{(1.6 \text{ s})^2} = 36.1 \text{ ft/s}^2.$$

What is the uncertainty in her answer?

The uncertainty in her answer can be found by using the rules just developed. To this end, we need to know the fractional uncertainties in each of the factors in the expression $g = 2h/t^2$ used to calculate g. The factor 2 has no uncertainty. The fractional uncertainties in h and t are

$$\frac{\delta h}{h} = \frac{0.3}{46.2} = 0.7\%$$

and

$$\frac{\delta t}{t} = \frac{0.1}{1.6} = 6.3\%.$$

According to the rule (3.10), the fractional uncertainty of t^2 is twice that of t. Therefore, applying the rule (3.8) for products and quotients to the formula $g = 2h/t^2$, we find the fractional uncertainty

$$\frac{\delta g}{g} = \frac{\delta h}{h} + 2\frac{\delta t}{t}$$

$$= 0.7\% + 2 \times (6.3\%) = 13.3\%, \qquad (3.11)$$

and hence the uncertainty

$$\delta g = (36.1 \text{ ft/s}^2) \times \frac{13.3}{100} = 4.80 \text{ ft/s}^2.$$

Thus, our student's final answer (properly rounded) is

$$g = 36 \pm 5 \text{ ft/s}^2.$$

This example illustrates how simple the estimation of uncertainties can often be. It also illustrates how error analysis tells you not only the size of uncertainties but also how to reduce them. In this example, (3.11) shows that the largest contribution comes from the measurement of the time. If we want a more precise value of g, then the measurement of t must be improved; any attempt to improve the measurement of h will be wasted effort.

Finally, the accepted value of g is 32 ft/s^2, which lies within our student's margins of error. Thus, she can conclude that her measurement, although not especially accurate, is perfectly consistent with the known value of g.

3.5   Independent Uncertainties in a Sum

The rules presented thus far can be summarized quickly: When measured quantities are added or subtracted, the *uncertainties add;* when measured quantities are multiplied or divided, the *fractional uncertainties add.* In this and the next section, I discuss how, under certain conditions, the uncertainties calculated by using these rules may be unnecessarily large. Specifically, you will see that if the original uncertainties are *independent* and *random,* a more realistic (and smaller) estimate of the final uncertainty is given by similar rules in which the uncertainties (or fractional uncertainties) are *added in quadrature* (a procedure defined shortly).

Let us first consider computing the sum, $q = x + y$, of two numbers x and y that have been measured in the standard form

$$(\text{measured value of } x) = x_{\text{best}} \pm \delta x,$$

with a similar expression for y. The argument used in the last section was as follows: First, the best estimate for $q = x + y$ is obviously $q_{\text{best}} = x_{\text{best}} + y_{\text{best}}$. Second, since the highest probable values for x and y are $x_{\text{best}} + \delta x$ and $y_{\text{best}} + \delta y$, the highest probable value for q is

$$x_{\text{best}} + y_{\text{best}} + \delta x + \delta y. \tag{3.12}$$

Similarly, the lowest probable value of q is

$$x_{\text{best}} + y_{\text{best}} - \delta x - \delta y.$$

Therefore, we concluded, the value of q probably lies between these two numbers, and the uncertainty in q is

$$\delta q \approx \delta x + \delta y.$$

To see why this formula is likely to overestimate δq, let us consider how the actual value of q could equal the highest extreme (3.12). Obviously, this occurs if we have underestimated x by the full amount δx *and* underestimated y by the full δy, obviously, a fairly unlikely event. If x and y are measured independently and our errors are random in nature, we have a 50% chance that an *underestimate* of x is accompanied by an *overestimate* of y, or *vice versa*. Clearly, then, the probability we will underestimate both x and y by the full amounts δx and δy is fairly small. Therefore, the value $\delta q \approx \delta x + \delta y$ overstates our probable error.

What constitutes a better estimate of δq? The answer depends on precisely what we mean by uncertainties (that is, what we mean by the statement that q is "probably" somewhere between $q_{\text{best}} - \delta q$ and $q_{\text{best}} + \delta q$). It also depends on the statistical laws governing our errors in measurement. Chapter 5 discusses the normal, or Gauss, distribution, which describes measurements subject to random uncertainties. It shows that if the measurements of x and y are made independently and are both governed by the normal distribution, then the uncertainty in $q = x + y$ is given by

$$\delta q = \sqrt{(\delta x)^2 + (\delta y)^2}. \tag{3.13}$$

When we combine two numbers by squaring them, adding the squares, and taking the square root, as in (3.13), the numbers are said to be *added in quadrature*. Thus, the rule embodied in (3.13) can be stated as follows: If the measurements of x and y are independent and subject only to random uncertainties, then the uncertainty δq in the calculated value of $q = x + y$ is the *sum in quadrature* or *quadratic sum* of the uncertainties δx and δy.

Compare the new expression (3.13) for the uncertainty in $q = x + y$ with our old expression,

$$\delta q \approx \delta x + \delta y. \tag{3.14}$$

First, the new expression (3.13) is always smaller than the old (3.14), as we can see from a simple geometrical argument: For any two positive numbers a and b, the numbers a, b, and $\sqrt{a^2 + b^2}$ are the three sides of a right-angled triangle (Figure 3.2). Because the length of any side of a triangle is always less than the sum of the

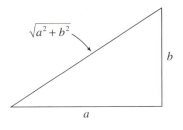

Figure 3.2. Because any side of a triangle is less than the sum of the other two sides, the inequality $\sqrt{a^2 + b^2} < a + b$ is always true.

other two sides, it follows that $\sqrt{a^2 + b^2} < a + b$ and hence that (3.13) is always less than (3.14).

Because expression (3.13) for the uncertainty in $q = x + y$ is always smaller than (3.14), you should always use (3.13) *when* it is applicable. It is, however, *not* always applicable. Expression (3.13) reflects the possibility that an overestimate of x can be offset by an underestimate of y or vice versa, but there are measurements for which this cancellation is not possible.

Suppose, for example, that $q = x + y$ is the sum of two lengths x and y measured with the same steel tape. Suppose further that the main source of uncertainty is our fear that the tape was designed for use at a temperature different from the present temperature. If we don't know this temperature (and don't have a reliable tape for comparison), we have to recognize that our tape may be longer or shorter than its calibrated length and hence may yield readings under or over the correct length. This uncertainty can be easily allowed for.[4] The point, however, is that if the tape is too long, then we *underestimate both x and y*; and if the tape is too short, we *overestimate both x and y*. Thus, there is no possibility for the cancellations that justified using the sum in quadrature to compute the uncertainty in $q = x + y$.

I will prove later (in Chapter 9) that, whether or not our errors are independent and random, the uncertainty in $q = x + y$ is *certainly no larger* than the simple sum $\delta x + \delta y$:

$$\delta q \leq \delta x + \delta y. \tag{3.15}$$

That is, our old expression (3.14) for δq is actually an *upper bound* that holds in all cases. If we have any reason to suspect the errors in x and y are *not* independent and random (as in the example of the steel tape measure), we are not justified in using the quadratic sum (3.13) for δq. On the other hand, the bound (3.15) guarantees that δq is certainly no worse than $\delta x + \delta y$, and our safest course is to use the old rule

$$\delta q \approx \delta x + \delta y.$$

Often, whether uncertainties are added in quadrature or directly makes little difference. For example, suppose that x and y are lengths both measured with uncertainties $\delta x = \delta y = 2$ mm. If we are sure these uncertainties are independent and random, we would estimate the error in $x + y$ to be the sum in quadrature,

$$\sqrt{(\delta x)^2 + (\delta y)^2} = \sqrt{4 + 4}\ \text{mm} = 2.8\ \text{mm} \approx 3\ \text{mm},$$

but if we suspect that the uncertainties may not be independent, we would have to use the ordinary sum,

$$\delta x + \delta y \approx (2 + 2)\ \text{mm} = 4\ \text{mm}.$$

In many experiments, the estimation of uncertainties is so crude that the difference between these two answers (3 mm and 4 mm) is unimportant. On the other hand, sometimes the sum in quadrature is significantly smaller than the ordinary sum. Also, rather surprisingly, the sum in quadrature is sometimes easier to compute than the ordinary sum. Examples of these effects are given in the next section.

---

[4] Suppose, for example, that the tape has a coefficient of expansion $\alpha = 10^{-5}$ per degree and that we decide that the difference between its calibration temperature and the present temperature is unlikely to be more than 10 degrees. The tape is then unlikely to be more than 10^{-4}, or 0.01%, away from its correct length, and our uncertainty is therefore 0.01%.

Quick Check 3.5. Suppose you measure the volumes of water in two beakers as

$$V_1 = 130 \pm 6 \text{ ml} \quad \text{and} \quad V_2 = 65 \pm 4 \text{ ml}$$

and then carefully pour the contents of the first into the second. What is your prediction for the total volume $V = V_1 + V_2$ with its uncertainty, δV, assuming the original uncertainties are independent and random? What would you give for δV if you suspected the original uncertainties were not independent?

3.6   More About Independent Uncertainties

In the previous section, I discussed how independent random uncertainties in two quantities x and y propagate to cause an uncertainty in the sum $x + y$. We saw that for this type of uncertainty the two errors should be added in quadrature. We can naturally consider the corresponding problem for differences, products, and quotients. As we will see in Section 5.6, in all cases our previous rules (3.4) and (3.8) are modified only in that the sums of errors (or fractional errors) are replaced by quadratic sums. Further, the old expressions (3.4) and (3.8) will be proven to be upper bounds that always hold whether or not the uncertainties are independent and random. Thus, the final versions of our two main rules are as follows:

Uncertainty in Sums and Differences

Suppose that $x, \ldots, w$ are measured with uncertainties δx, $\ldots, \delta w$ and the measured values used to compute

$$q = x + \cdots + z - (u + \cdots + w).$$

If the uncertainties in $x, \ldots, w$ are known to be *independent and random,* then the uncertainty in q is the quadratic sum

$$\delta q = \sqrt{(\delta x)^2 + \cdots + (\delta z)^2 + (\delta u)^2 + \cdots + (\delta w)^2} \quad (3.16)$$

of the original uncertainties. In any case, δq is never larger than their ordinary sum,

$$\delta q \leq \delta x + \cdots + \delta z + \delta u + \cdots + \delta w. \quad (3.17)$$

---

and

Uncertainties in Products and Quotients

Suppose that $x, \ldots, w$ are measured with uncertainties $\delta x, \ldots, \delta w$, and the measured values are used to compute

$$q = \frac{x \times \cdots \times z}{u \times \cdots \times w}.$$

If the uncertainties in $x, \ldots, w$ are *independent and random*, then the fractional uncertainty in q is the sum in quadrature of the original fractional uncertainties,

$$\frac{\delta q}{|q|} = \sqrt{\left(\frac{\delta x}{x}\right)^2 + \cdots + \left(\frac{\delta z}{z}\right)^2 + \left(\frac{\delta u}{u}\right)^2 + \cdots + \left(\frac{\delta w}{w}\right)^2}. \tag{3.18}$$

In any case, it is never larger than their ordinary sum,

$$\frac{\delta q}{|q|} \leq \frac{\delta x}{|x|} + \cdots + \frac{\delta z}{|z|} + \frac{\delta u}{|u|} + \cdots + \frac{\delta w}{|w|}. \tag{3.19}$$

Notice that I have not yet justified the use of addition in quadrature for independent random uncertainties. I have argued only that when the various uncertainties are independent and random, there is a good chance of partial cancellations of errors and that the resulting uncertainty (or fractional uncertainty) should be smaller than the simple sum of the original uncertainties (or fractional uncertainties); the sum in quadrature does have this property. I give a proper justification of its use in Chapter 5. The bounds (3.17) and (3.19) are proved in Chapter 9.

Example: Straight Addition vs Addition in Quadrature

As discussed, sometimes there is no significant difference between uncertainties computed by addition in quadrature and those computed by straight addition. Often, however, there is a significant difference, and—surprisingly enough—the sum in quadrature is often much simpler to compute. To see how this situation can arise, consider the following example.

Suppose we want to find the efficiency of a D.C. electric motor by using it to lift a mass m through a height h. The work accomplished is mgh, and the electric energy delivered to the motor is VIt, where V is the applied voltage, I the current, and t the time for which the motor runs. The efficiency is then

$$\text{efficiency, } e = \frac{\text{work done by motor}}{\text{energy delivered to motor}} = \frac{mgh}{VIt}.$$

Let us suppose that $m, h, V,$ and I can all be measured with 1% accuracy,

$$\text{(fractional uncertainty for } m, h, V, \text{ and } I) = 1\%,$$

and that the time t has an uncertainty of 5%,

$$(\text{fractional uncertainty for } t) = 5\%.$$

(Of course, g is known with negligible uncertainty.) If we now compute the efficiency e, then according to our old rule ("fractional errors add"), we have an uncertainty

$$\frac{\delta e}{e} \approx \frac{\delta m}{m} + \frac{\delta h}{h} + \frac{\delta V}{V} + \frac{\delta I}{I} + \frac{\delta t}{t}$$
$$= (1 + 1 + 1 + 1 + 5)\% = 9\%.$$

On the other hand, if we are confident that the various uncertainties are independent and random, then we can compute $\delta e/e$ by the quadratic sum to give

$$\frac{\delta e}{e} = \sqrt{\left(\frac{\delta m}{m}\right)^2 + \left(\frac{\delta h}{h}\right)^2 + \left(\frac{\delta V}{V}\right)^2 + \left(\frac{\delta I}{I}\right)^2 + \left(\frac{\delta t}{t}\right)^2}$$
$$= \sqrt{(1\%)^2 + (1\%)^2 + (1\%)^2 + (1\%)^2 + (5\%)^2}$$
$$= \sqrt{29}\% \approx 5\%.$$

Clearly, the quadratic sum leads to a significantly smaller estimate for δe. Furthermore, to one significant figure, the uncertainties in m, h, V, and I *make no contribution at all* to the uncertainty in e computed in this way; that is, to one significant figure, we have found (in this example)

$$\frac{\delta e}{e} = \frac{\delta t}{t}.$$

This striking simplification is easily understood. When numbers are added in quadrature, they are squared first and then summed. The process of squaring greatly exaggerates the importance of the larger numbers. Thus, if one number is 5 times any of the others (as in our example), its square is 25 times that of the others, and we can usually neglect the others entirely.

This example illustrates how combining errors in quadrature is usually better and often easier than computing them by straight addition. The example also illustrates the type of problem in which the errors *are* independent and for which addition in quadrature is justified. (For the moment I take for granted that the errors are random and will discuss this more difficult point in Chapter 4.) The five quantities measured (m, h, V, I, and t) are physically distinct quantities with different units and are measured by entirely different processes. For the sources of error in any quantity to be correlated with those in any other is almost inconceivable. Therefore, the errors can reasonably be treated as independent and combined in quadrature.

Quick Check 3.6.   Suppose you measure three numbers as follows:

$$x = 200 \pm 2, \quad y = 50 \pm 2, \quad z = 20 \pm 1,$$

where the three uncertainties are independent and random. What would you give for the values of $q = x + y - z$ and $r = xy/z$ with their uncertainties?

3.7 Arbitrary Functions of One Variable

You have now seen how uncertainties, both independent and otherwise, propagate through sums, differences, products, and quotients. However, many calculations require more complicated operations, such as computation of a sine, cosine, or square root, and you will need to know how uncertainties propagate in these cases.

As an example, imagine finding the refractive index n of glass by measuring the critical angle θ. We know from elementary optics that $n = 1/\sin\theta$. Therefore, if we can measure the angle θ, we can easily calculate the refractive index n, but we must then decide what uncertainty δn in $n = 1/\sin\theta$ results from the uncertainty $\delta\theta$ in our measurement of θ.

More generally, suppose we have measured a quantity x in the standard form $x_{best} \pm \delta x$ and want to calculate some known function $q(x)$, such as $q(x) = 1/\sin x$ or $q(x) = \sqrt{x}$. A simple way to think about this calculation is to draw a graph of $q(x)$ as in Figure 3.3. The best estimate for $q(x)$ is, of course, $q_{best} = q(x_{best})$, and the values x_{best} and q_{best} are shown connected by the heavy lines in Figure 3.3.

To decide on the uncertainty δq, we employ the usual argument. The largest probable value of x is $x_{best} + \delta x$; using the graph, we can immediately find the largest probable value of q, which is shown as q_{max}. Similarly, we can draw in the smallest probable value, q_{min}, as shown. If the uncertainty δx is small (as we always suppose it is), then the section of graph involved in this construction is approximately straight, and q_{max} and q_{min} are easily seen to be equally spaced on either side of q_{best}. The uncertainty δq can then be taken from the graph as either of the lengths shown, and we have found the value of q in the standard form $q_{best} \pm \delta q$.

Occasionally, uncertainties are calculated from a graph as just described. (See Problems 3.26 and 3.30 for examples.) Usually, however, the function $q(x)$ is known

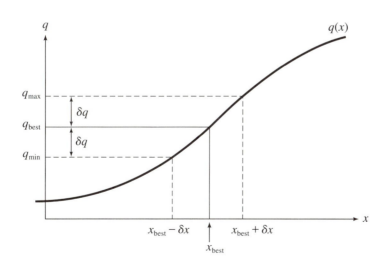

Figure 3.3. Graph of $q(x)$ vs x. If x is measured as $x_{best} \pm \delta x$, then the best estimate for $q(x)$ is $q_{best} = q(x_{best})$. The largest and smallest probable values of $q(x)$ correspond to the values $x_{best} \pm \delta x$ of x.

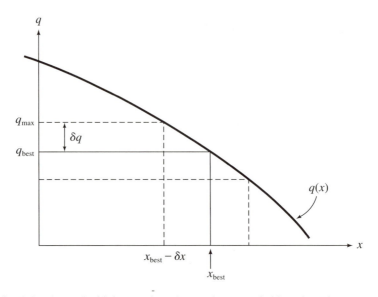

Figure 3.4. If the slope of $q(x)$ is negative, the maximum probable value of q corresponds to the minimum value of x, and vice versa.

explicitly—$q(x) = \sin x$ or $q(x) = \sqrt{x}$, for example—and the uncertainty δq can be calculated analytically. From Figure 3.3, we see that

$$\delta q = q(x_{\text{best}} + \delta x) - q(x_{\text{best}}). \tag{3.20}$$

Now, a fundamental approximation of calculus asserts that, for any function $q(x)$ and any sufficiently small increment u,

$$q(x + u) - q(x) = \frac{dq}{dx} u.$$

Thus, provided the uncertainty δx is small (as we always assume it is), we can rewrite the difference in (3.20) to give

$$\delta q = \frac{dq}{dx} \delta x. \tag{3.21}$$

Thus, to find the uncertainty δq, we just calculate the derivative dq/dx and multiply by the uncertainty δx.

The rule (3.21) is not quite in its final form. It was derived for a function, like that of Figure 3.3, whose slope is positive. Figure 3.4 shows a function with negative slope. Here, the maximum probable value q_{max} obviously corresponds to the minimum value of x, so that

$$\delta q = -\frac{dq}{dx} \delta x. \tag{3.22}$$

Because dq/dx is negative, we can write $-dq/dx$ as $|dq/dx|$, and we have the following general rule.

Uncertainty in Any Function of One Variable

If x is measured with uncertainty δx and is used to calculate the function $q(x)$, then the uncertainty δq is

$$\delta q = \left| \frac{dq}{dx} \right| \delta x. \tag{3.23}$$

This rule usually allows us to find δq quickly and easily. Occasionally, if $q(x)$ is very complicated, evaluating its derivative may be a nuisance, and going back to (3.20) is sometimes easier, as we discuss in Problem 3.32. Particularly if you have programmed your calculator or computer to find $q(x)$, then finding $q(x_{\text{best}} + \delta x)$ and $q(x_{\text{best}})$ and their difference may be easier than differentiating $q(x)$ explicitly.

Example: Uncertainty in a Cosine

As a simple application of the rule (3.23), suppose we have measured an angle θ as

$$\theta = 20 \pm 3°$$

and that we wish to find $\cos \theta$. Our best estimate of $\cos \theta$ is, of course, $\cos 20° = 0.94$, and according to (3.23), the uncertainty is

$$\delta(\cos \theta) = \left| \frac{d \cos \theta}{d\theta} \right| \delta \theta$$
$$= |\sin \theta| \, \delta \theta \text{ (in rad).} \tag{3.24}$$

We have indicated that $\delta \theta$ must be expressed in radians, because the derivative of $\cos \theta$ is $-\sin \theta$ only if θ is expressed in radians. Therefore, we rewrite $\delta \theta = 3°$ as $\delta \theta = 0.05$ rad; then (3.24) gives

$$\delta(\cos \theta) = (\sin 20°) \times 0.05$$
$$= 0.34 \times 0.05 = 0.02.$$

Thus, our final answer is

$$\cos \theta = 0.94 \pm 0.02.$$

Quick Check 3.7. Suppose you measure x as 3.0 ± 0.1 and then calculate $q = e^x$. What is your answer, with its uncertainty? (Remember that the derivative of e^x is e^x.)

As another example of the rule (3.23), we can rederive and generalize a result found in Section 3.4. Suppose we measure the quantity x and then calculate the power $q(x) = x^n$, where n is any known, fixed number, positive or negative. According to (3.23), the resulting uncertainty in q is

$$\delta q \; = \; \left| \frac{dq}{dx} \right| \delta x = |nx^{n-1}|\delta x.$$

If we divide both sides of this equation by $|q| = |x^n|$, we find that

$$\frac{\delta q}{|q|} \; = \; |n| \frac{\delta x}{|x|} \; ; \tag{3.25}$$

that is, the fractional uncertainty in $q = x^n$ is $|n|$ times that in x. This result (3.25) is just the rule (3.10) found earlier, except that the result here is more general, because n can now be any number. For example, if $n = 1/2$, then $q = \sqrt{x}$, and

$$\frac{\delta q}{|q|} \; = \; \frac{1}{2} \frac{\delta x}{|x|} \; ;$$

that is, the fractional uncertainty in $\sqrt{x}$ is *half* that in x itself. Similarly, the fractional uncertainty in $1/x = x^{-1}$ is the same as that in x itself.

The result (3.25) is just a special case of the rule (3.23). It is sufficiently important, however, to deserve separate statement as the following general rule.

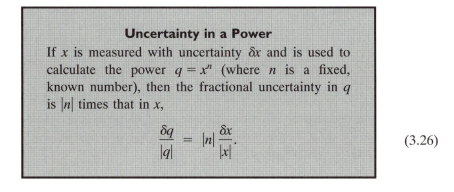

Uncertainty in a Power
If x is measured with uncertainty δx and is used to calculate the power $q = x^n$ (where n is a fixed, known number), then the fractional uncertainty in q is $|n|$ times that in x,

$$\frac{\delta q}{|q|} \; = \; |n| \frac{\delta x}{|x|}. \tag{3.26}$$

Quick Check 3.8. If you measure x as 100 ± 6, what should you report for $\sqrt{x}$, with its uncertainty?

3.8 Propagation Step by Step

We now have enough tools to handle almost any problem in the propagation of errors. Any calculation can be broken down into a sequence of steps, each involving just one of the following types of operation: (1) sums and differences; (2) products and quotients; and (3) computation of a function of one variable, such as x^n, $\sin x$,

e^x, or ln x. For example, we could calculate

$$q = x(y - z \sin u) \tag{3.27}$$

from the measured quantities x, y, z, and u in the following steps: Compute the *function* sin u, then the *product* of z and sin u, next the *difference* of y and z sin u, and finally the *product* of x and $(y - z \sin u)$.

We know how uncertainties propagate through each of these separate operations. Thus, provided the various quantities involved are independent, we can calculate the uncertainty in the final answer by proceeding in steps from the uncertainties in the original measurement. For example, if the quantities x, y, z, and u in (3.27) have been measured with corresponding uncertainties δx, . . . , δu, we could calculate the uncertainty in q as follows. First, find the uncertainty in the function sin u; knowing this, find the uncertainty in the product z sin u, and then that in the difference $y - z \sin u$; finally, find the uncertainty in the complete product (3.27).

Quick Check 3.9.   Suppose you measure three numbers as follows:

$$x = 200 \pm 2, \quad y = 50 \pm 2, \quad z = 40 \pm 2,$$

where the three uncertainties are independent and random. Use step-by-step propagation to find the quantity $q = x/(y - z)$ with its uncertainty. [First find the uncertainty in the difference $y - z$ and then the quotient $x/(y - z)$.]

Before I discuss some examples of this step-by-step calculation of errors, let me emphasize three general points. First, because uncertainties in sums or differences involve absolute uncertainties (such as δx) whereas those in products or quotients involve fractional uncertainties (such as $\delta x/|x|$), the calculations will require some facility in passing from absolute to fractional uncertainties and vice versa, as demonstrated below.

Second, an important simplifying feature of all these calculations is that (as repeatedly emphasized) uncertainties are seldom needed to more than one significant figure. Hence, much of the calculation can be done rapidly in your head, and many smaller uncertainties can be completely neglected. In a typical experiment involving several trials, you may need to do a careful calculation on paper of all error propagations for the first trial. After that, you will often find that all trials are sufficiently similar that no further calculation is needed or, at worst, that for subsequent trials the calculations of the first trial can be modified in your head.

Finally, you need to be aware that you will sometimes encounter functions $q(x)$ whose uncertainty cannot be found reliably by the stepwise method advocated here. These functions always involve at least one variable that appears more than once. Suppose, for example, that in place of the function (3.27), we had to evaluate

$$q = y - x \sin y.$$

This function is the difference of two terms, y and $x \sin y$, but these two terms are definitely *not* independent because both depend on y. Thus, to estimate the uncertainty, we would have to treat the terms as dependent (that is, add their uncertainties directly, not in quadrature). Under some circumstances, this treatment may seriously overestimate the true uncertainty. Faced with a function like this, we must recognize that a stepwise calculation may give an uncertainty that is unnecessarily big, and the only satisfactory procedure is then to use the general formula to be developed in Section 3.11.

3.9   Examples

In this and the next section, I give three examples of the type of calculation encountered in introductory laboratories. None of these examples is especially complicated; in fact, few real problems are much more complicated than the ones described here.

Example: Measurement of g with a Simple Pendulum

As a first example, suppose that we measure g, the acceleration of gravity, using a simple pendulum. The period of such a pendulum is well known to be $T = 2\pi\sqrt{l/g}$, where l is the length of the pendulum. Thus, if l and T are measured, we can find g as

$$g = 4\pi^2 l/T^2. \tag{3.28}$$

This result gives g as the product or quotient of three factors, $4\pi^2$, l, and T^2. If the various uncertainties are independent and random, the fractional uncertainty in our answer is just the quadratic sum of the fractional uncertainties in these factors. The factor $4\pi^2$ has no uncertainty, and the fractional uncertainty in T^2 is twice that in T:

$$\frac{\delta(T^2)}{T^2} = 2\frac{\delta T}{T}.$$

Thus, the fractional uncertainty in our answer for g will be

$$\frac{\delta g}{g} = \sqrt{\left(\frac{\delta l}{l}\right)^2 + \left(2\frac{\delta T}{T}\right)^2}. \tag{3.29}$$

Suppose we measure the period T for one value of the length l and get the results[5]

$$l = 92.95 \pm 0.1 \text{ cm},$$
$$T = 1.936 \pm 0.004 \text{ s}.$$

---

[5] Although at first sight an uncertainty $\delta T = 0.004$ s may seem unrealistically small, you can easily achieve it by timing several oscillations. If you can measure with an accuracy of 0.1 s, as is certainly possible with a stopwatch, then by timing 25 oscillations you will find T within 0.004 s.

Our best estimate for g is easily found from (3.28) as

$$g_{\text{best}} = \frac{4\pi^2 \times (92.95 \text{ cm})}{(1.936 \text{ s})^2} = 979 \text{ cm/s}^2.$$

To find our uncertainty in g using (3.29), we need the fractional uncertainties in l and T. These are easily calculated (in the head) as

$$\frac{\delta l}{l} = 0.1\% \quad \text{and} \quad \frac{\delta T}{T} = 0.2\%.$$

Substituting into (3.29), we find

$$\frac{\delta g}{g} = \sqrt{(0.1)^2 + (2 \times 0.2)^2} \, \% = 0.4\%;$$

from which

$$\delta g = 0.004 \times 979 \text{ cm/s}^2 = 4 \text{ cm/s}^2.$$

Thus, based on these measurements, our final answer is

$$g = 979 \pm 4 \text{ cm/s}^2.$$

Having found the measured value of g and its uncertainty, we would naturally compare these values with the accepted value of g. If the latter has its usual value of 981 cm/s^2, the present value is entirely satisfactory.

If this experiment is repeated (as most such experiments should be) with different values of the parameters, the uncertainty calculations usually do not need to be repeated in complete detail. We can often easily convince ourselves that all uncertainties (in the answers for g) are close enough that no further calculations are needed; sometimes the uncertainty in a few representative values of g can be calculated and the remainder estimated by inspection. In any case, the best procedure is almost always to record the various values of l, T, and g and the corresponding uncertainties in a single table. (See Problem 3.40.)

Example: Refractive Index Using Snell's Law

If a ray of light passes from air into glass, the angles of incidence i and refraction r are defined as in Figure 3.5 and are related by Snell's law, $\sin i = n \sin r$, where n is the refractive index of the glass. Thus, if you measure the angles i and r, you

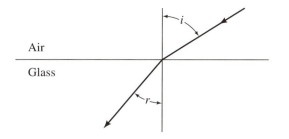

Figure 3.5. The angles of incidence i and refraction r when a ray of light passes from air into glass.

can calculate the refractive index n as

$$n = \frac{\sin i}{\sin r}. \tag{3.30}$$

The uncertainty in this answer is easily calculated. Because n is the quotient of $\sin i$ and $\sin r$, the fractional uncertainty in n is the quadratic sum of those in $\sin i$ and $\sin r$:

$$\frac{\delta n}{n} = \sqrt{\left(\frac{\delta \sin i}{\sin i}\right)^2 + \left(\frac{\delta \sin r}{\sin r}\right)^2}. \tag{3.31}$$

To find the fractional uncertainty in the sine of any angle θ, we note that

$$\delta \sin \theta = \left|\frac{d \sin \theta}{d\theta}\right| \delta\theta$$

$$= |\cos \theta| \, \delta\theta \text{ (in rad)}.$$

Thus, the fractional uncertainty is

$$\frac{\delta \sin \theta}{|\sin \theta|} = |\cot \theta| \, \delta\theta \text{ (in rad)}. \tag{3.32}$$

Suppose we now measure the angle r for a couple of values of i and get the results shown in the first two columns of Table 3.1 (with all measurements judged to be uncertain by $\pm 1°$, or 0.02 rad). The calculation of $n = \sin i / \sin r$ is easily carried out as shown in the next three columns of Table 3.1. The uncertainty in n can then be found as in the last three columns; the fractional uncertainties in $\sin i$ and $\sin r$ are calculated using (3.32), and finally the fractional uncertainty in n is found using (3.31).

Table 3.1. Finding the refractive index.

i (deg) all ± 1	r (deg) all ± 1	$\sin i$	$\sin r$	n	$\dfrac{\delta \sin i}{|\sin i|}$	$\dfrac{\delta \sin r}{|\sin r|}$	$\dfrac{\delta n}{n}$
20	13	0.342	0.225	1.52	5%	8%	9%
40	23.5	0.643	0.399	1.61	2%	4%	5%

Before making a series of measurements like the two shown in Table 3.1, you should think carefully how best to record the data and calculations. A tidy display like that in Table 3.1 makes the recording of data easier and reduces the danger of mistakes in calculation. It is also easier for the reader to follow and check.

If you repeat an experiment like this one several times, the error calculations can become tedious if you do them for each repetition. If you have a programmable calculator, you may decide to write a program to do the repetitive calculations automatically. You should recognize, however, that you almost never need to do the error calculations for all the repetitions; if you find the uncertainties in n corresponding to the smallest and largest values of i (and possibly a few intermediate values), then these uncertainties suffice for most purposes.

3.10   A More Complicated Example

The two examples just given are typical of many experiments in the introductory physics laboratory. A few experiments require more complicated calculations, however. As an example of such an experiment, I discuss here the measurement of the acceleration of a cart rolling down a slope.[6]

Example: Acceleration of a Cart Down a Slope

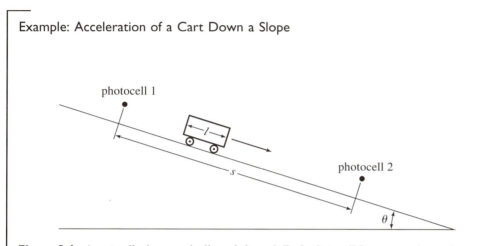

Figure 3.6. A cart rolls down an incline of slope θ. Each photocell is connected to a timer to measure the time for the cart to pass it.

Let us consider a cart rolling down an incline of slope θ as in Figure 3.6. The expected acceleration is $g \sin \theta$ and, if we measure θ, we can easily calculate the expected acceleration and its uncertainty (Problem 3.42). We can measure the actual acceleration a by timing the cart past two photocells as shown, each connected to a timer. If the cart has length l and takes time t_1 to pass the first photocell, its speed there is $v_1 = l/t_1$. In the same way, $v_2 = l/t_2$. (Strictly speaking, these speeds are the cart's *average* speeds while passing the two photocells. However, provided l is small, the difference between the average and instantaneous speeds is unimportant.) If the distance between the photocells is s, then the well-known formula $v_2{}^2 = v_1{}^2 + 2as$ implies that

$$a = \frac{v_2{}^2 - v_1{}^2}{2s}$$

$$= \left(\frac{l^2}{2s}\right)\left(\frac{1}{t_2{}^2} - \frac{1}{t_1{}^2}\right). \tag{3.33}$$

Using this formula and the measured values of l, s, t_1, and t_2, we can easily find the observed acceleration and its uncertainty.

---

[6]If you wish, you could omit this section without loss of continuity or return to study it in connection with Problem 3.42.

One set of data for this experiment, including uncertainties, was as follows (the numbers in parentheses are the corresponding percentage uncertainties, as you can easily check):

$$
\begin{aligned}
l &= 5.00 \pm 0.05 \text{ cm} \ (1\%) \\
s &= 100.0 \pm 0.2 \text{ cm} \ (0.2\%) \\
t_1 &= 0.054 \pm 0.001 \text{ s} \ (2\%) \\
t_2 &= 0.031 \pm 0.001 \text{ s} \ (3\%).
\end{aligned}
\tag{3.34}
$$

From these values, we can immediately calculate the first factor in (3.33) as $l^2/2s = 0.125$ cm. Because the fractional uncertainties in l and s are 1% and 0.2%, that in $l^2/2s$ is

$$
\begin{aligned}
\text{(fractional uncertainty in } l^2/2s) &= \sqrt{\left(2\frac{\delta l}{l}\right)^2 + \left(\frac{\delta s}{s}\right)^2} \\
&= \sqrt{(2 \times 1\%)^2 + (0.2\%)^2} = 2\%.
\end{aligned}
$$

(Note how the uncertainty in s makes no appreciable contribution and could have been ignored.) Therefore,

$$
l^2/2s = 0.125 \text{ cm} \pm 2\%.
\tag{3.35}
$$

To calculate the second factor in (3.33) and its uncertainty, we proceed in steps. Because the fractional uncertainty in t_1 is 2%, that in $1/t_1^2$ is 4%. Thus, since $t_1 = 0.054$ s,

$$
1/t_1^2 = 343 \pm 14 \text{ s}^{-2}.
$$

In the same way, the fractional uncertainty in $1/t_2^2$ is 6% and

$$
1/t_2^2 = 1041 \pm 62 \text{ s}^{-2}.
$$

Subtracting these (and combining the errors in quadrature), we find

$$
\frac{1}{t_2^2} - \frac{1}{t_1^2} = 698 \pm 64 \text{ s}^{-2} \quad \text{(or 9\%)}.
\tag{3.36}
$$

Finally, according to (3.33), the required acceleration is the product of (3.35) and (3.36). Multiplying these equations together (and combining the fractional uncertainties in quadrature), we obtain

$$
\begin{aligned}
a &= (0.125 \text{ cm} \pm 2\%) \times (698 \text{ s}^{-2} \pm 9\%) \\
&= 87.3 \text{ cm/s}^2 \pm 9\%
\end{aligned}
$$

or

$$
a = 87 \pm 8 \text{ cm/s}^2.
\tag{3.37}
$$

This answer could now be compared with the expected acceleration $g\sin\theta$, if the latter had been calculated.

When the calculations leading to (3.37) are studied carefully, several interesting features emerge. First, the 2% uncertainty in the factor $l^2/2s$ is completely swamped by the 9% uncertainty in $(1/t_2{}^2) - (1/t_1{}^2)$. If further calculations are needed for subsequent trials, the uncertainties in l and s can therefore be ignored (so long as a quick check shows they are still just as unimportant).

Another important feature of our calculation is the way in which the 2% and 3% uncertainties in t_1 and t_2 grow when we evaluate $1/t_1{}^2$, $1/t_2{}^2$, and the difference $(1/t_2{}^2) - (1/t_1{}^2)$, so that the final uncertainty is 9%. This growth results partly from taking squares and partly from taking the difference of large numbers. We could imagine extending the experiment to check the constancy of a by giving the cart an initial push, so that the speeds v_1 and v_2 are both larger. If we did, the times t_1 and t_2 would get smaller, and the effects just described would get worse (see Problem 3.42).

3.11   General Formula for Error Propagation[7]

So far, we have established three main rules for the propagation of errors: that for sums and differences, that for products and quotients, and that for arbitrary functions of one variable. In the past three sections, we have seen how the computation of a complicated function can often be broken into steps and the uncertainty in the function computed stepwise using our three simple rules.

In this final section, I give a single general formula from which all three of these rules can be derived and with which any problem in error propagation can be solved. Although this formula is often rather cumbersome to use, it is useful theoretically. Furthermore, there are some problems in which, instead of calculating the uncertainty in steps as in the past three sections, you will do better to calculate it in one step by means of the general formula.

To illustrate the kind of problem for which the one-step calculation is preferable, suppose that we measure three quantities x, y, and z and have to compute a function such as

$$q = \frac{x + y}{x + z} \tag{3.38}$$

in which a variable appears more than once (x in this case). If we were to calculate the uncertainty δq in steps, then we would first compute the uncertainties in the two sums $x + y$ and $x + z$, and then that in their quotient. Proceeding in this way, we would completely miss the possibility that errors in the numerator due to errors in x may, to some extent, cancel errors in the denominator due to errors in x. To understand how this cancellation can happen, suppose that x, y, and z are all positive numbers, and consider what happens if our measurement of x is subject to error. If we *overestimate* x, we *overestimate both* $x + y$ and $x + z$, and (to a large extent) these overestimates cancel one another when we calculate $(x + y)/(x + z)$. Similarly, an *underestimate* of x leads to *underestimates of both* $x + y$ and $x + z$, which again cancel when we form the quotient. In either case, an error in x is substantially

---

[7] You can postpone reading this section without a serious loss of continuity. The material covered here is not used again until Section 5.6.

canceled out of the quotient $(x + y)/(x + z)$, and our stepwise calculation completely misses these cancellations.

Whenever a function involves the same quantity more than once, as in (3.38), some errors may cancel themselves (an effect sometimes called *compensating errors*). If this cancellation is possible, then a stepwise calculation of the uncertainty may overestimate the final uncertainty. The only way to avoid this overestimation is to calculate the uncertainty in one step by using the method I will now develop.[8]

Let us suppose at first that we measure two quantities x and y and then calculate some function $q = q(x, y)$. This function could be as simple as $q = x + y$ or something more complicated such as $q = (x^3 + y) \sin(xy)$. For a function $q(x)$ of a *single* variable, we argued that if the best estimate for x is the number x_{best}, then the best estimate for $q(x)$ is $q(x_{best})$. Next, we argued that the extreme (that is, largest and smallest) probable values of x are $x_{best} \pm \delta x$ and that the corresponding extreme values of q are therefore

$$q(x_{best} \pm \delta x). \tag{3.39}$$

Finally, we used the approximation

$$q(x + u) \approx q(x) + \frac{dq}{dx} u \tag{3.40}$$

(for any small increment u) to rewrite the extreme probable values (3.39) as

$$q(x_{best}) \pm \left| \frac{dq}{dx} \right| \delta x, \tag{3.41}$$

where the absolute value is to allow for the possibility that dq/dx may be negative. The result (3.41) means that $\delta q \approx |dq/dx| \delta x$.

When q is a function of two variables, $q(x, y)$, the argument is similar. If x_{best} and y_{best} are the best estimates for x and y, we expect the best estimate for q to be

$$q_{best} = q(x_{best}, y_{best})$$

in the usual way. To estimate the uncertainty in this result, we need to generalize the approximation (3.40) for a function of two variables. The required generalization is

$$q(x + u, y + v) \approx q(x, y) + \frac{\partial q}{\partial x} u + \frac{\partial q}{\partial y} v, \tag{3.42}$$

where u and v are any small increments in x and y, and $\partial q/\partial x$ and $\partial q/\partial y$ are the so-called *partial derivatives* of q with respect to x and y. That is, $\partial q/\partial x$ is the result of differentiating q with respect to x while treating y as fixed, and vice versa for $\partial q/\partial y$. [For further discussion of partial derivatives and the approximation (3.42), see Problems 3.43 and 3.44.]

The extreme probable values for x and y are $x_{best} \pm \delta x$ and $y_{best} \pm \delta y$. If we insert these values into (3.42) and recall that $\partial q/\partial x$ and $\partial q/\partial y$ may be positive or

---

[8]Sometimes a function that involves a variable more than once can be rewritten in a different form that does not. For example, $q = xy - xz$ can be rewritten as $q = x(y - z)$. In the second form, the uncertainty δq can be calculated in steps without any danger of overestimation.

negative, we find, for the extreme values of q,

$$q(x_{\text{best}}, y_{\text{best}}) \pm \left(\left| \frac{\partial q}{\partial x} \right| \delta x + \left| \frac{\partial q}{\partial y} \right| \delta y \right).$$

This means that the uncertainty in $q(x, y)$ is

$$\delta q \approx \left| \frac{\partial q}{\partial x} \right| \delta x + \left| \frac{\partial q}{\partial y} \right| \delta y. \tag{3.43}$$

Before I discuss various generalizations of this new rule, let us apply it to rederive some familiar cases. Suppose, for instance, that

$$q(x, y) = x + y; \tag{3.44}$$

that is, q is just the sum of x and y. The partial derivatives are both one,

$$\frac{\partial q}{\partial x} = \frac{\partial q}{\partial y} = 1, \tag{3.45}$$

and so, according to (3.43),

$$\delta q \approx \delta x + \delta y. \tag{3.46}$$

This is just our original provisional rule that the uncertainty in $x + y$ is the sum of the uncertainties in x and y.

In much the same way, if q is the product $q = xy$, you can check that (3.43) implies the familiar rule that the fractional uncertainty in q is the sum of the fractional uncertainties in x and y (see Problem 3.45).

The rule (3.43) can be generalized in various ways. You will not be surprised to learn that when the uncertainties δx and δy are independent and random, the sum (3.43) can be replaced by a sum in quadrature. If the function q depends on more than two variables, then we simply add an extra term for each extra variable. In this way, we arrive at the following general rule (whose full justification will appear in Chapters 5 and 9).

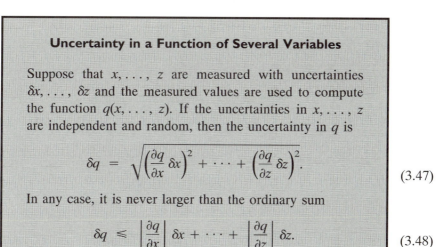

Uncertainty in a Function of Several Variables

Suppose that $x, \ldots, z$ are measured with uncertainties $\delta x, \ldots, \delta z$ and the measured values are used to compute the function $q(x, \ldots, z)$. If the uncertainties in $x, \ldots, z$ are independent and random, then the uncertainty in q is

$$\delta q = \sqrt{\left(\frac{\partial q}{\partial x} \delta x \right)^2 + \cdots + \left(\frac{\partial q}{\partial z} \delta z \right)^2}. \tag{3.47}$$

In any case, it is never larger than the ordinary sum

$$\delta q \leq \left| \frac{\partial q}{\partial x} \right| \delta x + \cdots + \left| \frac{\partial q}{\partial z} \right| \delta z. \tag{3.48}$$

Although the formulas (3.47) and (3.48) look fairly complicated, they are easy to understand if you think about them one term at a time. For example, suppose for a moment that among all the measured quantities, $x, y, \ldots, z$, only x is subject to any uncertainty. (That is, $\delta y = \ldots = \delta z = 0$.) Then (3.47) contains only one term and we would find

$$\delta q = \left| \frac{\partial q}{\partial x} \right| \delta x \quad (\text{if } \delta y = \cdots = \delta z = 0). \tag{3.49}$$

In other words, the term $|\partial q/\partial x| \delta x$ by itself is the uncertainty, or partial uncertainty, in q caused by the uncertainty in x alone. In the same way, $|\partial q/\partial y| \delta y$ is the partial uncertainty in q due to δy alone, and so on. Referring back to (3.47), we see that the total uncertainty in q is the quadratic sum of the partial uncertainties due to each of the separate uncertainties $\delta x, \delta y, \ldots, \delta z$ (provided the latter are independent). This is a good way to think about the result (3.47), and it suggests the simplest way to use (3.47) to calculate the total uncertainty in q: First, calculate the partial uncertainties in q due to $\delta x, \delta y, \ldots, \delta z$ separately, using (3.49) and its analogs for $y, \ldots, z$; then simply combine these separate uncertainties in quadrature to give the total uncertainty as in (3.47).

In the same way, whether or not the uncertainties $\delta x, \delta y, \ldots, \delta z$ are independent, the rule (3.48) says that the total uncertainty in q never exceeds the simple sum of the partial uncertainties due to each of $\delta x, \delta y, \ldots, \delta z$ separately.

Example: Using the General Formula (3.47)

To determine the quantity

$$q = x^2 y - xy^2,$$

a scientist measures x and y as follows:

$$x = 3.0 \pm 0.1 \quad \text{and} \quad y = 2.0 \pm 0.1.$$

What is his answer for q and its uncertainty, as given by (3.47)?

His best estimate for q is easily seen to be $q_{\text{best}} = 6.0$. To find δq, we follow the steps just outlined. The uncertainty in q due to δx alone, which we denote by δq_x, is given by (3.49) as

$$\delta q_x = (\text{error in } q \text{ due to } \delta x \text{ alone})$$

$$= \left| \frac{\partial q}{\partial x} \right| \delta x \tag{3.50}$$

$$= |2xy - y^2| \delta x = |12 - 4| \times 0.1 = 0.8.$$

Similarly, the uncertainty in q due to δy is

$$\delta q_y = (\text{error in } q \text{ due to } \delta y \text{ alone})$$

$$= \left| \frac{\partial q}{\partial y} \right| \delta y \tag{3.51}$$

$$= |x^2 - 2xy| \delta y = |9 - 12| \times 0.1 = 0.3.$$

Finally, according to (3.47), the total uncertainty in q is the quadratic sum of these two partial uncertainties:

$$\delta q = \sqrt{(\delta q_x)^2 + (\delta q_y)^2} \tag{3.52}$$

$$= \sqrt{(0.8)^2 + (0.3)^2} = 0.9.$$

Thus, the final answer for q is

$$q = 6.0 \pm 0.9.$$

The use of (3.47) or (3.48) to calculate uncertainties is reasonably straightforward if you follow the procedure used in this example; that is, first calculate each separate contribution to δq and only then combine them to give the total uncertainty. This procedure breaks the problem into calculations small enough that you have a good chance of getting them right. It has the further advantage that it lets you see which of the measurements $x, y, \ldots, z$ are the main contributors to the final uncertainty. (For instance, in the example above, the contribution $\delta q_y = 0.3$ was so small compared with $\delta q_x = 0.8$ that the former could almost be ignored.)

Generally speaking, when the stepwise propagation described in Sections 3.8 to 3.10 is possible, it is usually simpler than the general rules (3.47) or (3.48) discussed here. Nevertheless, you must recognize that if the function $q(x, \ldots, z)$ involves any variable more than once, there may be compensating errors; if so, a stepwise calculation may overestimate the final uncertainty, and calculating δq in one step using (3.47) or (3.48) is better.

Principal Definitions and Equations of Chapter 3

THE SQUARE-ROOT RULE FOR A COUNTING EXPERIMENT

If we observe the occurrences of an event that happens at random but with a definite average rate and we count ν occurrences in a time T, our estimate for the true average number is

$$(\text{average number of events in time } T) = \nu \pm \sqrt{\nu}. \qquad [\text{See (3.2)}]$$

RULES FOR ERROR PROPAGATION

The rules of error propagation refer to a situation in which we have found various quantities, $x, \ldots, w$ with uncertainties $\delta x, \ldots, \delta w$ and then use these values to calculate a quantity q. The uncertainties in $x, \ldots, w$ "propagate" through the calculation to cause an uncertainty in q as follows:

Sums and Differences: If

$$q = x + \cdots + z - (u + \cdots + w),$$

then

$$\delta q = \sqrt{(\delta x)^2 + \cdots + (\delta z)^2 + (\delta u)^2 + \cdots + (\delta w)^2}$$
(provided all errors are independent and random)

and

$$\delta q \leq \delta x + \cdots + \delta z + \delta u + \cdots + \delta w$$
(always).   [See (3.16) & (3.17)]

Products and Quotients: If

$$q = \frac{x \times \cdots \times z}{u \times \cdots \times w},$$

then

$$\frac{\delta q}{|q|} = \sqrt{\left(\frac{\delta x}{x}\right)^2 + \cdots + \left(\frac{\delta z}{z}\right)^2 + \left(\frac{\delta u}{u}\right)^2 + \cdots + \left(\frac{\delta w}{w}\right)^2}$$
(provided all errors are independent and random)

and

$$\frac{\delta q}{|q|} \leq \frac{\delta x}{|x|} + \cdots + \frac{\delta z}{|z|} + \frac{\delta u}{|u|} + \cdots + \frac{\delta w}{|w|}$$
(always).   [See (3.18) & (3.19)]

Measured Quantity Times Exact Number: If B is known exactly and

$$q = Bx,$$

then

$$\delta q = |B|\,\delta x \quad \text{or, equivalently,} \quad \frac{\delta q}{|q|} = \frac{\delta x}{|x|}. \qquad \text{[See (3.9)]}$$

Uncertainty in a Power: If n is an exact number and

$$q = x^n,$$

then

$$\frac{\delta q}{|q|} = |n|\frac{\delta x}{|x|}. \qquad \text{[See (3.26)]}$$

Uncertainty in a Function of One Variable: If $q = q(x)$ is any function of x, then

$$\delta q = \left|\frac{dq}{dx}\right|\,\delta x. \qquad \text{[See (3.23)]}$$

Sometimes, if $q(x)$ is complicated and if you have written a program to calculate $q(x)$ then, instead of differentiating $q(x)$, you may find it easier to use the equivalent formula,

$$\delta q = |q(x_{\text{best}} + \delta x) - q(x_{\text{best}})|. \qquad \text{[See Problem 3.32]}$$

General Formula for Error Propagation: If $q = q(x, \ldots, z)$ is any function of $x, \ldots, z$, then

$$\delta q = \sqrt{\left(\frac{\partial q}{\partial x}\delta x\right)^2 + \cdots + \left(\frac{\partial q}{\partial z}\delta z\right)^2}$$

(provided all errors are independent and random)

and

$$\delta q \leq \left|\frac{\partial q}{\partial x}\right|\delta x + \cdots + \left|\frac{\partial q}{\partial z}\right|\delta z$$

(always).   [See (3.47) & (3.48)]

Problems for Chapter 3

For Section 3.2: The Square-Root Rule for a Counting Experiment

3.1. ★   To measure the activity of a radioactive sample, two students count the alpha particles it emits. Student A watches for 3 minutes and counts 28 particles; Student B watches for 30 minutes and counts 310 particles. **(a)** What should Student A report for the average number emitted in 3 minutes, with his uncertainty? **(b)** What should Student B report for the average number emitted in 30 minutes, with her uncertainty? **(c)** What are the fractional uncertainties in the two measurements? Comment.

3.2. ★   A nuclear physicist studies the particles ejected by a beam of radioactive nuclei. According to a proposed theory, the average rates at which particles are ejected in the forward and backward directions should be equal. To test this theory, he counts the total number ejected forward and backward in a certain 10-hour interval and finds 998 forward and 1,037 backward. **(a)** What are the uncertainties associated with these numbers? **(b)** Do these results cast any doubt on the theory that the average rates should be equal?

3.3. ★   Most of the ideas of error analysis have important applications in many different fields. This applicability is especially true for the square-root rule (3.2) for counting experiments, as the following example illustrates. The normal average incidence of a certain kind of cancer has been established as 2 cases per 10,000 people per year. The suspicion has been aired that a certain town (population 20,000) suffers a high incidence of this cancer because of a nearby chemical dump. To test this claim, a reporter investigates the town's records for the past 4 years and finds 20 cases of the cancer. He calculates that the expected number is 16 (check this) and concludes that the observed rate is 25% more than expected. Is he justified in claiming that this result proves that the town has a higher than normal rate for this cancer?

3.4. ★★   As a sample of radioactive atoms decays, the number of atoms steadily diminishes and the sample's radioactivity decreases in proportion. To study this effect, a nuclear physicist monitors the particles ejected by a radioactive sample for 2 hours. She counts the number of particles emitted in a 1-minute period and repeats the measurement at half-hour intervals, with the following results:

Time elapsed, t (hours):	0.0	0.5	1.0	1.5	2.0
Number counted, ν, in 1 min:	214	134	101	61	54

(a) Plot the number counted against elapsed time, including error bars to show the uncertainty in the numbers. (Neglect any uncertainty in the elapsed time.) **(b)** Theory predicts that the number of emitted particles should diminish exponentially as $\nu = \nu_o \exp(-rt)$, where (in this case) $\nu_o = 200$ and $r = 0.693$ h^{-1}. On the same graph, plot this expected curve and comment on how well the data seem to fit the theoretical prediction.

For Section 3.3: Sums and Differences; Products and Quotients

3.5. ★   Using the provisional rules (3.4) and (3.8), compute the following:
    (a) $(5 \pm 1) + (8 \pm 2) - (10 \pm 4)$
    (b) $(5 \pm 1) \times (8 \pm 2)$
    (c) $(10 \pm 1)/(20 \pm 2)$
    (d) $(30 \pm 1) \times (50 \pm 1)/(5.0 \pm 0.1)$

3.6. ★   Using the provisional rules (3.4) and (3.8), compute the following:
    (a) $(3.5 \pm 0.1) + (8.0 \pm 0.2) - (5.0 \pm 0.4)$
    (b) $(3.5 \pm 0.1) \times (8.0 \pm 0.2)$
    (c) $(8.0 \pm 0.2)/(5.0 \pm 0.4)$
    (d) $(3.5 \pm 0.1) \times (8.0 \pm 0.2)/(5.0 \pm 0.4)$

3.7. ★   A student makes the following measurements:

$$a = 5 \pm 1 \text{ cm}, \quad b = 18 \pm 2 \text{ cm}, \quad c = 12 \pm 1 \text{ cm},$$
$$t = 3.0 \pm 0.5 \text{ s}, \quad m = 18 \pm 1 \text{ gram}$$

Using the provisional rules (3.4) and (3.8), compute the following quantities with their uncertainties and percentage uncertainties: **(a)** $a + b + c$, **(b)** $a + b - c$, **(c)** ct, and **(d)** mb/t.

3.8. ★★   The binomial theorem states that for any number n and any x with $|x| < 1$,

$$(1 + x)^n = 1 + nx + \frac{n(n - 1)}{1 \cdot 2} x^2 + \frac{n(n - 1)(n - 2)}{1 \cdot 2 \cdot 3} x^3 + \cdots$$

(a) Show that if n is a positive integer, this infinite series terminates (that is, has only a finite number of nonzero terms). Write the series down explicitly for the cases $n = 2$ and $n = 3$. **(b)** Write down the binomial series for the case $n = -1$. This case gives an infinite series for $1/(1 + x)$, but when x is small, you get a good approximation if you keep just the first two terms:

$$\frac{1}{1 + x} \approx 1 - x,$$

as quoted in (3.6). Calculate both sides of this approximation for each of the values $x = 0.5, 0.1$, and 0.01, and in each case find the percentage by which the approximation $(1 - x)$ differs from the exact value of $1/(1 + x)$.

For Section 3.4: Two Important Special Cases

3.9. ★   I measure the diameter of a circular disc as $d = 6.0 \pm 0.1$ cm and use this value to calculate the circumference $c = \pi d$ and radius $r = d/2$. What are my answers? [The rule (3.9) for "measured quantity $\times$ exact number" applies to both of these calculations. In particular, you can write r as $d \times 1/2$, where the number $1/2$ is, of course, exact.]

3.10. ★   I have a set of callipers that can measure thicknesses of a few inches with an uncertainty of ± 0.005 inches. I measure the thickness of a deck of 52 cards and get 0.590 in. **(a)** If I now calculate the thickness of 1 card, what is my answer (including its uncertainty)? **(b)** I can improve this result by measuring several decks together. If I want to know the thickness of 1 card with an uncertainty of only 0.00002 in, how many decks do I need to measure together?

3.11. ★   With a good stopwatch and some practice, you can measure times ranging from approximately 1 second up to many minutes with an uncertainty of 0.1 second or so. Suppose that we wish to find the period τ of a pendulum with $\tau \approx 0.5$ s. If we time 1 oscillation, we have an uncertainty of approximately 20%; but by timing several oscillations together, we can do much better, as the following questions illustrate:

(a) If we measure the total time for 5 oscillations and get 2.4 ± 0.1 s, what is our final answer for τ, with its absolute and percent uncertainties? [Remember the rule (3.9).]

(b) What if we measure 20 oscillations and get 9.4 ± 0.1 s?

(c) Could the uncertainty in τ be improved indefinitely by timing more oscillations?

3.12. ★   If x has been measured as 4.0 ± 0.1 cm, what should I report for x^2 and x^3? Give percent and absolute uncertainties, as determined by the rule (3.10) for a power.

3.13. ★   If I have measured the radius of a sphere as $r = 2.0 \pm 0.1$ m, what should I report for the sphere's volume?

3.14. ★   A visitor to a medieval castle measures the depth of a well by dropping a stone and timing its fall. She finds the time to fall is $t = 3.0 \pm 0.5$ sec and calculates the depth as $d = \frac{1}{2}gt^2$. What is her conclusion, if she takes $g = 9.80$ m/s^2 with negligible uncertainty?

3.15. ★★   Two students are asked to measure the rate of emission of alpha particles from a certain radioactive sample. Student A watches for 2 minutes and counts 32 particles. Student B watches for 1 hour and counts 786 particles. (The sample decays slowly enough that the expected rate of emission can be assumed to be constant during the measurements.) **(a)** What is the uncertainty in Student A's result, 32, for the number of particles emitted in 2 minutes? **(b)** What is the uncertainty in Student B's result, 786, for the number of particles emitted in 1 hour? **(c)** Each student now divides his count by his number of minutes to find the *rate* of emission in particles per minute. Assuming the times, 2 min and 60 min, have negligible uncertainty, what are the two students' answers for the rate, with their uncertainties? Comment.

For Section 3.5: Independent Uncertainties in a Sum

3.16. ★   A student measures five lengths:

$$a = 50 \pm 5, \quad b = 30 \pm 3, \quad c = 60 \pm 2, \quad d = 40 \pm 1, \quad e = 5.8 \pm 0.3$$

(all in cm) and calculates the four sums $a + b$, $a + c$, $a + d$, $a + e$. Assuming the original errors were independent and random, find the uncertainties in her four answers [rule (3.13), "errors add in quadrature"]. If she has reason to think the original errors were *not* independent, what would she have to give for her final uncertainties [rule (3.14), "errors add directly"]? Assuming the uncertainties are needed with only one significant figure, identify those cases in which the second uncertainty (that in b, c, d, e) can be entirely ignored. If you decide to do the additions in quadrature on a calculator, note that the conversion from rectangular to polar coordinates automatically calculates $\sqrt{x^2 + y^2}$ for given x and y.

3.17. ★   Evaluate each of the following:
    (a) $(5.6 \pm 0.7) + (3.70 \pm 0.03)$
    (b) $(5.6 \pm 0.7) + (2.3 \pm 0.1)$
    (c) $(5.6 \pm 0.7) + (4.1 \pm 0.2)$
    (d) $(5.6 \pm 0.7) + (1.9 \pm 0.3)$

For each sum, consider both the case that the original uncertainties are independent and random ("errors add in quadrature") and that they are not ("errors add directly"). Assuming the uncertainties are needed with only one significant figure, identify those cases in which the second of the original uncertainties can be ignored entirely. If you decide to do the additions in quadrature on a calculator, note that the conversion from rectangular to polar coordinates automatically calculates $\sqrt{x^2 + y^2}$ for given x and y.

For Section 3.6: More About Independent Uncertainties

3.18. ★   If you have not yet done it, do Problem 3.7 (assuming that the original uncertainties are *not* independent), and repeat each calculation assuming that the original uncertainties *are* independent and random. Arrange your answers in a table so that you can compare the two different methods of propagating errors.

3.19. ★   If you have not yet done it, do Problem 3.5 (assuming that the original uncertainties are *not* independent) and repeat each calculation assuming that the original uncertainties *are* independent and random. Arrange your answers in a table so that you can compare the two different methods of propagating errors.

3.20. ★   If you have not yet done it, do Problem 3.6 (assuming that the original uncertainties are *not* independent) and repeat each calculation assuming that the original uncertainties *are* independent and random. Arrange your answers in a table so that you can compare the two different methods of propagating errors.

3.21. ★   **(a)** To find the velocity of a cart on a horizontal air track, a student measures the distance d it travels and the time taken t as

$$d = 5.10 \pm 0.01 \text{ m} \quad \text{and} \quad t = 6.02 \pm 0.02 \text{ s}.$$

What is his result for $v = d/t$, with its uncertainty? **(b)** If he measures the cart's mass as $m = 0.711 \pm 0.002$ kg, what would be his answer for the momentum $p = mv = md/t$? (Assume all errors are random and independent.)

3.22. ★   A student is studying the properties of a resistor. She measures the current flowing through the resistor and the voltage across it as

$$I = 2.10 \pm 0.02 \text{ amps} \quad \text{and} \quad V = 1.02 \pm 0.01 \text{ volts}.$$

(a) What should be her calculated value for the power delivered to the resistor, $P = IV$, with its uncertainty? **(b)** What for the resistance $R = V/I$? (Assume the original uncertainties are independent. With I in amps and V in volts, the power P comes out in watts and the resistance R in ohms.)

3.23. ★   In an experiment on the conservation of angular momentum, a student needs to find the angular momentum L of a uniform disc of mass M and radius R as it rotates with angular velocity ω. She makes the following measurements:

$$M = 1.10 \pm 0.01 \text{ kg},$$
$$R = 0.250 \pm 0.005 \text{ m},$$
$$\omega = 21.5 \pm 0.4 \text{ rad/s}$$

and then calculates L as $L = \frac{1}{2}MR^2\omega$. (The factor $\frac{1}{2}MR^2$ is just the moment of inertia of the uniform disc.) What is her answer for L with its uncertainty? (Consider the three original uncertainties independent and remember that the fractional uncertainty in R^2 is twice that in R.)

3.24. ★★   In his famous experiment with electrons, J.J. Thomson measured the "charge-to-mass ratio" $r = e/m$, where e is the electron's charge and m its mass. A modern classroom version of this experiment finds the ratio r by accelerating electrons through a voltage V and then bending them in a magnetic field. The ratio $r = e/m$ is given by the formula

$$r = \frac{125}{32\mu_o^2 N^2} \frac{D^2 V}{d^2 I^2}. \tag{3.53}$$

In this equation, μ_o is the permeability constant of the vacuum (equal to $4\pi \times 10^{-7} \text{N/A}^2$ exactly) and N is the number of turns in the coil that produces the magnetic field; D is the diameter of the field coils, V is the voltage that accelerates the electrons, d is the diameter of the electrons' curved path, and I is the current in the field coils. A student makes the following measurements:

$$N = 72 \text{ (exactly)}$$
$$D = 661 \pm 2 \text{ mm}$$
$$V = 45.0 \pm 0.2 \text{ volts}$$
$$d = 91.4 \pm 0.5 \text{ mm}$$
$$I = 2.48 \pm 0.04 \text{ amps}$$

(a) Find the student's answer for the charge-to-mass ratio of the electron, with its uncertainty. [Assume all uncertainties are independent and random. Note that the first factor in (3.53) is known exactly and can thus be treated as a single known constant, K. The second factor is a product and quotient of four numbers, D^2, V, d^2, and I^2, so the fractional uncertainty in the final answer is given by the rule (3.18). Remember that the fractional uncertainty in D^2 is twice that in D, and so on.] (b) How well does this answer agree with the accepted value $r = 1.759 \times 10^{11}$ C/kg? (Note that you don't actually need to understand the theory of this experiment to do the problem. Nor do you need to worry about the units; if you use SI units for all the input quantities, the answer automatically comes out in the units given.)

3.25. ★★ We know from the rule (3.10) for uncertainties in a power that if $q = x^2$, the fractional uncertainty in q is twice that in x;

$$\frac{\delta q}{q} = 2\frac{\delta x}{|x|}.$$

Consider the following (fallacious) argument. We can regard x^2 as x times x; so

$$q = x \times x;$$

therefore, by the rule (3.18),

$$\frac{\delta q}{q} = \sqrt{\left(\frac{\delta x}{x}\right)^2 + \left(\frac{\delta x}{x}\right)^2} = \sqrt{2}\,\frac{\delta x}{|x|}.$$

This conclusion is wrong. In a few sentences, explain why.

For Section 3.7: Arbitrary Functions of One Variable

3.26. ★ In nuclear physics, the energy of a subatomic article can be measured in various ways. One way is to measure how quickly the particle is stopped by an obstacle such as a piece of lead and then to use published graphs of energy versus stopping rate. Figure 3.7 shows such a graph for photons (the particles of light) in lead. The vertical axis shows the photons' energy E in MeV (millions of electron volts), and the horizontal axis shows the corresponding absorption coefficient μ in cm^2/g. (The precise definition of this coefficient need not concern us here; μ is simply a suitable measure of how quickly the photon is stopped in the lead.) From this graph, you can obviously find the energy E of a photon as soon as you know its absorption coefficient μ.

(a) A student observes a beam of photons (all with the same energy, E) and finds that their absorption coefficient in lead is $\mu = 0.10 \pm 0.01$ cm^2/gram. Using the graph, find the energy E and the uncertainty δE. (You may find it helpful to draw on the graph the lines connecting the various points of interest, as done in Figure 3.3.) (b) What answer would the student have found if he had measured $\mu = 0.22 \pm 0.01$ cm^2/gram?

3.27. ★ A student finds the refractive index n of a piece of glass by measuring the critical angle θ for light passing from the glass into air as $\theta = 41 \pm 1°$. The relation

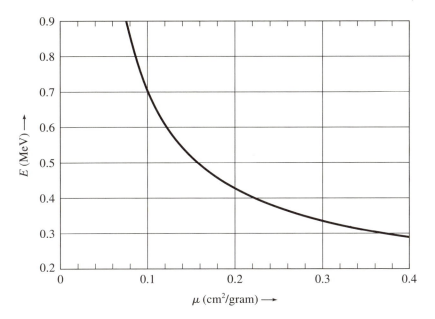

Figure 3.7. Energy E against absorption coefficient μ for photons in lead; for Problem 3.26.

between these is known to be $n = 1/\sin\theta$. Find the student's answer for n and use the rule (3.23) to find its uncertainty. (Don't forget to express $\delta\theta$ in radians.)

3.28. ★ (a) According to theory, the period T of a simple pendulum is $T = 2\pi\sqrt{L/g}$, where L is the length of the pendulum. If L is measured as $L = 1.40 \pm 0.01$ m, what is the predicted value of T? **(b)** Would you say that a measured value of $T = 2.39 \pm 0.01$ s is consistent with the theoretical prediction of part (a)?

3.29. ★ (a) An experiment to measure Planck's constant h gives it in the form $h = K\lambda^{1/3}$ where K is a constant known exactly and λ is the measured wavelength emitted by a hydrogen lamp. If a student has measured λ with a fractional uncertainty she estimates as 0.3%, what will be the fractional uncertainty in her answer for h? Comment. **(b)** If the student's best estimate for h is 6.644×10^{-34} J·s, is her result in satisfactory agreement with the accepted value of 6.626×10^{-34} J·s?

3.30. ★★ A spectrometer is a device for separating the different wavelengths in a beam of light and measuring the wavelengths. It deflects the different wavelengths through different angles θ, and, if the relation between the angle θ and wavelength λ is known, the experimenter can find λ by measuring θ. Careful measurements with a certain spectrometer have established the calibration curve shown in Figure 3.8; this figure is simply a graph of λ (in nanometers, or nm) against θ, obtained by measuring θ for several accurately known wavelengths λ. A student directs a narrow beam of light from a hydrogen lamp through this spectrometer and finds that the

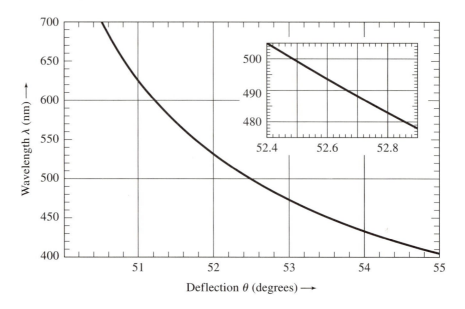

Figure 3.8. Calibration curve of wavelength λ against deflection θ for a spectrometer; for Problem 3.30.

light consists of just three well-defined wavelengths; that is, he sees three narrow beams (one red, one turquoise, and one violet) emerging at three different angles. He measures these angles as

$$\theta_1 = 51.0 \pm 0.2°$$
$$\theta_2 = 52.6 \pm 0.2°$$
$$\theta_3 = 54.0 \pm 0.2°$$

(a) Use the calibration curve of Figure 3.8 to find the corresponding wavelengths λ_1, λ_2, and λ_3 with their uncertainties. **(b)** According to theory, these wavelengths should be 656, 486, and 434 nm. Are the student's measurements in satisfactory agreement with these theoretical values? **(c)** If the spectrometer has a vernier scale to read the angles, the angles can be measured with an uncertainty of 0.05° or even less. Let us suppose the three measurements above have uncertainties of $\pm 0.05°$. Given this new, smaller uncertainty in the angles and *without drawing any more lines on the graph*, use your answers from part (a) to find the new uncertainties in the three wavelengths, explaining clearly how you do it. (Hint: the calibration curve is nearly straight in the vicinity of any one measurement.) **(d)** To take advantage of more accurate measurements, an experimenter may need to enlarge the calibration curve. The inset in Figure 3.8 is an enlargement of the vicinity of the angle θ_2. Use this graph to find the wavelength λ_2 if θ_2 has been measured as $52.72 \pm 0.05°$; check that your prediction for the uncertainty of λ_2 in part (c) was correct.

3.31. ★★ **(a)** An angle θ is measured as $125 \pm 2°$, and this value is used to compute $\sin\theta$. Using the rule (3.23), calculate $\sin\theta$ and its uncertainty. **(b)** If a is measured as $a_{\text{best}} \pm \delta a$, and this value used to compute $f(a) = e^a$, what are f_{best} and

δf? If $a = 3.0 \pm 0.1$, what are e^a and its uncertainty? **(c)** Repeat part (b) for the function $f(a) = \ln a$.

3.32. ★★★ The rule (3.23), $\delta q = |dq/dx|\delta x$, usually allows the uncertainty in a function $q(x)$ to be found quickly and easily. Occasionally, if $q(x)$ is very complicated, evaluating its derivative may be a nuisance, and going back to (3.20), from which (3.23) was derived, is sometimes easier. Note, however, that (3.20) was derived for a function whose slope was positive; if the slope is negative, the signs need to be reversed, and the general form of (3.20) is

$$\delta q = |q(x_{best} + \delta x) - q(x_{best})|. \tag{3.54}$$

Particularly if you have programmed your calculator or computer to find $q(x)$, then finding $q(x_{best} + \delta x)$ and $q(x_{best})$ and their difference will be easy.
(a) If you have a computer or programmable calculator, write a program to calculate the function

$$q(x) = \frac{(1 + x^2)^3}{x^2 + \cot x}.$$

Use this program to find $q(x)$ if $x = 0.75 \pm 0.1$, using the new rule (3.54) to find δq. **(b)** If you have the courage, differentiate $q(x)$ and check your value of δq using the rule (3.23).

3.33. ★★★ Do Problem 3.32 but use the function

$$q(x) = (1 - x^2) \cos\left(\frac{x + 2}{x^3}\right)$$

and the measured value $x = 1.70 \pm 0.02$.

For Section 3.8: Propagation Step by Step

3.34 ★ Use step-by-step propagation to find the following quantities (assuming that all given uncertainties are independent and random):
  (a) $(20 \pm 1) + [(5.0 \pm 0.4) \times (3.0 \pm 0.2)]$
  (b) $(20 \pm 1)/[(5.0 \pm 0.1) - (3.0 \pm 0.1)]$
  (c) $(1.5 \pm 0.1) - 2 \sin(30 \pm 6°)$
[In part (c), the number 2 is exact.]

3.35. ★ Use step-by-step propagation to find the following quantities (assuming that all given uncertainties are independent and random):
  (a) $(20 \pm 1) + [(50 \pm 1)/(5.0 \pm 0.2)]$
  (b) $(20 \pm 1) \times [(30 \pm 1) - (24 \pm 1)]$
  (c) $(2.0 \pm 0.1) \times \tan(45 \pm 3°)$

3.36. ★ Calculate the following quantities in steps as described in Section 3.8. Assume all uncertainties are independent and random.
  (a) $(12 \pm 1) \times [(25 \pm 3) - (10 \pm 1)]$
  (b) $\sqrt{16 \pm 4} + (3.0 \pm 0.1)^3(2.0 \pm 0.1)$
  (c) $(20 \pm 2)e^{-(1.0 \pm 0.1)}$

3.37. ★ **(a)** To find the acceleration of a glider moving down a sloping air track, I measure its velocities (v_1 and v_2) at two points and the time t it takes between them, as follows:

$$v_1 = 0.21 \pm 0.05, \quad v_2 = 0.85 \pm 0.05$$

(both in m/s) and

$$t = 8.0 \pm 0.1 \text{ s.}$$

Assuming all uncertainties are independent and random, what should I report for the acceleration, $a = (v_2 - v_1)/t$ and its uncertainty? **(b)** I have calculated theoretically that the acceleration should be 0.13 ± 0.01 m/s^2. Does my measurement agree with this prediction?

3.38. ★ **(a)**   As in Problem 3.37, I measure the velocities, v_1 and v_2, of a glider at two points on a sloping air track with the results given there. Instead of measuring the time between the two points, I measure the distance as

$$d = 3.740 \pm 0.002 \text{ m.}$$

If I now calculate the acceleration as $a = (v_2{}^2 - v_1{}^2)/2d$, what should be my answer with its uncertainty? **(b)** How well does it agree with my theoretical prediction that $a = 0.13 \pm 0.01$ m/s^2?

3.39. ★★ **(a)**   The glider on a horizontal air track is attached to a spring that causes it to oscillate back and forth. The total energy of the system is $E = \frac{1}{2}mv^2 + \frac{1}{2}kx^2$, where m is the glider's mass, v is its velocity, k is the spring's force constant, and x is the extension of the spring from equilibrium. A student makes the following measurements:

$$m = 0.230 \pm 0.001 \text{ kg}, \quad v = 0.89 \pm 0.01 \text{ m/s},$$
$$k = 1.03 \pm 0.01 \text{ N/m}, \quad x = 0.551 \pm 0.005 \text{ m.}$$

What is her answer for the total energy E? **(b)** She next measures the position x_{max} of the glider at the extreme end of its oscillation, where $v = 0$, as

$$x_{max} = 0.698 \pm 0.002 \text{ m.}$$

What is her value for the energy at the end point? **(c)** Are her results consistent with conservation of energy, which requires that these two energies should be the same?

For Section 3.9: Examples

3.40. ★★   Review the discussion of the simple pendulum in Section 3.9. In a real experiment, one should measure the period T for several different lengths l and hence obtain several different values of g for comparison. With a little thought, you can organize all data and calculations so that they appear in a single convenient tabulation, as in Table 3.2. Using Table 3.2 (or some other arrangement that you prefer), calculate g and its uncertainty δg for the four pairs of data shown. Are your answers consistent with the accepted value, 980 cm/s^2? Comment on the variation of δg as l gets smaller. (The answers given for the first pair of data will let you check your method of calculation.)

Table 3.2. Finding g with a pendulum; for Problem 3.40.

l (cm) all ± 0.1	T (sec) all ± 0.001	g (cm/s^2)	$\delta l/l$ (%)	$\delta T/T$ (%)	$\delta g/g$ (%)	answer $g \pm \delta g$
93.8	1.944	980	0.1	0.05	0.14	980 ± 1.4
70.3	1.681					
45.7	1.358					
21.2	0.922					

3.41. ★★ Review the measurement of the refractive index of glass in Section 3.9. Using a table similar to Table 3.1, calculate the refractive index n and its fractional uncertainty for the data in Table 3.3. Are your answers consistent with the manufacturer's claim that $n = 1.50$? Comment on the variation in the uncertainties. (All angles are in degrees, i is the angle of incidence, r that of refraction.)

Table 3.3. Refractive index data (in degrees); for Problem 3.41.

i (all ± 1)	10	20	30	50	70
r (all ± 1)	7	13	20	29	38

For Section 3.10: A More Complicated Example

3.42. ★★★ Review the experiment in Section 3.10, in which a cart is rolled down an incline of slope θ. **(a)** If the cart's wheels are smooth and light, the expected acceleration is $g \sin \theta$. If θ is measured as 5.4 ± 0.1 degrees, what are the expected acceleration and its uncertainty? **(b)** If the experiment is repeated giving the cart various pushes at the top of the slope, the data and all calculations can be recorded as usual, in a single tabulation like Table 3.4. Using Equation (3.33) for the acceler-

Table 3.4. Acceleration data; for Problem 3.42.

t_1 (s) all ± 0.001	t_2 (s) all ± 0.001	$\dfrac{1}{t_1^2}$	$\dfrac{1}{t_2^2}$	$\dfrac{1}{t_2^2} - \dfrac{1}{t_1^2}$	a (cm/s^2)
$0.054 \pm 2\%$	$0.031 \pm 3\%$	343 ± 14	1040 ± 62	698 ± 64	87 ± 8
0.038	0.027				
0.025	0.020				

ation (and the same values $l^2/2s = 0.125$ cm $\pm 2\%$ as before), calculate a and δa for the data shown. Are the results consistent with the expected constancy of a and with the expected value $g \sin \theta$ of part (a)? Would pushing the cart harder to check the constancy of a at even higher speeds be worthwhile? Explain.

For Section 3.11: General Formula for Error Propagation

3.43. ★   The partial derivative $\partial q/\partial x$ of $q(x, y)$ is obtained by differentiating q with respect to x while treating y as a constant. Write down the partial derivatives $\partial q/\partial x$ and $\partial q/\partial y$ for the three functions:

$$\textbf{(a)} \ q(x, y) \ = \ x + y, \quad \textbf{(b)} \ q(x, y) \ = \ xy, \quad \textbf{(c)} \ q(x, y) \ = \ x^2y^3.$$

3.44. ★★   The crucial approximation used in Section 3.11 relates the value of the function q at the point $(x + u, y + v)$ to that at the nearby point (x, y):

$$q(x + u, y + v) \ \approx \ q(x, y) + \frac{\partial q}{\partial x}u + \frac{\partial q}{\partial y}v \tag{3.55}$$

when u and v are small. Verify explicitly that this approximation is good for the three functions of Problem 3.43. That is, for each function, write both sides of Equation (3.55) exactly, and show that they are approximately equal when u and v are small. For example, if $q(x, y) \ = \ xy$, then the left side of Equation (3.55) is

$$(x + u)(y + v) \ = \ xy + uy + xv + uv.$$

As you will show, the right side of (3.55) is

$$xy + yu + xv.$$

If u and v are small, then uv can be neglected in the first expression, and the two expressions are approximately equal.

3.45. ★ (a)   For the function $q(x, y) = xy$, write the partial derivatives $\partial q/\partial x$ and $\partial q/\partial y$. Suppose we measure x and y with uncertainties δx and δy and then calculate $q(x, y)$. Using the general rules (3.47) and (3.48), write the uncertainty δq both for the case when δx and δy are independent and random, and for the case when they are not. Divide through by $|q| = |xy|$, and show that you recover the simple rules (3.18) and (3.19) for the fractional uncertainty in a product. **(b)** Repeat part (a) for the function $q(x, y) = x^ny^m$, where n and m are known fixed numbers. **(c)** What do Equations (3.47) and (3.48) become when $q(x)$ depends on only one variable?

3.46. ★★   If you measure two independent variables as

$$x \ = \ 6.0 \pm 0.1 \quad \text{and} \quad y \ = \ 3.0 \pm 0.1,$$

and use these values to calculate $q = xy + x^2/y$, what will be your answer and its uncertainty? [You must use the general rule (3.47) to find δq. To simplify your calculation, do it by first finding the two separate contributions δq_x and δq_y as defined in (3.50) and (3.51) and then combining them in quadrature.]

3.47. ★★   The Atwood machine consists of two masses M and m (with $M > m$) attached to the ends of a light string that passes over a light, frictionless pulley. When the masses are released, the mass M is easily shown to accelerate down with an acceleration

$$a \ = \ g\frac{M - m}{M + m}.$$

Suppose that M and m are measured as $M = 100 \pm 1$ and $m = 50 \pm 1$, both in grams. Use the general rule (3.47) to derive a formula for the uncertainty in the expected acceleration δa in terms of the masses and their uncertainties and then find δa for the given numbers.

3.48. ★★★ If we measure three independent quantities x, y, and z and then calculate a function such as $q = (x + y)/(x + z)$, then, as discussed at the beginning of Section 3.11, a stepwise calculation of the uncertainty in q may overestimate the uncertainty δq. **(a)** Consider the measured values $x = 20 \pm 1$, $y = 2$, and $z = 0$, and for simplicity, suppose that δy and δz are negligible. Calculate the uncertainty δq correctly using the general rule (3.47) and compare your result with what you would get if you were to calculate δq in steps. **(b)** Do the same for the values $x = 20 \pm 1$, $y = -40$, and $z = 0$. Explain any differences between parts (a) and (b).

3.49. ★★★ If an object is placed at a distance p from a lens and an image is formed at a distance q from the lens, the lens's focal length can be found as

$$f = \frac{pq}{p + q}. \qquad (3.56)$$

[This equation follows from the "lens equation," $1/f = (1/p) + (1/q)$.] **(a)** Use the general rule (3.47) to derive a formula for the uncertainty δf in terms of p, q, and their uncertainties. **(b)** Starting from (3.56) directly, you cannot find δf in steps because p and q both appear in numerator and denominator. Show, however, that f can be rewritten as

$$f = \frac{1}{(1/p) + (1/q)}.$$

Starting from this form, you *can* evaluate δf in steps. Do so, and verify that you get the same answer as in part (a).

3.50. ★★★ Suppose you measure three independent variables as

$$x = 10 \pm 2, \quad y = 7 \pm 1, \quad \theta = 40 \pm 3°,$$

and use these values to compute

$$q = \frac{x + 2}{x + y \cos(4\theta)}.$$

What should be your answer for q and its uncertainty? Note that you cannot do the error propagation in steps here because the variable x appears in both numerator and denominator; therefore, you must use the general rule (3.47).

Chapter 4

Statistical Analysis of Random Uncertainties

We have seen that one of the best ways to assess the reliability of a measurement is to repeat it several times and examine the different values obtained. In this chapter and Chapter 5, I describe statistical methods for analyzing measurements in this way.

As noted before, not all types of experimental uncertainty can be assessed by statistical analysis based on repeated measurements. For this reason, uncertainties are classified into two groups: the *random* uncertainties, which *can* be treated statistically, and the *systematic* uncertainties, which *cannot*. This distinction is described in Section 4.1. Most of the remainder of this chapter is devoted to random uncertainties. Section 4.2 introduces, without formal justification, two important definitions related to a series of measured values $x_1, \ldots, x_N$, all of some single quantity x. First, I define the *average* or *mean* $\bar{x}$ of $x_1, \ldots, x_N$. Under suitable conditions, $\bar{x}$ is the best estimate of x based on the measured values $x_1, \ldots, x_N$. I then define the *standard deviation* of $x_1, \ldots, x_N$, which is denoted σ_x and characterizes the average uncertainty in the separate measured values $x_1, \ldots, x_N$. Section 4.3 gives an example of the use of the standard deviation.

Section 4.4 introduces the important notion of the *standard deviation of the mean*. This parameter is denoted $\sigma_{\bar{x}}$ and characterizes the uncertainty in the mean $\bar{x}$ as the best estimate for x. Section 4.5 gives examples of the standard deviation of the mean. Finally, in Section 4.6, I return to the vexing problem of systematic errors.

Nowhere in this chapter do I attempt a complete justification of the methods described. The main aim is to introduce the basic formulas and describe how they are used. In Chapter 5, I give proper justifications, based on the important idea of the normal distribution curve.

The relation of the material of this chapter (statistical analysis) to the material of Chapter 3 (error propagation) deserves mention. From a practical point of view, these two topics can be viewed as separate, though related, branches of error analysis (somewhat as algebra and geometry are separate, though related, branches of mathematics). Both topics need to be mastered, because most experiments require the use of both.

In a few kinds of experiments, the roles of error propagation and of statistical analysis are complementary. That is, the experiment can be analyzed using either error propagation or statistical methods. Consider an example: Suppose you decide to measure the acceleration of gravity, g, by measuring the period, T, and the length, l, of a simple pendulum. Since $T = 2\pi\sqrt{l/g}$, you can find g as $g = 4\pi^2 l/T^2$. You might decide to repeat this experiment using several different values of l and measuring the corresponding period T for each. In this way, you would arrive at several values for g. To find the uncertainty in these values of g, you could proceed in either of two ways. If you can estimate realistically the uncertainties in your measurements of l and T, you could propagate these uncertainties to find the uncertainties in your values of g. Alternatively, given your several values of g, you could analyze them statistically; in particular, their *standard deviation* will be a good measure of their uncertainty. Unfortunately, you do not truly have a choice of how to find the uncertainty. If the uncertainty can be found in these two ways, you really ought to do so *both* ways to check that they do give, at least approximately, the same answer.

4.1   Random and Systematic Errors

Experimental uncertainties that can be revealed by repeating the measurements are called *random* errors; those that cannot be revealed in this way are called *systematic*. To illustrate this distinction, let us consider some examples. Suppose first that we time a revolution of a steadily rotating turntable. One source of error will be our reaction time in starting and stopping the watch. If our reaction time were always exactly the same, these two delays would cancel one another. In practice, however, our reaction time will vary. We may delay more in starting, and so underestimate the time of a revolution; or we may delay more in stopping, and so overestimate the time. Since either possibility is equally likely, the sign of the effect is *random*. If we repeat the measurement several times, we will sometimes overestimate and sometimes underestimate. Thus, our variable reaction time will show up as a variation of the answers found. By analyzing the spread in results statistically, we can get a very reliable estimate of this kind of error.

On the other hand, if our stopwatch is running consistently slow, then all our times will be underestimates, and no amount of repetition (with the same watch) will reveal this source of error. This kind of error is called *systematic*, because it always pushes our result in the same direction. (If the watch runs slow, we always underestimate; if the watch runs fast, we always overestimate.) Systematic errors cannot be discovered by the kind of statistical analysis contemplated here.

As a second example of random versus systematic errors, suppose we have to measure some well-defined length with a ruler. One source of uncertainty will be the need to interpolate between scale markings; and this uncertainty is probably random. (When interpolating, we are probably just as likely to overestimate as to underestimate.) But there is also the possibility that our ruler has become distorted; and this source of uncertainty would probably be systematic. (If the ruler has stretched, we always underestimate; if it has shrunk, we always overestimate.)

Just as in these two examples, almost all measurements are subject to both random and systematic uncertainties. You should have no difficulty finding more examples. In particular, notice that common sources of random uncertainties are

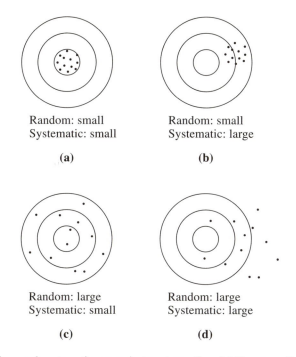

Random: small
Systematic: small

(a)

Random: small
Systematic: large

(b)

Random: large
Systematic: small

(c)

Random: large
Systematic: large

(d)

Figure 4.1. Random and systematic errors in target practice. **(a)** Because all shots arrived close to one another, we can tell the random errors are small. Because the distribution of shots is centered on the center of the target, the systematic errors are also small. **(b)** The random errors are still small, but the systematic ones are much larger—the shots are "systematically" off-center toward the right. **(c)** Here, the random errors are large, but the systematic ones are small—the shots are widely scattered but not systematically off-center. **(d)** Here, both random and systematic errors are large.

small errors of judgment by the observer (as when interpolating), small disturbances of the apparatus (such as mechanical vibrations), problems of definition, and several others. Perhaps the most obvious cause of systematic error is the miscalibration of instruments, such as the watch that runs slow, the ruler that has been stretched, or a meter that is improperly zeroed.

To get a better feel for the difference between random and systematic errors, consider the analogy shown in Figure 4.1. Here the "experiment" is a series of shots fired at a target; accurate "measurements" are shots that arrive close to the center. Random errors are caused by anything that makes the shots arrive at randomly different points. For example, the marksman may have an unsteady hand, or fluctuating atmospheric conditions between the marksman and the target may distort the view of the target in a random way. Systematic errors arise if anything makes the shots arrive off-center in one "systematic" direction, for instance, if the gun's sights are misaligned. Note from Figure 4.1 how the results change according to the various combinations of small or large random or systematic errors.

Although Figure 4.1 is an excellent illustration of the effects of random and systematic errors, it is, nonetheless, misleading in one important respect. Because

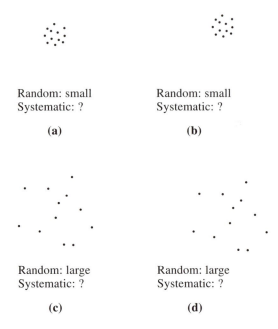

Random: small
Systematic: ?

(a)

Random: small
Systematic: ?

(b)

Random: large
Systematic: ?

(c)

Random: large
Systematic: ?

(d)

Figure 4.2. The same experiment as in Figure 4.1 redrawn without showing the position of the target. This situation corresponds closely to the one in most real experiments, in which we do not know the true value of the quantity being measured. Here, we can still assess the random errors easily but cannot tell anything about the systematic ones.

each of the four pictures shows the position of the target, we can tell at a glance whether a particular shot was accurate or not. In particular, the difference between the top two pictures is immediately evident. The shots in the left picture cluster around the target's center, whereas those in the right picture cluster around a point well off-center; clearly, therefore, the marksman responsible for the left picture had little systematic error, but the one responsible for the right picture had a lot more. Knowing the position of the target in Figure 4.1 corresponds, in a laboratory measurement, to knowing the true value of the measured quantity, and in the vast majority of real measurements, we do *not* know this true value. (If we knew the true value, we would usually not bother to measure it.)

To improve the analogy of Figure 4.1 with most real experiments, we need to redraw it without the rings that show the position of the target, as in Figure 4.2. In these pictures, identifying the random errors is still easy. (The top two pictures still obviously have smaller random errors than the bottom two.) Determining which marksman had larger systematic errors, however, is *impossible* based on Figure 4.2. This situation is exactly what prevails in most real experiments; by examining the distribution of measured values, we can easily assess the random errors but get no guidance concerning the systematic errors.

The distinction between random and systematic errors is not always clear-cut, and a problem that causes random errors in one experiment may produce systematic errors in another. For example, if you position your head first to one side and then to another to read a typical meter (such as an ordinary clock), the reading on the meter changes. This effect, called *parallax,* means that a meter can be read correctly only if you position yourself directly in front of it. No matter how careful you are, you cannot always position your eye *exactly* in front of the meter; consequently, your measurements will have a small uncertainty due to parallax, and this uncertainty will probably be random. On the other hand, a careless experimenter who places a meter to one side of his seat and forgets to worry about parallax will introduce a systematic error into all his readings. Thus, the same effect, parallax, can produce random uncertainties in one case, and systematic uncertainties in another.

The treatment of random errors is different from that of systematic errors. The statistical methods described in the following sections give a reliable estimate of the random uncertainties, and, as we shall see, provide a well-defined procedure for reducing them. For the reasons just discussed, systematic uncertainties are usually hard to evaluate and even to detect. The experienced scientist has to learn to anticipate the possible sources of systematic error and to make sure that all systematic errors are much less than the required precision. Doing so will involve, for example, checking the meters against accepted standards and correcting them or buying better ones if necessary. Unfortunately, in the first-year physics laboratory, such checks are rarely possible, so the treatment of systematic errors is often awkward. This concept is discussed further in Section 4.6. For now, I will discuss experiments in which all sources of systematic error have been identified and made much smaller than the required precision.

4.2 The Mean and Standard Deviation

Suppose we need to measure some quantity x, and we have identified all sources of systematic error and reduced them to a negligible level. Because all remaining sources of uncertainty are random, we should be able to detect them by repeating the measurement several times. We might, for example, make the measurement five times and find the results

$$71, 72, 72, 73, 71 \qquad (4.1)$$

(where, for convenience, we have omitted any units).

The first question we address is this: Given the five measured values (4.1), what should we take for our best estimate x_{best} of the quantity x? Reasonably, our best estimate would seem to be the *average* or *mean* $\bar{x}$ of the five values found, and in Chapter 5, I will prove that this choice is normally best. Thus,

$$\begin{aligned} x_{\text{best}} &= \bar{x} \\ &= \frac{71 + 72 + 72 + 73 + 71}{5} \\ &= 71.8. \end{aligned} \qquad (4.2)$$

Here, the second line is simply the definition of the mean $\bar{x}$ for the numbers at hand.[1]

More generally, suppose we make N measurements of the quantity x (all using the same equipment and procedures) and find the N values

$$x_1, x_2, \ldots, x_N. \tag{4.3}$$

Once again, the best estimate for x is usually the average of $x_1, \ldots, x_N$. That is,

$$x_{\text{best}} = \bar{x}, \tag{4.4}$$

where

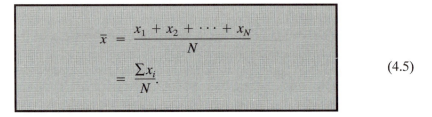

$$\bar{x} = \frac{x_1 + x_2 + \cdots + x_N}{N}$$

$$= \frac{\sum x_i}{N}. \tag{4.5}$$

In the last line, I have introduced the useful sigma notation, according to which

$$\sum_{i=1}^{N} x_i = \sum_i x_i = \sum x_i = x_1 + x_2 + \cdots + x_N;$$

the second and third expressions here are common abbreviations, which I will use when there is no danger of confusion.

The concept of the average or mean is almost certainly familiar to most readers. Our next concept, that of the *standard deviation,* is probably less so. The standard deviation of the measurements $x_1, \ldots, x_N$ is an estimate of the *average uncertainty of the measurements* $x_1, \ldots, x_N$ and is determined as follows.

Given that the mean $\bar{x}$ is our best estimate of the quantity x, it is natural to consider the difference $x_i - \bar{x} = d_i$. This difference, often called the *deviation* (or residual) of x_i from $\bar{x}$, tells us *how much the ith measurement x_i differs from the average $\bar{x}$.* If the deviations $d_i = x_i - \bar{x}$ are all very small, our measurements are all close together and presumably very precise. If some of the deviations are large, our measurements are obviously not so precise.

To be sure you understand the idea of the deviation, let us calculate the deviations for the set of five measurements reported in (4.1). These deviations can be listed as shown in Table 4.1. Notice that the deviations are not (of course) all the same size; d_i is small if the ith measurement x_i happens to be close to $\bar{x}$, but d_i is large if x_i is far from $\bar{x}$. Notice also that some of the d_i are positive and some negative because some of the x_i are bound to be higher than the average $\bar{x}$, and some are bound to be lower.

To estimate the average reliability of the measurements $x_1, \ldots, x_5$, we might naturally try averaging the deviations d_i. Unfortunately, as a glance at Table 4.1 shows, the average of the deviations is zero. In fact, this average will be zero for

---

[1] In this age of pocket calculators, it is worth pointing out that an average such as (4.2) is easily calculated in your head. Because all the numbers are in the seventies, the same must be true of the average. All that remains is to average the numbers 1, 2, 2, 3, 1 in the units place. These numbers obviously average to $9/5 = 1.8$, and our answer is $\bar{x} = 71.8$.

Table 4.1. Calculation of deviations.

Trial number i	Measured value x_i	Deviation $d_i = x_i - \bar{x}$
1	71	−0.8
2	72	0.2
3	72	0.2
4	73	1.2
5	71	−0.8

$$\Sigma x_i = 359 \qquad \Sigma d_i = 0.0$$

$$\text{mean, } \bar{x} = \Sigma x_i / N = 359/5 = 71.8$$

any set of measurements $x_1, \ldots, x_N$ because the definition of the average $\bar{x}$ ensures that $d_i = x_i - \bar{x}$ is sometimes positive and sometimes negative in just such a way that $\bar{d}$ is zero (see Problem 4.4). Obviously, then, the average of the deviations is not a useful way to characterize the reliability of the measurements $x_1, \ldots, x_N$.

The best way to avoid this annoyance is to *square* all the deviations, which will create a set of *positive* numbers, and then average these numbers.[2] If we then take the square root of the result, we obtain a quantity with the same units as x itself. This number is called the *standard deviation* of $x_1, \ldots, x_N$, and is denoted σ_x:

$$\sigma_x = \sqrt{\frac{1}{N} \sum_{i=1}^{N} (d_i)^2} = \sqrt{\frac{1}{N} \sum_{i=1}^{N} (x_i - \bar{x})^2}. \tag{4.6}$$

With this definition, the standard deviation can be described as the *root mean square* (or RMS) deviation of the measurements $x_1, \ldots, x_N$. It proves to be a useful way to characterize the reliability of the measurements. [As we will discuss shortly, the definition (4.6) is sometimes modified by replacing the denominator N by $N - 1$.]

To calculate the standard deviation σ_x as defined by (4.6), we must compute the deviations d_i, square them, average these squares, and then take the square root of the result. For the data of Table 4.1, we start this calculation in Table 4.2.

Table 4.2. Calculation of the standard deviation.

Trial number i	Measured value x_i	Deviation $d_i = x_i - \bar{x}$	Deviation squared d_i^2
1	71	−0.8	0.64
2	72	0.2	0.04
3	72	0.2	0.04
4	73	1.2	1.44
5	71	−0.8	0.64

$$\Sigma x_i = 359 \qquad \Sigma d_i = 0.0 \qquad \Sigma d_i^2 = 2.80$$

$$\bar{x} = 359/5 = 71.8$$

---

[2] Another possibility would be to take the absolute values $|d_i|$ and average them, but the average of the d_i^2 proves more useful. The average of the $|d_i|$ is sometimes (misleadingly) called the *average deviation*.

Summing the numbers d_i^2 in the fourth column of Table 4.2 and dividing by 5, we obtain the quantity σ_x^2 (often called the *variance* of the measurements),

$$\sigma_x^2 = \frac{1}{N}\sum d_i^2 = \frac{2.80}{5} = 0.56. \qquad (4.7)$$

Taking the square root, we find the standard deviation

$$\sigma_x \approx 0.7. \qquad (4.8)$$

Thus the average uncertainty of the five measurements 71, 72, 72, 73, 71 is approximately 0.7.

Unfortunately, the standard deviation has an alternative definition. There are theoretical arguments for replacing the factor N in (4.6) by $(N-1)$ and defining the standard deviation σ_x of $x_1, \ldots, x_N$ as

$$\sigma_x = \sqrt{\frac{1}{N-1}\sum d_i^2} = \sqrt{\frac{1}{N-1}\sum (x_i - \bar{x})^2}. \qquad (4.9)$$

I will not try here to prove that definition (4.9) of σ_x is better than (4.6), except to say that the new "improved" definition is obviously a little larger than the old one (4.6) and that (4.9) corrects a tendency for (4.6) to understate the uncertainty in the measurements $x_1, \ldots, x_N$, especially if the number of measurements N is small. This tendency can be understood by considering the extreme (and absurd) case that $N = 1$ (that is, we make only one measurement). Here, the average $\bar{x}$ is equal to our one reading x_1, and the one deviation is automatically zero. Therefore, the definition (4.6) gives the absurd result $\sigma_x = 0$. On the other hand, the definition (4.9) gives $0/0$; that is, with definition (4.9), σ_x is undefined, which correctly reflects our total ignorance of the uncertainty after just one measurement. The definition (4.6) is sometimes called the *population standard deviation* and (4.9) the *sample standard deviation.*

The difference between the two definitions (4.6) and (4.9) is almost always numerically insignificant. You should always repeat a measurement many times (at least five, and preferably many more). Even if you make only five measurements ($N = 5$), the difference between $\sqrt{N} = 2.2$ and $\sqrt{N-1} = 2$ is, for most purposes, insignificant. For example, if we recalculate the standard deviation (4.8) using the improved definition (4.9), we obtain $\sigma_x = 0.8$ instead of $\sigma_x = 0.7$, not a very important difference. Nevertheless, you need to be aware of both definitions. In the physics laboratory, using the more conservative (that is, larger) definition (4.9) is almost always best, but in any case, your laboratory report should state clearly which definition you are using so that your readers can check the calculations for themselves.

Quick Check 4.1.   You measure the time for a cart to roll down the same length of track four times and get the following results:

21, 24, 25, 22

(in seconds). Find the average time and the standard deviation as given by the improved definition (4.9).

To understand the notion of the standard deviation, you must be able to calculate it yourself for simple cases such as that in Quick Check 4.1. Most scientific calculators, however, have a built-in function to do the calculation automatically, and you will certainly want to use this function for real experiments that involve numerous measurements. If you are not sure how to use your calculator to obtain standard deviations, take the time to learn, and then use the function to check your answer to Quick Check 4.1. Some calculators give you a choice of the definitions (4.6) or (4.9); some use just (4.9). Make sure you know what yours does.

4.3   The Standard Deviation as the Uncertainty in a Single Measurement

Recall the claim that the standard deviation σ_x characterizes the average uncertainty of the measurements $x_1, \ldots, x_N$ from which it was calculated. In Chapter 5, I will justify this claim by proving the following more precise statement. If you measure the same quantity x many times, always using the same method, and if all your sources of uncertainty are small and random, then your results will be distributed around the true value x_{true} in accordance with the so-called normal, or bell-shaped, curve. In particular, *approximately 68% of your results*[3] *will fall within a distance σ_x on either side of x_{true}*; that is, 68% of your measurements will fall in the range $x_{\text{true}} \pm \sigma_x$.

In other words, if you make a *single* measurement (using the same method), the *probability* is 68% that your result will be within σ_x of the correct value. Thus, we can adopt σ_x to mean exactly what we have been calling "uncertainty." If you make one measurement of x, the uncertainty associated with this measurement can be taken to be

$$\delta x = \sigma_x;$$

with this choice, you can be 68% confident that the measurement is within δx of the correct answer.

To illustrate the application of these ideas, suppose we are given a box of similar springs and told to measure their spring constants k. We might measure the spring constants by loading each spring and observing the resulting extension or, perhaps better, by suspending a mass from each spring and timing its oscillations. Whatever method we choose, we need to know k and its uncertainty δk for each spring, but it would be hopelessly time-consuming to repeat our measurements many times for each spring. Instead we reason as follows: If we measure k for the first

---

[3] As we will see, the exact number is 68.27 . . . %, but stating this kind of number so precisely is obviously absurd. In fact, it is often best to think of this number as "about two thirds."

spring several (say, 5 or 10) times, then the mean of these measurements should give a good estimate of k for the first spring. More important for now, the standard deviation σ_k of these 5 or 10 measurements provides us with an estimate of the uncertainty in our method for measuring k. Provided our springs are all reasonably similar and we use the same method to measure each one, we can reasonably expect the same uncertainty in each measurement.[4] Thus, for each subsequent spring we need to make only one measurement, and we can immediately state that the uncertainty δk is the standard deviation σ_k measured for the first spring, with a 68% confidence that our answer is within σ_k of the correct value.

To illustrate these ideas numerically, we can imagine making 10 measurements on the first spring and obtaining the following measured values of k (in newtons/meter):

$$86, 85, 84, 89, 85, 89, 87, 85, 82, 85. \tag{4.10}$$

From these values, we can immediately calculate $\bar{k} = 85.7$ N/m and, using the definition (4.9),

$$\sigma_k = 2.16 \text{ N/m} \tag{4.11}$$

$$\approx 2 \text{ N/m}. \tag{4.12}$$

The uncertainty in any one measurement of k is therefore approximately 2 N/m. If we now measure the second spring once and obtain the answer $k = 71$ N/m, we can without further ado take $\delta k = \sigma_k = 2$ N/m and state with 68% confidence that k lies in the range

$$(k \text{ for second spring}) = 71 \pm 2 \text{ N/m}. \tag{4.13}$$

4.4   The Standard Deviation of the Mean

If $x_1, \ldots, x_N$ are the results of N measurements of the same quantity x, then, as we have seen, our best estimate for the quantity x is their mean $\bar{x}$. We have also seen that the standard deviation σ_x characterizes the average uncertainty of the separate measurements $x_1, \ldots, x_N$. Our answer $x_{\text{best}} = \bar{x}$, however, represents a judicious combination of all N measurements, and we have every reason to think it will be more reliable than any one of the measurements taken alone. In Chapter 5, I will prove that the uncertainty in the final answer $x_{\text{best}} = \bar{x}$ is given by the standard deviation σ_x *divided by* $\sqrt{N}$. This quantity is called the *standard deviation of the mean,* or SDOM, and is denoted $\sigma_{\bar{x}}$:

$$\sigma_{\bar{x}} = \sigma_x/\sqrt{N}. \tag{4.14}$$

(Other common names are *standard error* and *standard error of the mean.*) Thus, based on the N measured values $x_1, \ldots, x_N$, we can state our final answer for the

---

[4] If some springs are very different from the first, our uncertainty in measuring them may be different. Thus, if the springs differ a lot, we would need to check our uncertainty by making several measurements for each of two or three different springs.

value of x as

$$(\text{value of } x) = x_{\text{best}} \pm \delta x,$$

where $x_{\text{best}} = \bar{x}$, the mean of $x_1, \ldots, x_N$, and δx is the standard deviation of the mean,

$$\delta x = \sigma_{\bar{x}} = \sigma_x/\sqrt{N}. \tag{4.15}$$

As an example, we can consider the 10 measurements reported in (4.10) of the spring constant k of one spring. As we saw, the mean of these values is $\bar{k} = 85.7$ N/m, and the standard deviation is $\sigma_k = 2.2$ N/m. Therefore, the standard deviation of the mean is

$$\sigma_{\bar{k}} = \sigma_k/\sqrt{10} = 0.7 \text{ N/m}, \tag{4.16}$$

and our final answer, based on these 10 measurements, would be that the spring has

$$k = 85.7 \pm 0.7 \text{ newtons/meter}. \tag{4.17}$$

When you give an answer like this, you must state clearly what the numbers are—namely, the mean and the standard deviation of the mean—so your readers can judge their significance for themselves.

An important feature of the standard deviation of the mean, $\sigma_{\bar{x}} = \sigma_x/\sqrt{N}$, is the factor $\sqrt{N}$ in the denominator. The standard deviation σ_x represents the average uncertainty in the individual measurements $x_1, \ldots, x_N$. Thus, if we were to make some more measurements (using the same technique), the standard deviation σ_x would not change appreciably. On the other hand, the standard deviation of the mean, $\sigma_x/\sqrt{N}$, would slowly decrease as we increase N. This decrease is just what we would expect. If we make more measurements before computing an average, we would naturally expect the final result to be more reliable, and this improved reliability is just what the denominator $\sqrt{N}$ in (4.15) guarantees. This conclusion provides one obvious way to improve the precision of our measurements.

Unfortunately, the factor $\sqrt{N}$ grows rather slowly as we increase N. For example, if we wish to improve our precision by a factor of 10 simply by increasing the number of measurements N, we will have to increase N by a factor of 100—a daunting prospect, to say the least! Furthermore, we are for the moment neglecting systematic errors, and these are *not* reduced by increasing the number of measurements. Thus, in practice, if you want to increase your precision appreciably, you will probably do better to improve your technique than to rely merely on increased numbers of measurements.

Quick Check 4.2.   A student makes five measurements of e, the magnitude of the electron's charge, as follows:

$$15, 17, 18, 14, 16,$$

all in units of 10^{-20} coulombs. Find her best estimate for e (as given by the mean) and its uncertainty (as given by the SDOM).

4.5   Examples

In this section, I discuss two examples of simple experiments that make use of the ideas of the past three sections.

Example: Area of a Rectangle

As a first, simple application of the standard deviation of the mean, imagine that we have to measure very accurately the area A of a rectangular plate approximately 2.5 cm $\times$ 5 cm. We first find the best available measuring device, which might be a vernier caliper, and then make several measurements of the length l and breadth b of the plate. To allow for irregularities in the sides, we make our measurements at several different positions, and to allow for small defects in the instrument, we use several different calipers (if available). We might make 10 measurements each of l and b and obtain the results shown in Table 4.3.

Table 4.3. Length and breadth (in mm).

	Measured values	Mean	SD	SDOM
l	24.25, 24.26, 24.22, 24.28, 24.24 24.25, 24.22, 24.26, 24.23, 24.24	$\bar{l} = 24.245$	$\sigma_l = 0.019$	$\sigma_{\bar{l}} = 0.006$
b	50.36, 50.35, 50.41, 50.37, 50.36 50.32, 50.39, 50.38, 50.36, 50.38	$\bar{b} = 50.368$	$\sigma_b = 0.024$	$\sigma_{\bar{b}} = 0.008$

Using the 10 observed values of l, you can quickly calculate the mean $\bar{l}$, the standard deviation σ_l, and the standard deviation of the mean $\sigma_{\bar{l}}$, as shown in the columns labeled mean, SD, and SDOM. In the same way you can calculate $\bar{b}$, σ_b, and $\sigma_{\bar{b}}$. Before doing any further calculations, you should examine these results to see if they seem reasonable. For example, the two standard deviations σ_l and σ_b are supposed to be the average uncertainty in the measurements of l and b. Because l and b were measured in exactly the same way, σ_l and σ_b should not differ significantly from each other or from what we judge to be a reasonable uncertainty for the measurements.

Having convinced yourself that the results so far are reasonable, you can quickly finish the calculations. The best estimate for the length is the mean $\bar{l}$ and the uncertainty is the SDOM $\sigma_{\bar{l}}$; so the final value for l is

$$l = 24.245 \pm 0.006 \text{ mm} \quad \text{(or 0.025\%)};$$

the number in parenthesis is the percentage uncertainty. Similarly, the value for b is

$$b = 50.368 \pm 0.008 \text{ mm} \quad \text{(or 0.016\%)}.$$

Finally, the best estimate for the area $A = lb$ is the product of these values, with a fractional uncertainty given by the quadratic sum of those in l and b (assuming the errors are independent):

$$A = (24.245 \text{ mm} \pm 0.025\%) \times (50.368 \text{ mm} \pm 0.016\%)$$
$$= 1221.17 \text{ mm}^2 \pm 0.03\%$$
$$= 1221.2 \pm 0.4 \text{ mm}^2. \tag{4.18}$$

To arrive at the answer (4.18) for A, we calculated the averages $\bar{l}$ and $\bar{b}$, each with an uncertainty equal to the standard deviation of its mean. We then calculated the area A as the product of $\bar{l}$ and $\bar{b}$ and found the uncertainty by propagation of errors. We could have proceeded differently. For instance, we could have multiplied the first measured value of l by the first value of b to give a first answer for A. Continuing in this way we could have calculated 10 answers for A and then have subjected these 10 answers to statistical analysis, calculating $\bar{A}$, σ_A, and finally $\sigma_{\bar{A}}$. If, however, the errors in l and b are independent and random, and if we make enough measurements, this alternative procedure will produce the same result as the first one.[5]

Example: Another Spring

As a second example, consider a case in which a statistical analysis cannot be applied to the direct measurements but can to the final answers. Suppose we wish to measure the spring constant k of a spring by timing the oscillations of a mass m fixed to its end. We <u>know</u> from elementary mechanics that the period for such oscillations is $T = 2\pi\sqrt{m/k}$. Thus, by measuring T and m, we can find k as

$$k = 4\pi^2 m/T^2. \tag{4.19}$$

The simplest way to find k is to take a single, accurately known mass m and make several careful measurements of T. For various reasons, however, timing T for several *different* masses m may be more interesting. (For example, in this way, we could check that $T \propto \sqrt{m}$ as well as measure k.) We might then get a set of readings such as those in the first two lines of Table 4.4.

Table 4.4. Measurement of spring constant k.

Mass m (kg)	0.513	0.581	0.634	0.691	0.752	0.834	0.901	0.950
Period T (s)	1.24	1.33	1.36	1.44	1.50	1.59	1.65	1.69
$k = 4\pi^2 m/T^2$	13.17	12.97	etc.					

It obviously makes no sense to average the various different masses in the top line (or the times in the second line) because they are *not* different measurements of the same quantity. Nor can we learn anything about the uncertainty in our measurements by comparing the different values of m. On the other hand, we can com-

---

[5]The second procedure has a certain illogic because there is no particular reason to associate the first measurement of l with the first measurement of b. Indeed, we might have measured l eight times and b twelve times; then we couldn't pair off values. Thus, our first procedure is logically preferable.

bine each value of m with its corresponding period T and calculate k, as in the final line of Table 4.4. Our answers for k in the bottom line *are* all measurements of the same quantity and so can be subjected to statistical analysis. In particular, our best estimate for k is the mean, $\bar{k} = 13.16$ N/m, and our uncertainty is the standard deviation of the mean, $\sigma_{\bar{k}} = 0.06$ N/m (see Problem 4.20). Thus, the final answer, based on the data of Table 4.4, is

$$\text{spring constant } k = 13.16 \pm 0.06 \text{ N/m}. \tag{4.20}$$

If we had formed reasonable estimates of the uncertainties in our original measurements of m and T, we could also have estimated the uncertainty in k by using error propagation, starting from these estimates for δm and δT. In this case, it would be a good idea to compare the final uncertainties in k obtained by the two methods.

4.6 Systematic Errors

In the past few sections, I have been taking for granted that all systematic errors were reduced to a negligible level before serious measurements began. Here, I take up again the disagreeable possibility of appreciable systematic errors. In the example just discussed, we may have been measuring m with a balance that read consistently high or low, or our timer may have been running consistently fast or slow. Neither of these systematic errors will show up in the comparison of our various answers for the spring constant k. As a result, the standard deviation of the mean $\sigma_{\bar{k}}$ can be regarded as the *random component* δk_{ran} of the uncertainty δk but is certainly not the total uncertainty δk. Our problem is to decide how to estimate the *systematic component* δk_{sys} and then how to combine δk_{ran} and δk_{sys} to give the complete uncertainty δk.

No simple theory tells us what to do about systematic errors. In fact, the only theory of systematic errors is that they must be identified and reduced until they are much less than the required precision. In a teaching laboratory, however, this goal is often not attainable. Students often cannot check a meter against a better one to correct it, much less buy a new meter to replace an inadequate one. For this reason, some teaching laboratories establish a rule that, in the absence of more specific information, meters should be considered to have some definite systematic uncertainty. For example, the decision might be that all stopwatches have up to 0.5% systematic uncertainty, all balances up to 1%, all voltmeters and ammeters up to 3%, and so on.

Given rules of this kind, there are various possible ways to proceed. None can really be rigorously justified, and we describe just one approach here. (Problems 4.23 to 4.28 contain more examples.) In the last example in Section 4.5, the spring constant $k = 4\pi^2 m/T^2$ was found by measuring a series of values of m and the corresponding values of T. As we have seen, a statistical analysis of the various answers for k gives the random component of δk as

$$\delta k_{\text{ran}} = \sigma_{\bar{k}} = 0.06 \text{ N/m}. \tag{4.21}$$

Suppose now we have been told that the balance used to measure m and the clock used for T have systematic uncertainties up to 1% and 0.5%, respectively. We can then find the systematic component of δk by propagation of errors; the only question is whether to combine the errors in quadrature or directly. Because the errors in m and T are surely independent and some cancellation is therefore possible, using the quadratic sum is probably reasonable[6]; this choice gives

$$\frac{\delta k_{sys}}{k} = \sqrt{\left(\frac{\delta m_{sys}}{m}\right)^2 + \left(2\frac{\delta T_{sys}}{T}\right)^2} \tag{4.22}$$

$$= \sqrt{(1\%)^2 + (1\%)^2} = 1.4\% \tag{4.23}$$

and hence

$$\delta k_{sys} = k_{best} \times (1.4\%) \tag{4.24}$$

$$= (13.16 \text{ N/m}) \times 0.014 = 0.18 \text{ N/m}.$$

Now that we have estimates for both the random and systematic uncertainties in k, we must decide how to state our final conclusion for the spring constant k with its overall uncertainty. Because the method for combining δk_{ran} and δk_{sys} is not completely clear, many scientists leave the two components separate and state a final answer in the form

$$\text{(measured value of } k) = k_{best} \pm \delta k_{ran} \pm \delta k_{sys} \tag{4.25}$$

$$= 13.16 \pm 0.06 \pm 0.18 \text{ N/m}$$

(all of which should probably be rounded to one decimal place). Alternatively, a case can be made that δk_{ran} and δk_{sys} should be combined in quadrature, in which case we could state a single, total uncertainty

$$\delta k = \sqrt{(\delta k_{ran})^2 + (\delta k_{sys})^2} \tag{4.26}$$

$$= \sqrt{(0.06)^2 + (0.18)^2} = 0.19 \text{ N/m}$$

and replace the conclusion (4.25) by

$$\text{(measured value of } k) = k_{best} \pm \delta k$$

$$= 13.16 \pm 0.19 \text{ N/m}$$

or, probably better, 13.2 ± 0.2 N/m.

The expression (4.26) for δk cannot really be rigorously justified. Nor is the significance of the answer clear; for example, we probably cannot claim 68% confidence that the true answer lies in the range $\bar{k} \pm \delta k$. Nonetheless, the expression does at least provide a reasonable estimate of our total uncertainty, given that our apparatus has systematic uncertainties we could not eliminate. In particular, there is one important respect in which the answer (4.26) is realistic and instructive. We saw in Section 4.4 that the standard deviation of the mean $\sigma_{\bar{k}}$ approaches zero as the number of measurements N is increased. This result suggested that, if you have the

---

[6]Whether we should use the quadratic or ordinary sum really depends on what is meant by the statement that the balance has "up to 1% systematic uncertainty." If it means the error is *certainly* no more than 1% (and likewise for the clock), then direct addition is appropriate, and δk_{sys} is then *certainly* no more than 2%. On the other hand, perhaps an analysis of all balances in the laboratory has shown that they follow a normal distribution, with 68% of them better than 1% reliable (and likewise for the clocks). In this case, we can use addition in quadrature as in (4.22) with the usual significance of 68% confidence.

patience to make an enormous number of measurements, you can reduce the uncertainties indefinitely without having to improve your equipment or technique. We can now see that this suggestion is incorrect. Increasing N can reduce the *random* component $\delta k_{ran} = \sigma_{\bar{x}}$ indefinitely. But any given apparatus has *some* systematic uncertainty, which is *not* reduced as we increase N. From (4.26) we clearly see that little is gained from further reduction of δk_{ran}, once δk_{ran} is smaller than δk_{sys}. In particular, the total δk can never be made less than δk_{sys}. This fact simply confirms what we already guessed, that in practice a large reduction of the uncertainty requires improvements in techniques or equipment to reduce both the random and systematic errors in each single measurement.

As discussed in Chapter 2, a peculiar feature of the teaching laboratory is that you will probably be asked to measure quantities, such as the acceleration of gravity, for which an accurate, accepted value is already known. In this kind of experiment, the logic of the error analysis is a bit confusing. Probably the most honest course is to ignore the known accepted value until *after* you have done all calculations of your measured value, q_{best}, and its uncertainty. Then, of course, you must ask whether the accepted value lies inside (or at least close to) the range $q_{best} \pm \delta q$. If it does, you can simply record this agreement in your report. If the accepted value lies well outside the range $q_{best} \pm \delta q$, however, you have to examine the possible causes of the excessive discrepancy. For example, you might measure g, the acceleration of gravity, and get the results (all in m/s^2),

$$g_{best} = 9.97, \qquad (4.27)$$

with uncertainties

$$\delta g_{ran} = 0.02 \quad \text{and} \quad \delta g_{sys} = 0.03,$$

and hence a total uncertainty, as in (4.26), of

$$\delta g = 0.04.$$

Clearly, the accepted value of

$$g = 9.80 \text{ m/s}^2$$

lies far outside the measured range, 9.97 ± 0.04. (More specifically, the discrepancy is 0.17, which is four times the uncertainty.) This result is definitely *not* satisfactory and further analysis is required.

The first thing to check is the possibility that you made a downright mistake in calculating g_{best} or one of the uncertainties δg_{ran} and δg_{sys}. If you can convince yourself that all your calculations were correct, the next possibility is that the accepted value is wrong. In the case of $g = 9.80$ m/s^2 this possibility is rather unlikely, but it is entirely possible for plenty of other cases. For example, suppose you were measuring the density of air; because this is strongly dependent on the temperature and pressure, you could easily have looked up the wrong accepted value for this parameter.

Once you have eliminated these suspects, only one possibility is left: You must have overlooked some systematic error so that your value of δg_{sys} is too small. Ideally, you should try to find the culprit, but this search can be hard because of the many possibilities:

(1) Perhaps one of your meters had larger systematic errors than you had allowed for when you calculated δg_{sys}. You can investigate this possibility by determining how large a systematic error in your clock (or voltmeter, or whatever) would be needed to account for the offending discrepancy. If the needed error is not unreasonably large, you have one possible explanation of your difficulty.

(2) Another possible cause of systematic error is that you used an incorrect value for some parameter needed in your calculations. A celebrated example of this was Millikan's famous measurement of the electron's charge, e. Millikan's method depended on the viscosity of air, for which he used a value that was 0.4% too small. This discrepancy caused all of his values of e to be 0.6% too small, an error that was not noticed for nearly 20 years. This kind of mistake sometimes arises in a teaching laboratory when a student uses a value that has too few significant figures. For example, suppose you do an experiment with protons and you expect to have an accuracy better than 1%. If you take the proton's mass to be 1.7×10^{-27} kg (instead of the more exact 1.67×10^{-27} kg), you will have introduced a 2% systematic error, which will almost certainly frustrate your hope for 1% results.

(3) Much harder to analyze is the possibility of a flaw in the design of the experiment. For example, if you had measured g by dropping an object from a great height, air resistance could introduce an appreciable systematic error. [Note, however, that this error would not account for the large value of g in (4.27) because air resistance would cause an acceleration that was too *small*.] Similarly, if you try to measure the half-life of a radioactive material and your sample is contaminated with another material of shorter half-life, you will get an answer that is systematically too short.

Obviously, tracking down the source of systematic errors is difficult and has defied the best efforts of many great scientists. In all probability, your instructors are not going to penalize you too severely if you fail to do so. Nevertheless, they will expect an intelligent discussion of the problem and at least an honest admission that there appear to have been systematic errors that you were unable to identify.

Principal Definitions and Equations of Chapter 4

Suppose that we make N measurements, $x_1, x_2, \ldots, x_N$ of the same quantity x, all using the same method. Provided all uncertainties are random and small, we have the following results:

THE MEAN

The best estimate for x, based on these measurements, is their mean:

$$\bar{x} = \frac{1}{N} \sum_{i=1}^{N} x_i. \qquad \text{[See (4.5)]}$$

THE STANDARD DEVIATION

The average uncertainty of the individual measurements $x_1, x_2, \ldots, x_N$ is given by the standard deviation, or SD:

$$\sigma_x = \sqrt{\frac{1}{N-1} \Sigma (x_i - \bar{x})^2}. \qquad \text{[See (4.9)]}$$

This definition of the SD, often called the *sample* standard deviation, is the most appropriate for our purposes. The *population* standard deviation is obtained by replacing the factor $(N-1)$ in the denominator by N. You will usually want to calculate standard deviations using the built-in function on your calculator; be sure you know which definition it uses.

The detailed significance of the standard deviation σ_x is that approximately 68% of the measurements of x (using the same method) should lie within a distance σ_x of the true value. (This claim is justified in Section 5.4.) This result is what allows us to identify σ_x as the *uncertainty* in any one measurement of x,

$$\delta x = \sigma_x,$$

and, with this choice, we can be 68% confident that any one measurement will fall within σ_x of the correct answer.

THE STANDARD DEVIATION OF THE MEAN

As long as systematic uncertainties are negligible, the uncertainty in our best estimate for x (namely $\bar{x}$) is the standard deviation of the mean, or SDOM,

$$\sigma_{\bar{x}} = \frac{\sigma_x}{\sqrt{N}}. \qquad \text{[See (4.14)]}$$

If there *are* appreciable systematic errors, then $\sigma_{\bar{x}}$ gives the *random component* of the uncertainty in our best estimate for x:

$$\delta x_{\text{ran}} = \sigma_{\bar{x}}.$$

If you have some way to estimate the systematic component δx_{sys}, a reasonable (but not rigorously justified) expression for the total uncertainty is the quadratic sum of δx_{ran} and δx_{sys}:

$$\delta x_{\text{tot}} = \sqrt{(\delta x_{\text{ran}})^2 + (\delta x_{\text{sys}})^2}. \qquad \text{[See (4.26)]}$$

Problems for Chapter 4

For Section 4.2: The Mean and Standard Deviation

4.1. ★ You measure the time for a ball to drop from a second-floor window three times and get the results (in tenths of a second):

11, 13, 12.

(a) Find the mean and the standard deviation. For the latter, use both the "improved" definition (4.9) (the *sample* standard deviation) and the original definition (4.6) (the *population* standard deviation). Note how, even with only three measurements, the difference between the two definitions is not very big. (b) If you don't yet know how to calculate the mean and standard deviation using your calculator's built-in functions, take a few minutes to learn. Use your calculator to check your answers to part (a); in particular, find out which definition of the standard deviation your calculator uses.

4.2. ★   A student measures g, the acceleration of gravity, five times, with the results (all in m/s^2):

$$9.9, \ 9.6, \ 9.5, \ 9.7, \ 9.8.$$

(a) Find her mean and the standard deviation [as defined by (4.9)]. Do the calculation yourself using a layout similar to that in Table 4.2. (b) Check your value for the SD using the built-in function on your calculator. (If you don't yet know how to calculate the mean and standard deviation using your calculator's built-in functions, take a few minutes to learn.)

4.3. ★   Find the mean and standard deviation of the 10 measurements reported in (4.10). If you haven't done either of the two previous problems, be sure to do this calculation yourself, then check it using the built-in functions on your calculator.

4.4. ★   The mean of N quantities $x_1, \ldots, x_N$ is defined as their sum divided by N; that is, $\bar{x} = \Sigma x_i / N$. The deviation of x_i is the difference $d_i = x_i - \bar{x}$. Show clearly that the mean of the deviations $d_1, \ldots, d_N$ is always zero.

If you are not used to the Σ notation, you might want to do this problem both without and with the notation. For example, write out the sum $\Sigma(x_i - \bar{x})$ as $(x_1 - \bar{x}) + (x_2 - \bar{x}) + \ \ldots \ + (x_N - \bar{x})$ and regroup the terms.

4.5. ★★   (a) Computing the standard deviation σ_x of N measurements $x_1, \ldots, x_N$ of a single quantity x requires that you compute the sum $\Sigma(x_i - \bar{x})^2$. Prove that this sum can be rewritten as

$$\Sigma[(x_i - \bar{x})^2] \ = \ \Sigma[(x_i)^2] - \frac{1}{N}[\Sigma(x_i)]^2. \tag{4.28}$$

This problem is a good exercise in using the Σ notation. Many calculators use the result to compute the standard deviation for the following reason: To use the expression on the left, a calculator must keep track of all the data (which uses a lot of memory) to calculate $\bar{x}$ and then the sum indicated; to use the expression on the right, the machine needs only to keep a running total of $\Sigma(x_i^2)$ and $\Sigma(x_i)$, which uses much less memory. (b) Verify the identity (4.28) for the three measurements of Problem 4.1.

4.6. ★★   In Chapter 3, you learned that in a counting experiment, the uncertainty associated with a counted number is given by the "square-root rule" as the square root of that number. This rule can now be made more precise with the following statements (proved in Chapter 11): If we make several counts

$$\nu_1, \ \nu_2, \ \ldots, \ \nu_N$$

of the number ν of random events that occur in a time T, then: **(1)** the best estimate for the true average number that occur in time T is the mean $\bar{\nu} = \sum\nu_i/N$ of our measurements, and **(2)** the *standard deviation* of the observed numbers should be approximately equal to the *square root* of this same best estimate; that is, the uncertainty in each measurement is $\sqrt{\nu}$. In particular, if we make only one count ν, the best estimate is just ν and the uncertainty is the square root $\sqrt{\nu}$; this result is just the square-root rule of Chapter 3 with the additional information that the "uncertainty" is actually the standard deviation and gives the margins within which we can be approximately 68% confident the true answer lies. This problem and Problem 4.7 explore these ideas.

A nuclear physicist uses a Geiger counter to monitor the number of cosmic-ray particles arriving in his laboratory in any two-second interval. He counts this number 20 times with the following results:

$$10, \ 13, \ \ 8, \ 15, \ \ 8, \ 13, \ 14, \ 13, \ 19, \ \ 8,$$
$$13, \ 13, \ \ 7, \ \ 8, \ \ 6, \ \ 8, \ 11, \ 12, \ \ 8, \ \ 7.$$

(a) Find the mean and standard deviation of these numbers. **(b)** The latter should be approximately equal to the square root of the former. How well is this expectation borne out?

4.7. ★★★   Read the first paragraph of Problem 4.6 and then do the following problem: A health physicist is testing a new detector and places it near a weak radioactive sample. In five separate 10-second intervals, the detector counts the following numbers of radioactive emissions:

$$16, \ 21, \ 13, \ 12, \ 15.$$

(a) Find the mean and standard deviation of these five numbers. **(b)** Compare the standard deviation with its expected value, the square root of the average number. **(c)** Naturally, the two numbers in part (b) do not agree exactly, and we would like to have some way to assess their disagreement. This problem is, in fact, one of error propagation. We have measured the number ν. The expected standard deviation in this number is just $\sqrt{\nu}$, a simple function of ν. Thus, the uncertainty in the standard deviation can be found by error propagation. Show, in this way, that the uncertainty in the SD is 0.5. Do the numbers in part (b) agree within this uncertainty?

4.8. ★★★   Spreadsheet programs, such as Lotus 123 or Excel, provide an excellent way to find the standard deviation of a set of measurements (and to record and process many kinds of data in general). **(a)** If you have access to one of these programs, create a spreadsheet to calculate the SD of any set of 10 measurements using the layout of Table 4.2. The first column should list the trial number i, and the second column should be where the user will enter the data x_i. The mean $\bar{x}$ will be calculated by a formula that gives

(sum of data entries)/(number of data entries),

which you can place to the side of the main table. In the third column, put a formula to calculate the deviation, $x_i - \bar{x}$, and in the fourth, a formula to calculate the deviation squared. Finally, somewhere to the side of the main table, write a formula to find the SD as

$$SD = \sqrt{\frac{\text{sum of squared deviations}}{N-1}}.$$

(Most spreadsheet programs can calculate standard deviations automatically, but the point here is for you to create your own program to do it, not to use the built-in functions.) **(b)** Test your spreadsheet on the data of Equation (4.10). **(c)** (The hard part.) Most simple solutions to part (a) have two irritating drawbacks. First, as you enter each of the data, the spreadsheet will calculate answers for the mean and SD, but these answers will probably be incorrect (even for the data entered so far) until you have entered *all* the data. Second, your spreadsheet will probably need some modification before you can use it to find the mean and SD of a different number of data. If you can, modify your spreadsheet so that it does not suffer these defects and can find the mean and SD of *any number* of data (up to some convenient maximum, say 30). **(d)** Test your new spreadsheet using the data of Problem 4.6, and convince yourself that as you enter the data the program gives the correct answer at each stage.

For Section 4.3: The Standard Deviation as the Uncertainty in a Single Measurement

4.9. ★   A student measures the period of a pendulum three times and gets the answers 1.6, 1.8, and 1.7, all in seconds. What are the mean and standard deviation? [Use the improved definition (4.9) of the standard deviation.] If the student decides to make a fourth measurement, what is the probability that this new measurement will fall outside the range of 1.6 to 1.8 s? (The numbers here were chosen to "come out right." In Chapter 5, I will explain how to do this kind of problem even when the numbers don't come out right.)

4.10. ★   After several measurements of a quantity x, we expect to find that approximately 68% of the measurements fall within the range of $\bar{x} \pm \sigma_x$. (The exact number is 68.27 ... %, but the difference is usually insignificant.) If you have not already done so, check the mean and standard deviation of the 10 measurements in (4.10). How many of the measurements should we expect to fall within the range of $\bar{x} \pm \sigma_x$? How many do?

4.11. ★   A student has to measure several unknown charges Q on a capacitor by discharging it through a ballistic galvanometer. The discharge kicks the galvanometer's needle, whose resulting maximum swing tells the student the value of Q. Before making all her measurements, the student wants to know how reliably she can read the maximum displacement of the swinging needle. Therefore, she arranges to charge the capacitor to exactly the same voltage (and hence the same charge) five times and to measure the resulting charge. Her results (in microcoulombs) are:

$$1.2, 1.4, 1.6, 1.6, 1.2.$$

Based on these results, what would you suggest she take for the uncertainty δQ in her subsequent measurements of the charge Q, assuming she wants to be 68% confident that the correct value is within $\pm \delta Q$ of her measurement?

4.12. ★   To calibrate a prism spectrometer, a student sends light of 10 different known wavelengths λ through the spectrometer and measures the angle θ by which each beam is deflected. Using these results, he makes a calibration curve like Figure 3.8 showing λ as a function of θ. For the first value of λ, he measures θ six times and obtains these results (in degrees):

$$52.5, \ 52.3, \ 52.6, \ 52.5, \ 52.7, \ 52.4.$$

For each of the nine remaining values of λ, he measures the corresponding value of θ just once. What should he take for the uncertainty in each of these nine measurements of θ?

4.13. ★★   **(a)** Calculate the mean and standard deviation for the following 30 measurements of a time t (in seconds):

$$8.16, \ 8.14, \ 8.12, \ 8.16, \ 8.18, \ 8.10, \ 8.18, \ 8.18, \ 8.18, \ 8.24,$$
$$8.16, \ 8.14, \ 8.17, \ 8.18, \ 8.21, \ 8.12, \ 8.12, \ 8.17, \ 8.06, \ 8.10,$$
$$8.12, \ 8.10, \ 8.14, \ 8.09, \ 8.16, \ 8.16, \ 8.21, \ 8.14, \ 8.16, \ 8.13.$$

(You should certainly use the built-in functions on your calculator (or the spreadsheet you created in Problem 4.8 if you did), and you can save some button pushing if you drop all the leading 8s and shift the decimal point two places to the right before doing any calculation.) **(b)** We know that after several measurements, we can expect about 68% of the observed values to be within σ_t of $\bar{t}$ (that is, inside the range $\bar{t} \pm \sigma_t$). For the measurements of part (a), about how many would you expect to lie *outside* the range $\bar{t} \pm \sigma_t$? How many do? **(c)** In Chapter 5, I will show that we can also expect about 95% of the values to be within $2\sigma_t$ of $\bar{t}$ (that is, inside the range $\bar{t} \pm 2\sigma_t$). For the measurements of part (a), about how many would you expect to lie *outside* the range $\bar{t} \pm 2\sigma_t$? How many do?

4.14. ★★   **(a)** If you have not yet done it, do Problem 4.6. **(b)** About how many of the measurements of this problem should you expect to lie outside the range $\bar{\nu} \pm \sigma_\nu$? How many do? **(c)** In Chapter 5, I will show that we can expect about 95% of the measurements to be within $2\sigma_\nu$ of $\bar{\nu}$ (that is, inside the range $\bar{\nu} \pm 2\sigma_\nu$). For the measurements of part (a), about how many would you expect to lie *outside* the range $\bar{\nu} \pm 2\sigma_\nu$? How many do?

For Section 4.4: The Standard Deviation of the Mean

4.15. ★   Given the three measurements in Problem 4.1, what should you state for your best estimate for the time concerned and its uncertainty? (Your answer will illustrate how the mean can have more significant figures than the original measurements.)

4.16. ★   **(a)** Based on the five measurements of g reported in Problem 4.2, what should be the student's best estimate for g and its uncertainty? **(b)** How well does her result agree with the accepted value of 9.8 m/s^2?

4.17. ★   **(a)** Based on the 30 measurements in Problem 4.13, what would be your best estimate for the time involved and its uncertainty, assuming all uncertainties are random? **(b)** Comment on the number of significant digits in your best estimate, as compared with the number of significant digits in the data.

4.18. ★   After measuring the speed of sound u several times, a student concludes that the standard deviation σ_u of her measurements is $\sigma_u = 10$ m/s. If all uncertainties were truly random, she could get any desired precision by making enough measurements and averaging. **(a)** How many measurements are needed to give a final uncertainty of ± 3 m/s? **(b)** How many for a final uncertainty of only ± 0.5 m/s?

4.19. ★★★   **(a)** The data in Problem 4.6 are 20 measurements of the number of cosmic-ray particles counted by a Geiger counter in 2 seconds. By averaging these numbers, find the best estimate for the number of particles that arrive in 2 seconds and the uncertainty in that number (as given by the SDOM). **(b)** Another way to do this problem is as follows: If you add all the data, you will have the number counted in 40 seconds (with an uncertainty given by the square root of that number). If you now divide this result by 20, you should get the number of particles that arrive in 2 seconds and its uncertainty. Check that your answers in (b) and (c) agree, at least approximately. (Assume that that the 2-second time intervals are measured with negligible uncertainty.) **(c)** Prove that these two methods should, ideally, give the same answers. (Your proof should be general; that is, you should work in terms of algebraic symbols, not the specific numbers of Problem 4.6. Remember that the SD of any counted number should be the square root of that number.)

For Section 4.5: Examples

4.20. ★   Complete the calculations of the spring constant k in Table 4.4. Then compute $\bar{k}$ and its uncertainty (the SDOM, $\sigma_{\bar{k}}$).

4.21. ★   Table 4.3 records 10 measurements each of the length l and breadth b of a rectangle. These values were used to calculate the area $A = lb$. If the measurements were made in pairs (one of l and one of b), it would be natural to multiply each pair together to give a value of A—the first l times the first b to give a first value of A, and so on. Calculate the resulting 10 values of A, the mean $\bar{A}$, the SD σ_A, and the SDOM $\sigma_{\bar{A}}$. Compare the answers for $\bar{A}$ and $\sigma_{\bar{A}}$ with the answer (4.18) obtained by calculating the averages $\bar{l}$ and $\bar{b}$ and then taking A to be $\bar{l}\,\bar{b}$, with an uncertainty given by error propagation. (For a large number of measurements, the two methods should agree.)

4.22. ★★   This problem is an example of an experiment for which the error analysis can be done in either of two ways: by propagating the estimated errors in the original measurements or by doing a statistical analysis of the various answers. A student wants to measure g, the acceleration of gravity, using a simple pendulum, as described briefly in the introduction to this chapter. Because the period is known to be $T = 2\pi\sqrt{l/g}$, where l is the length of the pendulum, she can find g as $g = 4\pi^2 l/T^2$. She measures T for five different values of l and obtains the following results:

Length, l (cm):	57.3	61.1	73.2	83.7	95.0
Time, T (s):	1.521	1.567	1.718	1.835	1.952

(a) Copy the table of data and add a row in which you list her five computed values of g. **(b)** She estimates she can read the lengths l within about 0.3% (that is, two or three millimeters).[7] Similarly, she estimates that all of the times are within $\pm 0.2\%$. Use error propagation to find the uncertainty in her values for g. **(c)** Because her values of g are five measurements of the same quantity, we can analyze them statistically. In particular, their standard deviation should represent the uncertainty in any one of her answers. What is the SD, and how does it compare with the uncertainty found by error propagation in part (b)? [You should not expect the agreement to be *especially* good because we don't know the exact nature of her original estimated uncertainties (nor, probably, does she). Nevertheless, the two methods should agree *roughly*, and a large disagreement would be a clear signal that something had gone wrong.] **(d)** What is her final answer for g with its uncertainty? [Use the statistical analysis of part (c), and remember that the final uncertainty is the SDOM. How does her answer compare with the accepted value (in her laboratory) of 979.6 cm/s^2?

For Section 4.6: Systematic Errors

4.23. ★   A famous example of a systematic error occurred in Millikan's historic measurement of the electron's charge e. He worked on this experiment for several years and had reduced all random errors to a very low level, certainly less than 0.1%. Unfortunately, his answer for e depended on the viscosity of air (denoted η), and the value of η that he used was 0.4% too low. His value for e had the form $e = K\eta^{3/2}$, where K stands for a complicated expression involving several measured parameters but not η. Therefore, the systematic error in η caused a systematic error in e. Given that all other errors (random and systematic) were much less than 0.4%, what was his error in e? (This example is typical of many systematic errors. Until the errors are identified, nothing can be done about them. Once identified, the errors can be eliminated, in this case by using the right value of η.)

4.24. ★★   In some experiments, systematic errors can be caused by the neglect of an effect that is *not* (in the situation concerned) negligible, for example, neglect of heat losses from a badly insulated calorimeter or neglect of friction for a poorly lubricated cart. Here is another example: A student wants to measure the acceleration of gravity g by timing the fall of a wooden ball (3 or 4 inches across) dropped from four different windows in a tall building. He assumes that air resistance is negligible and that the distance fallen is given by $d = \frac{1}{2}gt^2$. Using a tape measure and an electric timer, he measures the distances and times of the four separate drops as follows:

| Distance, d (meters): | 15.43 | 17.37 | 19.62 | 21.68 |
| Time, t (seconds): | 1.804 | 1.915 | 2.043 | 2.149 |

(a) Copy these data and add a third row in which you put the corresponding accelerations, calculated as $g = 2d/t^2$. **(b)** Based on these results, what is his best estimate

---

[7] Although the percent uncertainties in the five measurements of l are probably not exactly the same, it is appropriate (and time saving) in many experiments to assume they are at least approximately so. In other words, instead of doing five separate error propagations, you may often appropriately do just one for a representative case and assume that all five cases are reasonably similar.

for g, assuming that all errors are random? Show that this answer is inconsistent with the accepted value of $g = 9.80$ m/s^2. **(c)** Having checked his calculations, tape measure, and timer, he concludes (correctly) that there must be some systematic error causing an acceleration different from 9.80 m/s^2, and he suggests that air resistance is probably the culprit. Give at least two arguments to support this suggestion. **(d)** Suggest a couple of ways he could modify the experiment to reduce the effect of this systematic error.

4.25. ★★ **(a)** A student measures the speed of sound as $u = f\lambda$, where f is the frequency shown on the dial of an audio oscillator, and λ is the wavelength measured by locating several maxima in a resonant air column. Because there are several measurements of λ, they can be analyzed statistically, and the student concludes that $\lambda = 11.2 \pm 0.5$ cm. Only one measurement has been taken of $f = 3,000$ Hz (the setting on the oscillator), and the student has no way to judge its reliability. The instructor says that the oscillator is "certainly 1% reliable"; therefore, the student allows for a 1% systematic error in f (but none in λ). What is the student's answer for u with its uncertainty? Is the possible 1% systematic error from the oscillator's calibration important? **(b)** If the student's measurement had been $\lambda = 11.2 \pm 0.1$ cm and the oscillator calibration had been 3% reliable, what would the answer have been? Is the systematic error important in this case?

4.26. ★★ A student wants to check the resistance of a resistor by measuring the voltage across it (V) and the resulting current through it (I) and then calculating the resistance as $R = V/I$. He measures four different values of V and the corresponding currents I, as follows:

Voltage, V (volts):	11.2	13.4	15.1	17.7
Current, I (amps):	4.67	5.46	6.28	7.22

(a) Calculate the four corresponding values of R (which will come out in ohms). What is his best estimate for R, and what is the random component of its uncertainty (δR_{ran})? **(b)** The resistor is rated at 2.50 ohms, which does not lie within the range $R_{best} \pm \delta R_{ran}$, so he considers the possibility that the voltmeter and ammeter suffer some systematic error. The laboratory technician states that many of the meters in the laboratory have up to 2% systematic error. Use error propagation to find the possible systematic error in R, and then combine the systematic and random errors to give the total uncertainty. (In both calculations, combine the errors in quadrature.) What is his final answer and how does it compare with the given value?

4.27. ★★ Some experiments require calibration of the equipment before the measurements can be made. Any random errors in the calibration will usually become systematic errors in the experiment itself. To illustrate this effect, consider an experiment on the Zeeman effect, in which a magnetic field causes a tiny shift in the frequency of light given out by an atom. This shift can be measured using a Fabry-Perot interferometer, which consists of two parallel reflecting surfaces a distance d apart (where d is typically a couple of millimeters). To use the interferometer, one must know the distance d; that is, one must calibrate the instrument by measuring d. A convenient way to make this measurement is to send light of an accurately known wavelength λ through the interferometer, which produces a series of interfer-

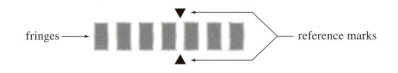

Figure 4.3. When light of one wavelength is sent through a Fabry-Perot interferometer, it produces a pattern of alternating light and dark fringes. If the air is pumped out of the interferometer, the whole pattern shifts sideways, and the number of complete fringes (light-dark-light) that pass the reference marks can be counted; for Problem 4.27.

ence fringes like those depicted in Figure 4.3. If all the air is then pumped out of the chamber that houses the interferometer, the interference pattern slowly shifts sideways, and the number of fringes that move past the reference marks is $N = 2(n - 1)d/\lambda$, where n denotes the refractive index of air. Because n and λ are known accurately, d can be found by counting N. Because N is not necessarily an integer, the fractions of complete fringes that pass the reference marks must be estimated, and this estimation introduces the only serious source of uncertainty.

(a) In one such experiment, a student measures N five times as follows:

$$\text{values of } N \ = \ 3.0, \ 3.5, \ 3.2, \ 3.0, \ 3.2.$$

What is her best estimate for N and its uncertainty? What is the resulting percent uncertainty in d? **(b)** The uncertainty you just found is purely random. Nevertheless, in all subsequent measurements using the interferometer, she will be using the same value of d she found in part (a), and any error in that value will cause a *systematic* error in her final answers. What percent value should she use for this systematic error in d?

4.28. ★★★ Systematic errors sometimes arise when the experimenter unwittingly measures the wrong quantity. Here is an example: A student tries to measure g using a pendulum made of a steel ball suspended by a light string. (See Figure 4.4.) He

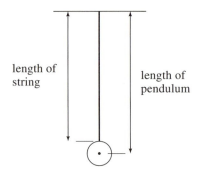

Figure 4.4. A pendulum consists of a metal ball suspended by a string. The effective length of the pendulum is the length of the string *plus* the radius of the ball; for Problem 4.28.

records five different lengths of the pendulum l and the corresponding periods T as follows:

Length, l (cm):   51.2   59.7   68.2   79.7   88.3
Period, T (s):   1.448 1.566 1.669 1.804 1.896.

(a) For each pair, he calculates g as $g = 4\pi^2 l/T^2$. He then calculates the mean of these five values, their SD, and their SDOM. Assuming all his errors are random, he takes the SDOM as his final uncertainty and quotes his answer in the standard form of mean $\pm$ SDOM. What is his answer for g? (b) He now compares his answer with the accepted value $g = 979.6$ cm/s^2 and is horrified to realize that his discrepancy is nearly 10 times larger than his uncertainty. Confirm this sad conclusion. (c) Having checked all his calculations, he concludes that he must have overlooked some systematic error. He is sure there was no problem with the measurement of the period T, so he asks himself the question: How large would a systematic error in the length l have to be so that the margins of the total error *just included* the accepted value 979.6 cm/s^2? Show that the answer is approximately 1.5%. (d) This result would mean that his length measurements suffered a systematic error of about a centimeter—a conclusion he first rejects as absurd. As he stares at the pendulum, however, he realizes that 1 cm is about the radius of the ball and that the lengths he recorded were the lengths *of the string*. Because the correct length of the pendulum is the distance from the pivot to the *center* of the ball (see Figure 4.4), his measurements were indeed systematically off by the radius of the ball. He therefore uses callipers to find the ball's diameter, which turns out to be 2.00 cm. Make the necessary corrections to his data and compute his final answer for g with its uncertainty.

Chapter 5

The Normal Distribution

This chapter continues our discussion of the statistical analysis of repeated measurements. Chapter 4 introduced the important ideas of the mean, the standard deviation, and the standard deviation of the mean; we saw their significance and some of their uses. This chapter supplies the theoretical justification for these statistical ideas and gives proofs of several results stated without proof in earlier chapters.

The first problem in discussing measurements repeated many times is to find a way to handle and display the values obtained. One convenient method is to use a *distribution* or *histogram,* as described in Section 5.1. Section 5.2 introduces the notion of the *limiting distribution,* the distribution of results that would be obtained if the number of measurements become infinitely large. In Section 5.3, I define the *normal distribution,* or *Gauss distribution,* which is the limiting distribution of results for any measurement subject to many small random errors.

Once the mathematical properties of the normal distribution are understood, we can proceed to prove several important results quite easily. Section 5.4 provides proof that, as anticipated in Chapter 4, about 68% of all measurements (all of one quantity and all using the same technique) should lie within one standard deviation of the true value. Section 5.5 proves the result, used back in Chapter 1, that if we make N measurements $x_1, x_2, \ldots, x_N$ of some quantity x, then our best estimate x_{best} based on these values is the mean $\bar{x} = \sum x_i/N$. Section 5.6 justifies the use of addition in quadrature when propagating errors that are independent and random. In Section 5.7, I prove that the uncertainty of the mean $\bar{x}$, when used as the best estimate of x, is given by the standard deviation of the mean $\sigma_{\bar{x}} = \sigma_x/\sqrt{N}$, as stated in Chapter 4. Finally, Section 5.8 discusses how to assign a numerical confidence to experimental results.

The mathematics used in this chapter is more advanced than used thus far. In particular, you will need to understand the basic ideas of integration—the integral as the area under a graph, changes of variables, and (occasionally) integration by parts. However, once you have worked through Section 5.3 on the normal distribution (going over calculations with a pencil and paper, if necessary) you should be able to follow the rest of the chapter without much difficulty.

5.1   Histograms and Distributions

It should be clear that the serious statistical analysis of an experiment requires us to make many measurements. Thus, we first need to devise methods for recording and displaying large numbers of measured values. Suppose, for instance, we were to make 10 measurements of some length x. For example, x might be the distance from a lens to an image formed by the lens. We might obtain the values (all in cm)

$$26, 24, 26, 28, 23, 24, 25, 24, 26, 25. \tag{5.1}$$

Written this way, these 10 numbers convey fairly little information, and if we were to record many more measurements this way, the result would be a confusing jungle of numbers. Obviously, a better system is needed.

As a first step, we can reorganize the numbers (5.1) in ascending order,

$$23, 24, 24, 24, 25, 25, 26, 26, 26, 28. \tag{5.2}$$

Next, rather than recording the three readings 24, 24, 24, we can simply record that we obtained the value 24 three times; in other words, we can record the *different* values of x obtained, together with the *number* of times each value was found, as in Table 5.1.

Table 5.1. Measured lengths x and their numbers of occurrences.

Different values, x_k	23	24	25	26	27	28
Number of times found, n_k	1	3	2	3	0	1

Here, I have introduced the notation x_k ($k = 1, 2, \ldots$) to denote the various different values found: $x_1 = 23$, $x_2 = 24$, $x_3 = 25$, and so on. And n_k ($k = 1, 2, \ldots$) denotes the number of times the corresponding value x_k was found: $n_1 = 1$, $n_2 = 3$, and so on.

If we record measurements as in Table 5.1, we can rewrite the definition of the mean $\bar{x}$ in what proves to be a more convenient way. From our old definition, we know that

$$\bar{x} = \frac{\sum_i x_i}{N} = \frac{23 + 24 + 24 + 24 + 25 + \ldots + 28}{10}. \tag{5.3}$$

This equation is the same as

$$\bar{x} = \frac{23 + (24 \times 3) + (25 \times 2) + \ldots + 28}{10}$$

or in general

$$\bar{x} = \frac{\sum_k x_k n_k}{N}. \tag{5.4}$$

In the original form (5.3), we sum over *all* the measurements made; in (5.4) we sum over all *different* values obtained, multiplying each value by the number of times it occurred. These two sums are obviously the same, but the form (5.4) proves more useful when we make many measurements. A sum like that in (5.4) is sometimes called a *weighted sum;* each value x_k is *weighted* by the number of times it occurred, n_k. For later reference, note that if we add up all the numbers n_k, we obtain the total number of measurements made, N. That is,

$$\sum_k n_k = N. \tag{5.5}$$

(For example, for Table 5.1 this equation asserts that the sum of the numbers in the bottom line is 10.)

Quick Check 5.1. In his first two years at college, Joe takes 20 courses (all with the same number of credits) and earns 7 As, 4 Bs, 7 Cs, and 2 Fs. For the purpose of computing a grade point average (GPA), each letter grade is assigned a numerical score in the usual way, as follows:

Letter grade:	F	D	C	B	A
Score, s_k:	$s_1 = 0$	$s_2 = 1$	$s_3 = 2$	$s_4 = 3$	$s_5 = 4$

Set up a table like Table 5.1 showing the different possible scores s_k and the number of times n_k they were obtained. Use Equation (5.4) to compute Joe's GPA, $\bar{s}$.

The ideas of the past two paragraphs can be rephrased in a way that is often more convenient. Instead of saying that the result $x = 24$ was obtained three times, we can say that $x = 24$ was obtained in 3/10 of all our measurements. In other words, instead of using n_k, the *number* of times the result x_k occurred, we introduce the fraction

$$F_k = \frac{n_k}{N}, \tag{5.6}$$

which is the *fraction* of our N measurements that gave the result x_k. The fractions F_k are said to specify the *distribution* of our results because they describe how our measurements were *distributed* among the different possible values.

In terms of the fractions F_k, we can rewrite the formula (5.4) for the mean $\bar{x}$ in the compact form

$$\bar{x} = \sum_k x_k F_k. \tag{5.7}$$

That is, the mean $\bar{x}$ is just the weighted sum of all the different values x_k obtained, with each x_k weighted by the fraction of times it occurred, F_k.

The result (5.5) implies that

$$\sum_k F_k = 1. \tag{5.8}$$

That is, if we add up the fractions F_k for all possible results x_k, we must get 1. Any set of numbers whose sum is 1 is said to be *normalized*, and the relation (5.8) is therefore called the *normalization condition*.

The distribution of our measurements can be displayed graphically in a *histogram*, as in Figure 5.1. This figure is just a plot of F_k against x_k, in which the

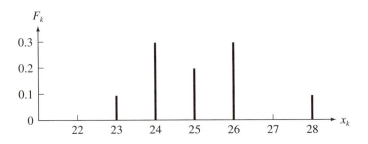

Figure 5.1. Histogram for 10 measurements of a length x. The vertical axis shows the fraction of times F_k that each value x_k was observed.

different measured values x_k are plotted along the horizontal axis and the fraction of times each x_k was obtained is indicated by the height of the vertical bar drawn above x_k. (We can also plot n_k against x_k, but for our purposes the plot of F_k against x_k is more convenient.) Data displayed in histograms like this one can be comprehended quickly and easily, as many writers for newspapers and magazines are aware.

A histogram like that in Figure 5.1 can be called a *bar histogram* because the distribution of results is indicated by the heights of the vertical bars above the x_k. This kind of histogram is appropriate whenever the values x_k are tidily spaced, with integer values. (For example, students' scores on an examination are usually integers and are displayed conveniently using a bar histogram.) Most measurements, however, do not provide tidy integer results because most physical quantities have a continuous range of possible values. For example, rather than the 10 lengths reported in Equation (5.1), you are much more likely to obtain 10 values like

$$26.4, \ 23.9, \ 25.1, \ 24.6, \ 22.7, \ 23.8, \ 25.1, \ 23.9, \ 25.3, \ 25.4. \tag{5.9}$$

A bar histogram of these 10 values would consist of 10 separate bars, all the same height, and would convey comparatively little information. Given measurements like those in (5.9), the best course is to divide the range of values into a convenient number of *intervals* or "bins," and to count how many values fall into each "bin." For example, we could count the number of the measurements (5.9) between $x = 22$ and 23, between $x = 23$ and 24, and so on. The results of counting in this way are shown in Table 5.2. (If a measurement happens to fall exactly on the boundary between two bins, you must decide where to place it. A simple and reasonable course is to assign half a measurement to each of the two bins.)

Table 5.2. The 10 measurements (5.9) grouped in bins.

Bin	22 to 23	23 to 24	24 to 25	25 to 26	26 to 27	27 to 28
Observations in bin	1	3	1	4	1	0

The results in Table 5.2 can be plotted in a form we can call a *bin histogram,* as shown in Figure 5.2. In this plot, the fraction of measurements that fall in each bin is indicated by the area of the rectangle drawn above the bin. Thus, the shaded

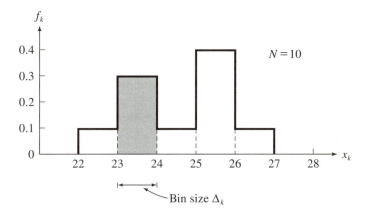

Figure 5.2. Bin histogram showing the fraction of the 10 measurements (5.9) of x that fall in the "bins" 22 to 23, 23 to 24, and so on. The area of the rectangle above each interval gives the fraction of measurements that fall in that interval. Thus, the area of the shaded rectangle is 0.3, indicating that $\frac{3}{10}$ of all measurements lie between 23 and 24.

rectangle above the interval from $x = 23$ to $x = 24$ has area $0.3 \times 1 = 0.3$, indicating that 3/10 of all the measurements fell in this interval. In general, we denote the width of the kth bin by Δ_k. (These widths are usually all the same, though they certainly don't have to be.) The height f_k of the rectangle drawn above this bin is chosen so that the area $f_k \Delta_k$ is

$$f_k \Delta_k = \text{fraction of measurements in } k\text{th bin.}$$

In other words, in a bin histogram the area $f_k \Delta_k$ of the kth rectangle has the same significance as the height F_k of the kth bar in a bar histogram.

Some care is needed in choosing the width Δ_k of the bins for a histogram. If the bins are made much too wide, then all the readings (or almost all) will fall in one bin, and the histogram will be an uninteresting single rectangle. If the bins are made too narrow, then few of them will contain more than one reading, and the histogram will consist of numerous narrow rectangles almost all of the same height.

Clearly, the bin width must be chosen so several readings fall in each of several bins. Thus, when the total number of measurements N is small, we have to choose our bins relatively wide, but if we increase N, then we can usually choose narrower bins.

Quick Check 5.2. A class of 20 students takes an exam, which is graded out of 50 points, and obtains the following results:

$$26, 33, 38, 41, 49, 28, 36, 38, 47, 41,$$
$$32, 37, 48, 44, 27, 32, 34, 44, 37, 30$$

(These scores were taken from an alphabetical list of the students.) On a piece of square-ruled paper, draw a bin histogram of the scores, using bin boundaries at 25, 30, 35, 40, 45, and 50. Label the vertical scale so that the area of each rectangle is the fraction of students in the corresponding bin.

5.2   Limiting Distributions

In most experiments, as the number of measurements increases, the histogram begins to take on a definite simple shape. This evolving shape is clearly visible in Figures 5.3 and 5.4, which show 100 and 1,000 measurements of the same quantity as in Figure 5.2. After 100 measurements, the histogram has become a single peak, which is approximately symmetrical. After 1,000 measurements, we have been able to halve the bin size, and the histogram has become quite smooth and regular. These three graphs illustrate an important property of most measurements. As the number of measurements approaches infinity, their distribution approaches some definite, continuous curve. When this happens, the continuous curve is called the *limiting distribution*.[1] Thus, for the measurements of Figures 5.2 through 5.4, the limiting

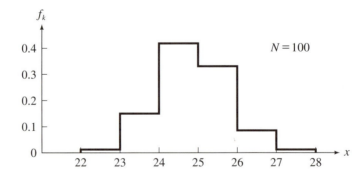

Figure 5.3. Histogram for 100 measurements of the same quantity as in Figure 5.2.

[1] Some common synonyms (or approximate synonyms) for the limiting distribution are: parent distribution, infinite parent distribution, universe distribution, and parent population.

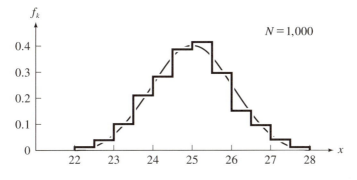

Figure 5.4. Histogram for 1,000 measurements of the same quantity as in Figure 5.3. The broken curve is the limiting distribution.

distribution appears to be close to the symmetric bell-shaped curve superimposed on Figure 5.4.

Note that the limiting distribution is a theoretical construct that can never itself be measured exactly. The more measurements we make, the closer our histogram approaches the limiting distribution. But only if we were to make an infinite number of measurements and use infinitesimally narrow bins would we actually obtain the limiting distribution itself. Nevertheless, there is good reason to believe that every measurement *does* have a limiting distribution to which our histogram approaches ever closer as we make more and more measurements.

A limiting distribution, such as the smooth curve in Figure 5.4, defines a function, which we call $f(x)$. The significance of this function is shown by Figure 5.5. As we make more and more measurements of the quantity x, our histogram will eventually be indistinguishable from the limiting curve $f(x)$. Therefore, the fraction of measurements that fall in any small interval x to $x + dx$ equals the area $f(x)\,dx$ of the shaded strip in Figure 5.5(a):

$$f(x)\,dx = \text{fraction of measurements that}$$
$$\text{fall between } x \text{ and } x + dx. \qquad (5.10)$$

More generally, the fraction of measurements that fall between any two values a and b is the total area under the graph between $x = a$ and $x = b$ (Figure 5.5b).

(a)   **(b)**

Figure 5.5. A limiting distribution $f(x)$. **(a)** After very many measurements, the fraction that falls between x and $x + dx$ is the area $f(x)\,dx$ of the narrow strip. **(b)** The fraction that falls between $x = a$ and $x = b$ is the shaded area.

This area is just the *definite integral* of $f(x)$. Thus, we have the important result that

$$\int_a^b f(x)\, dx = \text{fraction of measurements that} \atop \text{fall between } x = a \text{ and } x = b. \tag{5.11}$$

Understanding the meaning of the two statements (5.10) and (5.11) is important. Both tell us the fraction of measurements expected to lie in some interval after we make a *very large number of measurements*. Another, very useful, way to say this is that $f(x)\, dx$ is the *probability* that a single measurement of x will give an answer between x and $x + dx$,

$$f(x)\, dx = \text{probability that any one measurement} \atop \text{will give an answer between } x \text{ and } x + dx. \tag{5.12}$$

Similarly, the integral $\int_a^b f(x)\, dx$ tells us the probability that any one measurement will fall between $x = a$ and $x = b$. We have arrived at the following important conclusion: If we knew the limiting distribution $f(x)$ for the measurement of a given quantity x with a given apparatus, then we would know the probability of obtaining an answer in any interval $a \leqslant x \leqslant b$.

Because the total probability of obtaining an answer anywhere between $-\infty$ and $+\infty$ must be one, the limiting distribution $f(x)$ must satisfy

$$\int_{-\infty}^{\infty} f(x)\, dx = 1. \tag{5.13}$$

This identity is the natural analog of the normalization sum (5.8), $\sum_k F_k = 1$, and a function $f(x)$ satisfying (5.13) is said to be *normalized*.

The limits $\pm\infty$ in the integral (5.13) may seem puzzling. They do not mean that we really expect to obtain answers ranging all the way from $-\infty$ to ∞. Quite the contrary. In a real experiment, the measurements all fall in some fairly small finite interval. For example, the measurements of Figure 5.4 all lie between $x = 21$ and $x = 29$. Even after infinitely many measurements, the fraction lying outside $x = 21$ to $x = 29$ would be entirely negligible. In other words, $f(x)$ is essentially zero outside this range, and it makes no difference whether the integral (5.13) runs from $-\infty$ to $+\infty$ or 21 to 29. Because we generally don't know what these finite limits are, for convenience we leave them as $\pm\infty$.

If the measurement under consideration is very precise, all the values obtained will be close to the actual value of x, so the histogram of results, and hence the limiting distribution, will be narrowly peaked like the solid curve in Figure 5.6. If the measurement is of low precision, then the values found will be widely spread and the distribution will be broad and low like the dashed curve in Figure 5.6.

The limiting distribution $f(x)$ for measurement of a given quantity x using a given apparatus describes how results would be distributed after many, many measurements. Thus, if we knew $f(x)$, we could calculate the mean value $\bar{x}$ that would be found after many measurements. We saw in (5.7) that the mean of any number

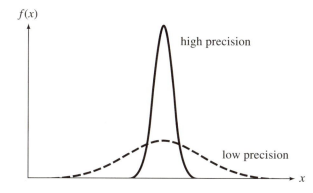

Figure 5.6. Two limiting distributions, one for a high-precision measurement, the other for a low-precision measurement.

of measurements is the sum of all different values x_k, each weighted by the fraction of times it is obtained:

$$\bar{x} = \sum_k x_k F_k. \tag{5.14}$$

In the present case, we have an enormous number of measurements with distribution $f(x)$. If we divide the whole range of values into small intervals x_k to $x_k + dx_k$, the fraction of values in each interval is $F_k = f(x_k) \, dx_k$ and in the limit that all intervals go to zero, (5.14) becomes

$$\bar{x} = \int_{-\infty}^{\infty} x f(x) \, dx. \tag{5.15}$$

Remember that this formula gives the mean $\bar{x}$ expected after infinitely many trials.

Similarly, we can calculate the standard deviation σ_x obtained after many measurements. Because we are concerned with the limit $N \to \infty$, it makes no difference which definition of σ_x we use, the original (4.6) or the "improved" (4.9) with N replaced by $N - 1$. In either case, when $N \to \infty$, σ_x^2 is the average of the squared deviation $(x - \bar{x})^2$. Thus, exactly the argument leading to (5.15) gives, after many trials,

$$\sigma_x^2 = \int_{-\infty}^{\infty} (x - \bar{x})^2 f(x) \, dx \tag{5.16}$$

(see Problem 5.10).

5.3   The Normal Distribution

Different types of measurements have different limiting distributions. Not all limiting distributions have the symmetric bell shape illustrated in Section 5.2. (For

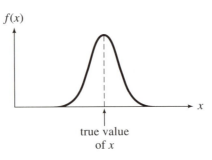

$f(x)$

true value
of x

Figure 5.7. The limiting distribution for a measurement subject to many small random errors. The distribution is bell-shaped and centered on the true value of the measured quantity x.

example, the binomial and Poisson distributions discussed in Chapters 10 and 11 are usually not symmetric.) Nevertheless, many measurements are found to have a symmetric bell-shaped curve for their limiting distribution. In fact, I will prove in Chapter 10 that if a measurement is subject to many small sources of random error and negligible systematic error, the measured values will be distributed in accordance with a bell-shaped curve and this curve will be centered on the true value of x, as in Figure 5.7. In the remainder of this chapter, I will confine my attention to measurements with this property.

If our measurements have appreciable systematic errors, we would *not* expect the limiting distribution to be centered on the true value. Random errors are equally likely to push our readings above or below the true value. If all errors are random, after many measurements the number of observations above the true value will be the same as that below it, and our distribution of results will therefore be centered on the true value. But a systematic error (such as that caused by a tape measure that is stretched or a clock that runs slow) pushes all values in one direction and so pushes the distribution of observed values off center from the true value. In this chapter, I will assume that the distribution is centered on the true value. This is equivalent to assuming that all systematic errors have been reduced to a negligible level.

I now turn briefly to a question we have avoided discussing so far: What is the "true value" of a physical quantity? This question is a hard one that has no satisfactory, simple answer. Because no measurement can exactly determine the true value of any continuous variable (a length, a time, etc.), whether the true value of such a quantity exists is not even clear. Nevertheless, I will make the convenient assumption that every physical quantity does have a true value.

We can think of the true value of a quantity as that value to which one approaches closer and closer as increasing numbers of measurements are made with increasing care. As such, the true value is an idealization similar to the mathematician's point with no size or line with no width; like the point or line, it is a useful idealization. I will often denote the true values of measured quantities $x, y, \ldots$, by their corresponding capital letters $X, Y, \ldots$. If the measurements of x are subject to many small random errors but negligible systematic errors, their distribution will be a symmetric bell-shaped curve centered on the true value X.

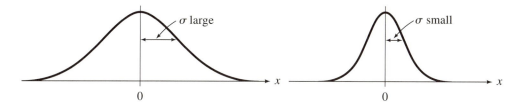

Figure 5.8. The Gauss function (5.17) is bell-shaped and centered on $x = 0$. The bell curve is wide if σ is large and narrow if σ is small. Although for now we will view σ as just a parameter that characterizes the bell curve's width, σ can be shown (as in Problem 5.13) to be the distance from the center of the curve to the point where the curvature changes sign. This distance is shown in the two graphs.

The mathematical function that describes the bell-shaped curve is called the *normal distribution,* or *Gauss function.*[2] The prototype of this function is

$$e^{-x^2/2\sigma^2},\tag{5.17}$$

where σ is a fixed parameter I will call the *width parameter.* It is important that you become familiar with the properties of this function.

When $x = 0$, the Gauss function (5.17) is equal to one. The function is symmetric about $x = 0$, because it has the same value for x and $-x$. As x moves away from zero in either direction, $x^2/2\sigma^2$ increases, quickly if σ is small, more slowly if σ is large. Therefore, as x moves away from the origin, the function (5.17) decreases toward zero. Thus, the general appearance of the Gauss function (5.17) is as shown in Figure 5.8. The graphs illustrate the name "width parameter" for σ because the bell shape is wide if σ is large and narrow if σ is small.

The Gauss function (5.17) is a bell-shaped curve centered on $x = 0$. To obtain a bell-shaped curve centered on some other point $x = X$, we merely replace x in (5.17) by $x - X$. Thus, the function

$$e^{-(x-X)^2/2\sigma^2}\tag{5.18}$$

has its maximum at $x = X$ and falls off symmetrically on either side of $x = X$, as in Figure 5.9.

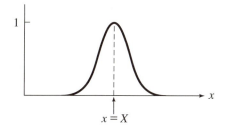

Figure 5.9. The Gauss function (5.18) is bell-shaped and centered on $x = X$.

---

[2] Other common names for the Gauss function are the Gaussian function (or just "Gaussian"), the normal density function, and the normal error function. The last of these names is rather unfortunate because the name "error function" is often used for the integral of the Gauss function (as discussed in Section 5.4).

The function (5.18) is not quite in its final form to describe a limiting distribution because any distribution must be *normalized;* that is, it must satisfy

$$\int_{-\infty}^{\infty} f(x)\, dx = 1. \tag{5.19}$$

To arrange this normalization, we set

$$f(x) = Ne^{-(x-X)^2/2\sigma^2}. \tag{5.20}$$

(Multiplication by the factor N does not change the shape, nor does it shift the maximum at $x = X$.) We must then choose the "normalization factor" N so that $f(x)$ is normalized as in (5.19). This involves some elementary manipulation of integrals, which I give in some detail:

$$\int_{-\infty}^{\infty} f(x)\, dx = \int_{-\infty}^{\infty} Ne^{-(x-X)^2/2\sigma^2}\, dx. \tag{5.21}$$

In evaluating this kind of integral, changing variables to simplify the integral is always a good idea. Thus, we can set $x - X = y$ (in which case $dx = dy$) and get

$$= N \int_{-\infty}^{\infty} e^{-y^2/2\sigma^2}\, dy. \tag{5.22}$$

Next, we can set $y/\sigma = z$ (in which case $dy = \sigma dz$) and get

$$= N\sigma \int_{-\infty}^{\infty} e^{-z^2/2}\, dz. \tag{5.23}$$

The remaining integral is one of the standard integrals of mathematical physics. It can be evaluated by elementary methods, but the details are not especially illuminating, so I will simply quote the result;[3]

$$\int_{-\infty}^{\infty} e^{-z^2/2}\, dz = \sqrt{2\pi}. \tag{5.24}$$

Returning to (5.21) and (5.23), we find that

$$\int_{-\infty}^{\infty} f(x)\, dx = N\sigma\sqrt{2\pi}.$$

Because this integral must equal 1, we must choose the normalization factor N to be

$$N = \frac{1}{\sigma\sqrt{2\pi}}.$$

With this choice for the normalization factor, we arrive at the final form for the Gauss, or normal, distribution function, which we denote by $G_{X,\sigma}(x)$:

---

[3] For a derivation, see, for example, H. D. Young, *Statistical Treatment of Experimental Data* (McGraw-Hill, 1962), Appendix D.

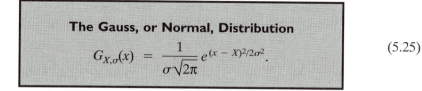

The Gauss, or Normal, Distribution

$$G_{X,\sigma}(x) = \frac{1}{\sigma\sqrt{2\pi}}e^{(x-X)^2/2\sigma^2}.$$ (5.25)

Notice that I have added subscripts X and σ to indicate the center and width of the distribution. The function $G_{X,\sigma}(x)$ describes the limiting distribution of results in a measurement of a quantity x whose true value is X, if the measurement is subject only to random errors. Measurements whose limiting distribution is given by the Gauss function (5.25) are said to be *normally distributed.*

The significance of the width parameter σ will be explored shortly. We have already seen that a small value of σ gives a sharply peaked distribution, which corresponds to a precise measurement, whereas a large value of σ gives a broad distribution, which corresponds to a low-precision measurement. Figure 5.10 shows two examples of Gauss distributions with different centers X and widths σ. Note

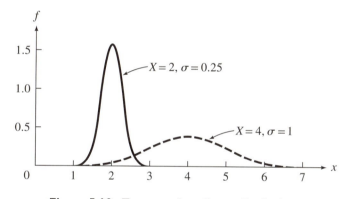

Figure 5.10. Two normal, or Gauss, distributions.

how the factor σ in the denominator of (5.25) guarantees that a narrower distribution (σ smaller) is automatically higher at its center, as it must be so that the total area under the curve equals 1.

Quick Check 5.3.   On square-ruled paper, sketch the Gauss function $G_{X,\sigma}(x)$ for $X = 10$ and $\sigma = 1$. Use your calculator to find the values at $x = 10$, 10.5, 11, 11.5, 12, and 12.5. You don't need to calculate the values for $x < 10$ because you know the function is symmetric about $x = 10$.

We saw in Section 5.2 that knowledge of the limiting distribution for a measurement lets us compute the average value $\bar{x}$ expected after numerous trials. According to (5.15), this expected average for the Gauss distribution $f(x) = G_{X,\sigma}(x)$ is

$$\bar{x} = \int_{-\infty}^{\infty} x\, G_{X,\sigma}(x)\, dx. \tag{5.26}$$

Before we evaluate this integral, we should note that the answer almost obviously will be X, because the symmetry of the Gauss function about X implies that the same number of results will fall any distance above X as will fall an equal distance below X. Thus, the average should be X.

We can calculate the integral (5.26) for the Gauss distribution as follows:

$$\bar{x} = \int_{-\infty}^{\infty} x\, G_{X,\sigma}(x)\, dx$$

$$= \frac{1}{\sigma\sqrt{2\pi}} \int_{-\infty}^{\infty} x\, e^{-(x-X)^2/2\sigma^2}\, dx. \tag{5.27}$$

If we make the change of variables $y = x - X$, then $dx = dy$ and $x = y + X$. Thus, the integral (5.27) becomes two terms,

$$\bar{x} = \frac{1}{\sigma\sqrt{2\pi}} \left(\int_{-\infty}^{\infty} y\, e^{-y^2/2\sigma^2}\, dy + X \int_{-\infty}^{\infty} e^{-y^2/2\sigma^2}\, dy \right). \tag{5.28}$$

The first integral here is exactly zero because the contribution from any point y is exactly canceled by that from the point $-y$. The second integral is the normalization integral encountered in (5.22) and has the value $\sigma\sqrt{2\pi}$. This integral cancels with the $\sigma\sqrt{2\pi}$ in the denominator and leaves the expected answer that

$$\bar{x} = X, \tag{5.29}$$

after many trials. In other words, if the measurements are distributed according to the Gauss distribution $G_{X,\sigma}(x)$, then, after numerous trials, the mean value $\bar{x}$ is the true value X, on which the Gauss function is centered.

The result (5.29) would be exactly true only if we could make infinitely many measurements. Its practical usefulness is that if we make a large (but finite) number of trials, our average will be *close* to X.

Another interesting quantity to compute is the *standard deviation* σ_x after a large number of trials. According to (5.16), this quantity is given by

$$\sigma_x^2 = \int_{-\infty}^{\infty} (x - \bar{x})^2 G_{X,\sigma}(x)\, dx. \tag{5.30}$$

This integral is evaluated easily. We replace $\bar{x}$ by X, make the substitutions $x - X = y$ and $y/\sigma = z$, and finally integrate by parts to obtain the result (see Problem 5.16)

$$\sigma_x^2 = \sigma^2 \tag{5.31}$$

after many trials. In other words, the width parameter σ of the Gauss function $G_{X,\sigma}(x)$ is just the standard deviation we would obtain after making many measurements. This is, of course, why the letter σ is used for the width parameter and explains why σ is often called the standard deviation of the Gauss distribution $G_{X,\sigma}(x)$. Strictly speaking, however, σ is the standard deviation expected only after *infinitely many* trials. If we make some finite number of measurements (10 or 20, say) of x, the observed standard deviation should be some approximation to σ, but we have no reason to think it will be *exactly* σ. Section 5.5 addresses what more can be said about the mean and standard deviation after a finite number of trials.

5.4   The Standard Deviation as 68% Confidence Limit

The limiting distribution $f(x)$ for measurement of some quantity x tells us the probability of obtaining any given value of x. Specifically, the integral

$$\int_a^b f(x)\,dx$$

is the probability that any one measurement gives an answer in the range $a \leqslant x \leqslant b$. If the limiting distribution is the Gauss function $G_{X,\sigma}(x)$, this integral can be evaluated. In particular, we can now calculate the probability (discussed in Chapter 4) that a measurement will fall within one standard deviation σ of the true value X. This probability is

$$Prob(\text{within } \sigma) \;=\; \int_{X-\sigma}^{X+\sigma} G_{X,\sigma}(x)\,dx \tag{5.32}$$

$$=\; \frac{1}{\sigma\sqrt{2\pi}} \int_{X-\sigma}^{X+\sigma} e^{-(x-X)^2/2\sigma^2}\,dx. \tag{5.33}$$

This integral is illustrated in Figure 5.11. It can be simplified in the now familiar

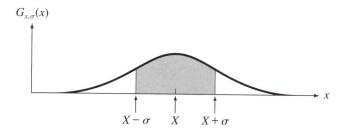

Figure 5.11. The shaded area between $X \pm \sigma$ is the probability of a measurement within one standard deviation of X.

way by substituting $(x - X)/\sigma = z$. With this substitution, $dx = \sigma dz$, and the limits of integration become $z = \pm 1$. Therefore,

$$Prob(\text{within } \sigma) \;=\; \frac{1}{\sqrt{2\pi}} \int_{-1}^{1} e^{-z^2/2}\,dz. \tag{5.34}$$

Before discussing the integral (5.34), let me remark that we could equally have found the probability for an answer within 2σ of X or within 1.5σ of X. More generally, we could calculate *Prob*(within $t\sigma$), which means "the probability for an

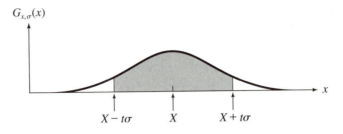

Figure 5.12. The shaded area between $X \pm t\sigma$ is the probability of a measurement within t standard deviations of X.

answer within $t\sigma$ of X," where t is any positive number. This probability is given by the area in Figure 5.12, and a calculation identical to that leading to (5.34) gives (as in Problem 5.22)

$$Prob(\text{within } t\sigma) = \frac{1}{\sqrt{2\pi}} \int_{-t}^{t} e^{-z^2/2} \, dz. \qquad (5.35)$$

The integral (5.35) is a standard integral of mathematical physics, and it is often called the *error function,* denoted erf(t), or the *normal error integral.* It cannot be evaluated analytically but is easily calculated on a computer (or even many calculators). Figure 5.13 shows this integral plotted as a function of t and tabulates a few values. A more complete tabulation can be found in Appendix A (see also Appendix B, which shows a different, but closely related, integral).

We first note from Figure 5.13 that the probability that a measurement will fall

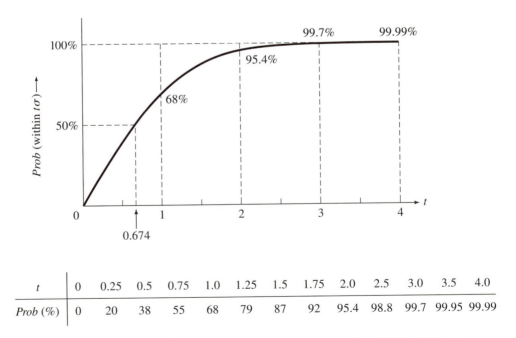

t	0	0.25	0.5	0.75	1.0	1.25	1.5	1.75	2.0	2.5	3.0	3.5	4.0
Prob (%)	0	20	38	55	68	79	87	92	95.4	98.8	99.7	99.95	99.99

Figure 5.13. The probability *Prob*(within $t\sigma$) that a measurement of x will fall within t standard deviations of the true value $x = X$. Two common names for this function are the *normal error integral* and the *error function,* erf(t).

within one standard deviation of the true answer is 68%, as anticipated in Chapter 4. If we quote the standard deviation as our uncertainty in such a measurement (that is, write $x = x_{best} \pm \delta x$, and take $\delta x = \sigma$), then we can be 68% confident that we are within σ of the correct answer.

We can also see in Figure 5.13 that the probability $Prob$(within $t\sigma$) rapidly approaches 100% as t increases. The probability that a measurement will fall inside 2σ is 95.4%; that for 3σ is 99.7%. To put these results another way, the probability that a measurement will fall *outside* one standard deviation is appreciable (32%), that it will lie outside 2σ is much smaller (4.6%), and that it will lie outside 3σ is extremely small (0.3%).

Of course, nothing is sacred about the number 68%; it just happens to be the confidence associated with the standard deviation σ. One alternative to the standard deviation is called the *probable error,* or PE, and is defined as that distance for which there is a 50% probability of a measurement between $X \pm$ PE. Figure 5.13 shows that (for a measurement that is normally distributed) the probable error is

$$\text{PE} \approx 0.67\sigma.$$

Some experimenters like to quote the PE as the uncertainty in their measurements. Nonetheless, the standard deviation σ is the most popular choice because its properties are so simple.

Quick Check 5.4.   The measurements of a certain distance x are distributed normally with $X = 10$ and $\sigma = 2$. What is the probability that a single measurement will lie between $x = 7$ and $x = 13$? What is the probability that it will lie *outside* the range from $x = 7$ to 13?

5.5   Justification of the Mean as Best Estimate

The past three sections have discussed the *limiting distribution f(x)*, the distribution obtained from an infinite number of measurements of a quantity x. If $f(x)$ were known, we could calculate the mean $\bar{x}$ and standard deviation σ obtained after infinitely many measurements, and (at least for the normal distribution) we would also know the true value X. Unfortunately, we never do know the limiting distribution. In practice, we have a finite number of measured values (5, 10, or perhaps 50),

$$x_1, x_2, \ldots, x_N,$$

and our problem is to arrive at *best estimates* of X and σ based on these N measured values.

If the measurements follow a normal distribution $G_{X,\sigma}(x)$ and if we knew the parameters X and σ, we could calculate the probability of obtaining the values $x_1, \ldots, x_N$ that were actually obtained. Thus, the probability of getting a reading near x_1, in a small interval dx_1, is

$$Prob(x \text{ between } x_1 \text{ and } x_1 + dx_1) = \frac{1}{\sigma\sqrt{2\pi}} e^{-(x_1-X)^2/2\sigma^2} dx_1.$$

In practice, we are not interested in the size of the interval dx_1 (or the factor $\sqrt{2\pi}$), so we abbreviate this equation to

$$Prob(x_1) \propto \frac{1}{\sigma} e^{-(x_1-X)^2/2\sigma^2}. \tag{5.36}$$

I will refer to (5.36) as the probability of getting the value x_1, although strictly speaking it is the probability of getting a value in an interval near x_1 as in the preceding equation.

The probability of obtaining the second reading x_2 is

$$Prob(x_2) \propto \frac{1}{\sigma} e^{-(x_2-X)^2/2\sigma^2}, \tag{5.37}$$

and we can similarly write down all the probabilities ending with

$$Prob(x_N) \propto \frac{1}{\sigma} e^{-(x_N-X)^2/2\sigma^2}. \tag{5.38}$$

Equations (5.36) through (5.38) give the probabilities of obtaining each of the readings $x_1, \ldots, x_N$, calculated in terms of the assumed limiting distribution $G_{X,\sigma}(x)$. The probability that we observe the whole set of N readings is just the product of these separate probabilities,[4]

$$Prob_{X,\sigma}(x_1, \ldots, x_N) = Prob(x_1) \times Prob(x_2) \times \ldots \times Prob(x_N)$$

or

$$Prob_{X,\sigma}(x_1, \ldots, x_N) \propto \frac{1}{\sigma^N} e^{-\Sigma(x_i - X)^2/2\sigma^2}. \tag{5.39}$$

Understanding the significance of the various quantities in (5.39) is most important. The numbers $x_1, \ldots, x_N$ are the actual results of N measurements; thus $x_1, \ldots, x_N$ are known, fixed numbers. The quantity $Prob_{X,\sigma}(x_1, \ldots, x_N)$ is the probability for obtaining the N results $x_1, \ldots, x_N$, calculated in terms of X and σ, the true value of x and the width parameter of its distribution. The numbers X and σ are *not* known; we want to find best estimates for X and σ based on the given observations $x_1, \ldots, x_N$. I have added subscripts X and σ to the probability (5.39) to emphasize that it depends on the (unknown) values of X and σ.

Because the actual values of X and σ are unknown, we might imagine guessing values X' and σ' and then using those guessed values to compute the probability $Prob_{X',\sigma'}(x_1, \ldots, x_N)$. If we next guessed two new values, X'' and σ'' and found that the corresponding probability $Prob_{X'',\sigma''}(x_1, \ldots, x_N)$ was larger, we would natu-

---

[4]We are using the well-known result that the probability for several independent events is the product of their separate probabilities. For example, the probability of throwing a "heads" with a coin is 1/2 and that of throwing a "six" with a die is 1/6. Therefore, the probability of throwing a "heads" *and* a "six" is $(1/2) \times (1/6) = 1/12$.

rally regard the new values X'' and σ'' as better estimates for X and σ. Continuing in this way, we could imagine hunting for the values of X and σ that make $Prob_{X,\sigma}(x_1, \ldots, x_N)$ as large as possible, and these values would be regarded as the best estimates for X and σ.

This plausible procedure for finding the best estimates for X and σ is called by statisticians the *principle of maximum likelihood*. It can be stated briefly as follows:

Given the N observed measurements $x_1, \ldots, x_N$, the best estimates for X and σ are those values for which the observed $x_1, \ldots, x_N$ are most likely. That is, the best estimates for X and σ are those values for which $Prob_{X,\sigma}(x_1, \ldots, x_N)$ is maximum, given that here

$$Prob_{X,\sigma}(x_1, \ldots, x_N) \propto \frac{1}{\sigma^N} e^{-\Sigma(x_i - X)^2/2\sigma^2}. \tag{5.40}$$

Using this principle, we can easily find the best estimate for the true value X. Obviously (5.40) is *maximum* if the sum in the exponent is *minimum*. Thus, the best estimate for X is that value of X for which

$$\sum_{i=1}^{N} (x_i - X)^2/\sigma^2 \tag{5.41}$$

is minimum. To locate this minimum, we differentiate with respect to X and set the derivative equal to zero, giving

$$\sum_{i=1}^{N} (x_i - X) = 0$$

or

$$(\text{best estimate for } X) = \frac{\Sigma x_i}{N}. \tag{5.42}$$

That is, the *best estimate* for the true value X is the mean of our N measurements, $\bar{x} = \Sigma x_i/N$, a result we have been assuming without proof since Chapter 1.

Finding the best estimate for σ, the width of the limiting distribution, is a little harder, because the probability (5.40) is a more complicated function of σ. We must differentiate (5.40) with respect to σ and set the derivative equal to zero. (I leave the details to you; see Problem 5.26.) This procedure gives the value of σ that maximizes (5.40) and that is therefore the best estimate for σ, as

$$(\text{best estimate for } \sigma) = \sqrt{\frac{1}{N} \sum_{i=1}^{N} (x_i - X)^2}. \tag{5.43}$$

The true value X is unknown. Thus, in practice, we have to replace X in (5.43) by our best estimate for X, namely the mean $\bar{x}$. This replacement yields the estimate

$$\sigma = \sqrt{\frac{1}{N} \sum_{i=1}^{N} (x_i - \bar{x})^2}. \tag{5.44}$$

In other words, our estimate for the width σ of the limiting distribution is the standard deviation of the N observed values, $x_1, \ldots, x_N$, as originally defined in (4.6).

You may have been surprised that the estimate (5.44) is the same as our original definition (4.6), using N, of the standard deviation, instead of our "improved" definition, using $N - 1$. In fact, in passing from the best estimate (5.43) to the expression (5.44), we have glossed over a rather elegant subtlety. The best estimate (5.43) involves the true value X, whereas in (5.44) we have replaced X by $\bar{x}$ (our best estimate for X). Now, these numbers are generally not the same, and you can easily see that the number (5.44) is *always less than,* or at most equal to, (5.43).[5] Thus, in passing from (5.43) to (5.44), we have consistently *underestimated* the width σ. Appendix E shows that this underestimation is corrected by replacing the denominator N in (5.44) by $N - 1$. That is, the best estimate for the width σ is precisely the "improved," or "sample," standard deviation of the measured values $x_1, x_2, \ldots, x_N$, with $(N - 1)$ in its denominator,

$$\text{(best estimate for } \sigma) = \sqrt{\frac{1}{N-1} \sum_{i=1}^{N} (x_i - \bar{x})^2}. \tag{5.45}$$

Now is a good time to review the rather complicated story that has unfolded so far. First, if the measurements of x are subject only to random errors, their limiting distribution is the Gauss function $G_{X,\sigma}(x)$ centered on the true value X and with width σ. The width σ is the 68% confidence limit, in that there is a 68% probability that any measurement will fall within a distance σ of the true value X. In practice, neither X nor σ is known. Instead, we know our N measured values $x_1, \ldots, x_N$, where N is as large as our time and patience allowed us to make it. Based on these N measured values, our best estimate of the true value X has been shown to be the mean $\bar{x} = \Sigma x_i/N$, and our best estimate of the width σ is the standard deviation σ_x of $x_1, \ldots, x_N$ as defined in (5.45).

Two further questions now arise. First, what is the uncertainty in $\bar{x}$ as an estimate of the true value of X? This question is discussed in Section 5.7, where the uncertainty in $\bar{x}$ is shown to be the *standard deviation of the mean,* or SDOM, as defined in Chapter 4. Second, what is the uncertainty in σ_x as an estimate of the true width σ? The formula for this "uncertainty in the uncertainty" or "standard deviation of the standard deviation" is derived in Appendix E; the result proved there is that the fractional uncertainty in σ_x is

$$\text{(fractional uncertainty in } \sigma_x) = \frac{1}{\sqrt{2(N-1)}}. \tag{5.46}$$

---

[5] If we regard (5.43) as a function of X, we have just seen that this function is minimum at $X = \bar{x}$. Thus (5.44) is always less than or equal to (5.43).

This result makes clear the need for numerous measurements before the uncertainty can be known reliably. For example, with just three measurements of a quantity ($N = 3$), the result (5.46) implies that the standard deviation is 50% uncertain!

Quick Check 5.5. To test the reliability of a ballistic galvanometer, a student discharges a capacitor (charged to a fixed voltage) through the galvanometer three times and measures the resulting charge q in microcoulombs. From his three measurements, he calculates the standard deviation to be $\sigma_q = 6 \ \mu C$. Use (5.46) to find the uncertainty $\delta\sigma_q$ in this parameter. The true value of the standard deviation could easily be as small as $\sigma_q - \delta\sigma_q$ or as large as $\sigma_q + \delta\sigma_q$. What are these two values for this experiment? Note well how unreliable σ_q is after only three measurements.

The results of the past two sections depend on the assumption that our measurements are normally distributed.[6] Although this assumption is reasonable, it is difficult to verify in practice and is sometimes not exactly true. This being the case, I should emphasize that, even when the distribution of measurements is *not* normal, it is almost always *approximately* normal, and you can safely use the ideas of this chapter, at least as good approximations.

5.6   Justification of Addition in Quadrature

Let us now return to the topic of Chapter 3, the propagation of errors. I stated there, without formal proof, that when errors are random and independent, they can be combined in quadrature according to certain standard rules, either the "simple rules" in (3.16) and (3.18), or the general rule in (3.47), which includes the "simple" rules as special cases. This use of addition in quadrature can now be justified.

The problem of error propagation arises when we measure one or more quantities $x, \ldots, z$, all with uncertainties, and then use our measured values to calculate some quantity $q(x, \ldots, z)$. The main question is, of course, to decide on the uncertainty in our answer for q. If the quantities $x, \ldots, z$ are subject only to random errors, they will be normally distributed with width parameters[7] $\sigma_x, \ldots, \sigma_z$, which we take to be the uncertainties associated with any single measurement of the corresponding quantities. The question to be decided now is this: Knowing the distributions of measurements of $x, \ldots, z$, what can we say about the distribution of values for q? In particular, what will be the width of the distribution of values of q? This question is answered in four steps, numbered I to IV.

---

[6] And that systematic errors have been reduced to a negligible level.

[7] When discussing several different measured quantities $x, \ldots, z$, I use subscripts $x, \ldots, z$ to distinguish the width parameters of their limiting distributions. Thus σ_x denotes the width of the Gauss distribution $G_{X,\sigma_x}(x)$ for the measurements of x, and so on.

I. MEASURED QUANTITY PLUS FIXED NUMBER

First, consider two simple, special cases. Suppose we measure a quantity x and proceed to calculate the quantity

$$q = x + A, \tag{5.47}$$

where A is a fixed number with no uncertainty (such as $A = 1$ or π). Suppose also that the measurements of x are normally distributed about the true value X, with

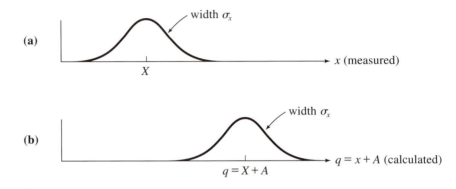

Figure 5.14. If the measured values of x are normally distributed with center $x = X$ and width σ_x, the calculated values of $q = x + A$ (with A fixed and known) will be normally distributed with center $q = X + A$ and the same width σ_x.

width σ_x, as in Figure 5.14(a). Then the probability of obtaining any value x (in a small interval dx) is $G_{X,\sigma_x}(x)\, dx$ or

$$(\text{probability of obtaining value } x) \propto e^{-(x-X)^2/2\sigma_x^2}. \tag{5.48}$$

Our problem is to deduce the probability of obtaining any value q of the quantity defined by (5.47). Now, from (5.47) we see that $x = q - A$ and hence that

$$(\text{probability of obtaining value } q) = (\text{probability of obtaining } x = q - A).$$

The second probability is given by (5.48), and so

$$(\text{probability of obtaining value } q) \propto e^{-[(q-A)-X]^2/2\sigma_x^2}$$
$$= e^{-[q-(X+A)]^2/2\sigma_x^2}. \tag{5.49}$$

The result (5.49) shows that the calculated values of q are normally distributed and centered on the value $X + A$, with width σ_x, as shown in Figure 5.14(b). In particular, the uncertainty in q is the same (namely, σ_x) as that in x, just as the rule (3.16) would have predicted.

II. MEASURED QUANTITY TIMES FIXED NUMBER

As a second simple example, suppose that we measure x and calculate the quantity

$$q = Bx,$$

where B is a fixed number (such as $B = 2$ or $B = \pi$). If the measurements of x are normally distributed, then, arguing exactly as before, we conclude that[8]

(probability of obtaining value q) $\propto$ (probability of obtaining $x = q/B$)

$$\propto \exp\left[-\left(\frac{q}{B} - X\right)^2 / 2\sigma_x^2\right]$$

$$= \exp[-(q - BX)^2 / 2B^2\sigma_x^2]. \qquad (5.50)$$

In other words, the values of $q = Bx$ will be normally distributed, with center at $q = BX$ and width $B\sigma_x$, as shown in Figure 5.15. In particular, the uncertainty in $q = Bx$ is B times that in x, just as our rule (3.18) would have predicted.

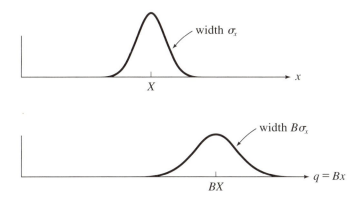

Figure 5.15. If the measured values of x are normally distributed with center $x = X$ and width σ_x, then the calculated values of $q = Bx$ (with B fixed and known) will be normally distributed with center BX and width $B\sigma_x$.

III. SUM OF TWO MEASURED QUANTITIES

As a first nontrivial example of error propagation, suppose we measure two independent quantities x and y and calculate their sum $x + y$. We suppose that the measurements of x and y are normally distributed about their true values X and Y, with widths σ_x and σ_y as in Figures 5.16(a) and (b), and we will try to find the distribution of the calculated values of $x + y$. We will find that the values of $x + y$

---

[8] Here I introduce the alternative notation $\exp(z)$ for the exponential function, $\exp(z) \equiv e^z$. When the exponent z becomes complicated, the "exp" notation is more convenient to type or print.

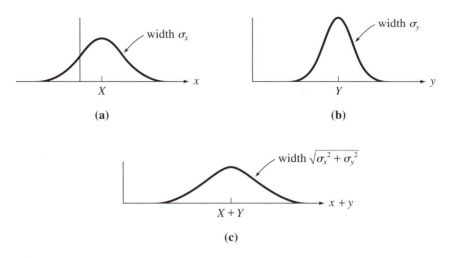

Figure 5.16. If the measurements of x and y are independent and normally distributed with centers X and Y and widths σ_x and σ_y, then the calculated values of $x+y$ are normally distributed with center $X + Y$ and width $\sqrt{\sigma_x^2 + \sigma_y^2}$.

are normally distributed, that their center is the true value $X + Y$, and that the width of their distribution is

$$\sqrt{\sigma_x^2 + \sigma_y^2},$$

as in Figure 5.16(c). In particular, this result justifies the rule of Chapter 3 that if x and y are subject to independent random uncertainties only, then the uncertainty in $x + y$ is the quadratic sum of the separate uncertainties in x and y.

To simplify our algebra, we assume at first that the true values X and Y are both zero. In this case, the probability of getting any particular value of x is

$$Prob(x) \propto \exp\left(\frac{-x^2}{2\sigma_x^2}\right) \tag{5.51}$$

and that of y is

$$Prob(y) \propto \exp\left(\frac{-y^2}{2\sigma_y^2}\right). \tag{5.52}$$

Our problem now is to calculate the probability of obtaining any particular value of $x + y$. We first observe that because x and y are independently measured, the probability of obtaining any given x *and* any given y is just the product of (5.51) and (5.52):

$$Prob(x, y) \propto \exp\left[-\frac{1}{2}\left(\frac{x^2}{\sigma_x^2} + \frac{y^2}{\sigma_y^2}\right)\right]. \tag{5.53}$$

Knowing the probability of obtaining any x *and* any y, we can now calculate the probability for any given value of $x + y$. The first step is to rewrite the exponent in (5.53) in terms of the variable of interest, $x + y$. This step can be done using the identity (which you can easily verify)

$$\frac{x^2}{A} + \frac{y^2}{B} = \frac{(x + y)^2}{A + B} + \frac{(Bx - Ay)^2}{AB(A + B)} \tag{5.54}$$

$$= \frac{(x + y)^2}{A + B} + z^2. \tag{5.55}$$

In the second line I have introduced the abbreviation z^2 for the second term on the right of (5.54) because its value does not interest us anyway.

If we substitute (5.55) into (5.53), replacing A with $\sigma_x{}^2$ and B with $\sigma_y{}^2$, we obtain

$$Prob(x, y) \propto \exp\left[-\frac{(x + y)^2}{2(\sigma_x{}^2 + \sigma_y{}^2)} - \frac{z^2}{2} \right]. \tag{5.56}$$

This probability for obtaining given values of x and y can just as well be viewed as the probability of obtaining given values of $x + y$ and z. Thus, we can rewrite (5.56) as

$$Prob(x + y, z) \propto \exp\left[\frac{-(x + y)^2}{2(\sigma_x{}^2 + \sigma_y{}^2)} \right] \exp\left[\frac{-z^2}{2} \right]. \tag{5.57}$$

Finally, what we want is the probability of obtaining a given value of $x + y$, *irrespective of the value of z*. This probability is obtained by summing, or rather integrating, (5.57) over all possible values of z; that is,

$$Prob(x + y) = \int_{-\infty}^{\infty} Prob(x + y, z)\, dz. \tag{5.58}$$

When we integrate (5.57) with respect to z, the factor $\exp(-z^2/2)$ integrates to $\sqrt{2\pi}$, and we find

$$Prob(x + y) \propto \exp\left[\frac{-(x + y)^2}{2(\sigma_x{}^2 + \sigma_y{}^2)} \right]. \tag{5.59}$$

This result shows that the values of $x + y$ are normally distributed with width $\sqrt{\sigma_x{}^2 + \sigma_y{}^2}$ as anticipated.

Our proof is complete for the case when the true values of x and y are both zero, $X = Y = 0$. If X and Y are nonzero, we can proceed as follows: We first write

$$x + y = (x - X) + (y - Y) + (X + Y). \tag{5.60}$$

Here, the first two terms are centered on zero, with widths σ_x and σ_y, by the result from step I. Therefore, by the result just proved [Equation (5.59)], the sum of the first two terms is normally distributed with width $\sqrt{\sigma_x{}^2 + \sigma_y{}^2}$. The third term in (5.60) is a fixed number; therefore, by the result of step I again, it simply shifts the center of the distribution to $(X + Y)$ but leaves the width unchanged. In other words, the values of $(x + y)$ as given by (5.60) are normally distributed about $(X + Y)$ with width $\sqrt{\sigma_x{}^2 + \sigma_y{}^2}$. This is the required result.

IV. THE GENERAL CASE

Having justified the error-propagation formula for the special case of a sum $x + y$, we can justify the general formula for error propagation surprisingly simply. Suppose we measure two independent quantities x and y whose observed values are normally distributed, and we now calculate some quantity $q(x, y)$ in terms of x and y. The distribution of values of $q(x, y)$ is found easily by using the results from steps I through III as follows:

First, the widths σ_x and σ_y (the uncertainties in x and y) must, as always, be small. This requirement means that we are concerned only with values of x close to X and y close to Y, and we can use the approximation (3.42) to write

$$q(x, y) \approx q(X, Y) + \left(\frac{\partial q}{\partial x}\right)(x - X) + \left(\frac{\partial q}{\partial y}\right)(y - Y). \tag{5.61}$$

This approximation is good because the only values of x and y that occur significantly often are close to X and Y. The two partial derivatives are evaluated at X and Y and are, therefore, fixed numbers.

The approximation (5.61) expresses the desired quantity $q(x, y)$ as the sum of three terms. The first term $q(X, Y)$ is a fixed number, so it merely shifts the distribution of answers. The second term is the fixed number $\partial q/\partial x$ times $(x - X)$, whose distribution has width σ_x, so the values of the second term are centered on zero, with width

$$\left(\frac{\partial q}{\partial x}\right)\sigma_x .$$

Similarly, the values of the third term are centered on zero with width

$$\left(\frac{\partial q}{\partial y}\right)\sigma_y .$$

Combining the three terms in (5.61) and invoking the results already established, we conclude that the values of $q(x, y)$ are normally distributed about the true value $q(X, Y)$ with width

$$\sigma_q = \sqrt{\left(\frac{\partial q}{\partial x}\sigma_x\right)^2 + \left(\frac{\partial q}{\partial y}\sigma_y\right)^2} . \tag{5.62}$$

If we identify the standard deviations σ_x and σ_y as the uncertainties in x and y, the result (5.62) is precisely the rule (3.47) for propagation of random errors, for the case when q is a function of just two variables, $q(x, y)$. If q depends on several variables, $q(x, y, \ldots, z)$, the preceding argument can be extended immediately to establish the general rule (3.47) for functions of several variables. Because the rules of Chapter 3 concerning propagation of random errors can be derived from (3.47), they are all now justified.

5.7   Standard Deviation of the Mean

One more important result, quoted in Chapter 4, remains to be proved. This result concerns the standard deviation of the mean $\sigma_{\bar{x}}$. I proved (in Section 5.5) that if we make N measurements $x_1, \ldots, x_N$ of a quantity x (that is normally distributed), the best estimate of the true value X is the mean $\bar{x}$ of $x_1, \ldots, x_N$. In Chapter 4, I stated that the uncertainty in this estimate is the standard deviation of the mean,

$$\sigma_{\bar{x}} = \sigma_x/\sqrt{N}. \tag{5.63}$$

Let us now prove this result. The proof is so surprisingly brief that you need to follow it very carefully.

Suppose that the measurements of x are normally distributed about the true value X with width parameter σ_x. We now want to know the reliability of *the average of the N measurements*. To answer this, we naturally imagine repeating our N measurements many times; that is, we imagine performing a sequence of experiments, in each of which we make N measurements and compute the average. We now want to find the distribution of these many determinations of the average of N measurements.

In each experiment, we measure N quantities $x_1, \ldots, x_N$ and then compute the function

$$\bar{x} = \frac{x_1 + \cdots + x_N}{N}. \tag{5.64}$$

Because the calculated quantity ($\bar{x}$) is a simple function of the measured quantities $x_1, \ldots, x_N$, we can now find the distribution of our answers for $\bar{x}$ by using the error-propagation formula. The only unusual feature of the function (5.64) is that all the measurements $x_1, \ldots, x_N$ happen to be measurements of the same quantity, with the same true value X and the same width σ_x.

We first observe that, because each of the measured quantities $x_1, \ldots, x_N$ is normally distributed, the same is true for the function $\bar{x}$ given by (5.64). Second, the true value for each of $x_1, \ldots, x_N$ is X; so the true value of $\bar{x}$ as given by (5.64) is

$$\frac{X + \cdots + X}{N} = X.$$

Thus, after making many determinations of the average $\bar{x}$ of N measurements, our many results for $\bar{x}$ will be normally distributed about the true value X. The only remaining (and most important) question is to find the width of our distribution of answers. According to the error-propagation formula (5.62), rewritten for N variables, this width is

$$\sigma_{\bar{x}} = \sqrt{\left(\frac{\partial \bar{x}}{\partial x_1} \sigma_{x_1}\right)^2 + \cdots + \left(\frac{\partial \bar{x}}{\partial x_N} \sigma_{x_N}\right)^2}. \tag{5.65}$$

Because $x_1, \ldots, x_N$ are all measurements of the same quantity x, their widths are all the same and are all equal to σ_x,

$$\sigma_{x_1} = \cdots = \sigma_{x_N} = \sigma_x.$$

We also see from (5.64) that all the partial derivatives in (5.65) are the same:

$$\frac{\partial \bar{x}}{\partial x_1} = \cdots = \frac{\partial \bar{x}}{\partial x_N} = \frac{1}{N}.$$

Therefore, (5.65) reduces to

$$\sigma_{\bar{x}} = \sqrt{\left(\frac{1}{N}\sigma_x\right)^2 + \cdots + \left(\frac{1}{N}\sigma_x\right)^2}$$

$$= \sqrt{N\frac{\sigma_x^2}{N^2}} = \frac{\sigma_x}{\sqrt{N}}, \tag{5.66}$$

as required.

We have arrived at the desired result (5.66) so quickly that we probably need to pause and review its significance. We imagined a large number of experiments, in each of which we made N measurements of x and then computed the average $\bar{x}$ of those N measurements. We have shown that, after repeating this experiment many times, our many answers for $\bar{x}$ will be normally distributed, that they will be centered on the true value X, and that the width of their distribution is $\sigma_{\bar{x}} = \sigma_x/\sqrt{N}$, as shown in Figure 5.17 for $N = 10$. This width $\sigma_{\bar{x}}$ is the 68% confidence limit for our experiment. If we find the mean of N measurements *once,* we can be 68% confident that our answer lies within a distance $\sigma_{\bar{x}}$ of the true value X. This result is exactly what should be signified by the *uncertainty in the mean.* It also explains clearly why that uncertainty is called the standard deviation of the mean.

With this simple and elegant proof, all the results quoted in earlier chapters concerning random uncertainties have now been justified.

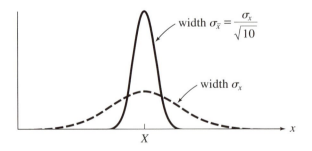

Figure 5.17. The individual measurements of x are normally distributed about X with width σ_x (dashed curve). If we use the same equipment to make many determinations of the average of 10 measurements, the results $\bar{x}$ will be normally distributed about X with width $\sigma_{\bar{x}} = \sigma_x/\sqrt{10}$ (solid curve).

5.8   Acceptability of a Measured Answer

We can now return to two questions first raised but not completely answered in Chapter 2. First, what is meant by the now-familiar statement that we are "reasonably confident" that a measured quantity lies in the range $x_{best} \pm \delta x$? Second, when we compare our value x_{best} with an expected value x_{exp} (the latter based on some theory, or just someone else's measurement), how do we decide whether or not the agreement between our value and the expected value is acceptable?

For the first question, the answer should by now be clear: If we measure a quantity x several times (as we usually would), the mean $\bar{x}$ of our measurements is the best estimate for x, and the standard deviation of the mean $\sigma_{\bar{x}}$ is a good measure of its uncertainty. We would report the conclusion that

$$\text{(value of } x) = \bar{x} \pm \sigma_{\bar{x}},$$

meaning that, based on our observations, we expect 68% of any measurements of x, made in the same way, to fall in the range $\bar{x} \pm \sigma_{\bar{x}}$.

We *could* choose to characterize our uncertainty differently. For example, we might choose to state our conclusion as

$$\text{(value of } x) = \bar{x} \pm 2\sigma_{\bar{x}};$$

here, we would be stating the range in which we expect 95% of all comparable measurements to fall. Clearly, the essential point in stating any measured value is to state a range (or uncertainty) and *the confidence level corresponding to that range.* The most popular choice is to give the standard deviation of the answer, with its familiar significance as the 68% confidence limit.

As emphasized in Chapter 2, almost all experimental conclusions involve the comparison of two or more numbers. With our statistical theory, we can now give a quantitative significance to many such comparisons. Here, I will consider just one type of experiment, in which we arrive at a number and compare our result with some known, expected answer. Notice that this general description fits many interesting experiments. For instance, in an experiment to check conservation of momentum, we may measure initial and final momenta, p and p', to verify that $p = p'$ (within uncertainties), but we can equally well regard this experiment as finding a value for $(p - p')$ to be compared with the expected answer of zero. More generally, when we want to compare any two measurements that are supposedly the same, we can form their difference and compare it with the expected answer of zero. Any experiment that involves measuring a quantity (such as g, the acceleration of gravity), for which an accurate accepted value is known, is also of this type, and the expected answer is the known accepted value.

Let us suppose that a student measures some quantity x (such as the difference of two momenta that are supposedly equal) in the form

$$\text{(value of } x) = x_{best} \pm \sigma,$$

where σ denotes the standard deviation of his answer (which would be the SDOM if x_{best} was the mean of several measurements). He now wants to compare this answer with the expected answer x_{exp}.

In Chapter 2, I argued that if the discrepancy $|x_{best} - x_{exp}|$ is less than (or only slightly more than) σ, then the agreement is satisfactory, but if $|x_{best} - x_{exp}|$ is much greater than σ, it is not satisfactory. These criteria are correct as far as they go, but they give no quantitative measure of how good or bad the agreement is. They also do not tell us where to draw the boundary of acceptability. Would a discrepancy of 1.5σ have been satisfactory? Or 2σ?

We can now answer these questions if we assume that our student's measurement was governed by a normal distribution (as is certainly reasonable). We start by making two working hypotheses about this distribution:

(1) The distribution is centered on the expected answer x_{exp}.

(2) The width parameter of the distribution is equal to the student's estimate σ.

Hypothesis (1) is, of course, what the student hopes is true. It assumes that all systematic errors were reduced to a negligible level (so that the distribution was centered on the true value) and that the true value was indeed x_{exp} (that is, that the reasons for expecting x_{exp} were correct). Hypothesis (2) is an approximation because σ must have been an estimate of the standard deviation, but it is a reasonable approximation if the number of measurements on which σ is based is large.[9] Taken together, our two hypotheses are the same as assuming that the student's procedures and calculations were essentially correct.

We must now decide whether the student's value x_{best} was a reasonable one to obtain if our hypotheses were correct. If the answer is yes, there is no reason to doubt the hypotheses, and all is well; if the answer is no, the hypotheses must be doubted, and the student must examine the possibilities of mistakes in the measurements or calculations, of undetected systematic errors, and of the expected answer x_{exp} being incorrect.

We first determine the discrepancy, $|x_{best} - x_{exp}|$, and then

$$t = \frac{|x_{best} - x_{exp}|}{\sigma}, \tag{5.67}$$

the number of standard deviations by which x_{best} differs from x_{exp}. (Here, σ denotes the standard deviation appropriate to x_{best}; if x_{best} is the mean of several measurements, then σ is the standard deviation of the mean.) Next, from the table of the normal error integral in Appendix A, we can find the probability (given our hypotheses) of obtaining an answer that differs from x_{exp} by t or more standard deviations. This probability is

$$Prob(\text{outside } t\sigma) = 1 - Prob(\text{within } t\sigma). \tag{5.68}$$

If this probability is large, the discrepancy $|x_{best} - x_{exp}|$ is perfectly reasonable, and the result x_{best} is acceptable; if the probability in (5.68) is "unreasonably small," the

---

[9] We are going to judge the reasonableness of our measurement, x_{best}, by comparing $|x_{best} - x_{exp}|$ with σ, our *estimate* of the width of the normal distribution concerned. If the number of measurements on which σ was based is *small*, this estimate may be fairly unreliable, and the confidence levels will be correspondingly inaccurate (although still a useful rough guide). With a small number of measurements, the accurate calculation of confidence limits requires use of the so-called "student's t distribution," which allows for the probable variations in our estimate σ of the width. See H. L. Alder and E. B. Roessler, *Introduction to Probability and Statistics,* 6th ed. (W. H. Freeman, 1977), Chapter 10.

discrepancy must be judged *significant* (that is, unacceptable), and our unlucky student must try to find out what has gone wrong.

Suppose, for example, the discrepancy $|x_{best} - x_{exp}|$ is one standard deviation. The probability of a discrepancy this large or larger is the familiar 32%. Clearly, a discrepancy of one standard deviation is quite likely to occur and is, therefore, insignificant. At the opposite extreme, the probability *Prob*(outside 3σ) is just 0.3%, and, if our hypotheses are correct, a discrepancy of 3σ is most unlikely. Turning this statement around, if our student's discrepancy is 3σ, our hypotheses were most unlikely to be correct.

The boundary between acceptability and unacceptability depends on the level below which we judge a discrepancy is unreasonably improbable. This level is a matter of opinion, to be decided by the experimenter. Many scientists regard 5% a fair boundary for "unreasonable improbability." If we accept this choice, then a discrepancy of 2σ would be just unacceptable, because *Prob*(outside 2σ) = 4.6%. In fact, from the table in Appendix A, we see that any discrepancy greater than 1.96σ is unacceptable at this 5% level, and discrepancies this large are sometimes called *"significant."* Similarly, at the 1% level, any discrepancy greater than 2.58σ would be unacceptable, and discrepancies this large are sometimes called *"highly significant."*

Quick Check 5.6.   A student measures the electron charge e and notes that her answer is 2.4 standard deviations away from the accepted value. Is this discrepancy significant at the 5% level? What about the 2% level? Or 1%?

We still do not have a clear-cut answer that a certain measured value x_{best} is, or is not, acceptable. Our theory of the normal distribution, however, has given us a clear, quantitative measure of the reasonableness of any particular answer, which is the best we can hope for.

Most physicists do not spend a lot of time debating precisely where the boundary of acceptability lies. If the discrepancy is appreciably less than 2σ (1.8σ, say), then by almost any standard the result would be judged acceptable. If the discrepancy is appreciably more than 2.5σ, then by any standard it is unacceptable. If the discrepancy falls in the gray region between about 1.9σ and about 2.6σ, the experiment is simply inconclusive. If the experiment is sufficiently important (as a test of a new theory, for example), it needs to be repeated (preferably with improved techniques) until a conclusive result *is* obtained.

Of course, some experiments are more complicated and require correspondingly more involved analyses. Most of the basic principles, however, have been illustrated by the simple, important case discussed here. Further examples can be found in Part II.

Principal Definitions and Equations of Chapter 5

LIMITING DISTRIBUTIONS

If $f(x)$ is the limiting distribution for measurement of a continuous variable x, then

$$f(x)\,dx = \text{probability that any one measurement will give an answer between } x \text{ and } x + dx,$$

and

$$\int_a^b f(x)\,dx = \text{probability that any one measurement will give an answer between } x = a \text{ and } x = b. \quad [\text{See (5.12)}]$$

The normalization condition is

$$\int_{-\infty}^{\infty} f(x)\,dx = 1. \quad [\text{See (5.13)}]$$

The mean value of x expected after many measurements is

$$\bar{x} = \int_{-\infty}^{\infty} x\,f(x)\,dx. \quad [\text{See (5.15)}]$$

THE GAUSS, OR NORMAL, DISTRIBUTION

If the measurements of x are subject to many small random errors but negligible systematic error, their limiting distribution will be the *normal,* or *Gauss, distribution:*

$$G_{X,\sigma}(x) = \frac{1}{\sigma\sqrt{2\pi}}\,e^{-(x-X)^2/2\sigma^2}, \quad [\text{See (5.25)}]$$

where

$$X = \text{true value of } x$$
$$= \text{center of distribution}$$
$$= \text{mean value after many measurements,}$$

and

$$\sigma = \text{width parameter of distribution}$$
$$= \text{standard deviation after many measurements.}$$

The probability of a single measurement falling within t standard deviations of X is

$$Prob(\text{within } t\sigma) = \frac{1}{\sqrt{2\pi}}\int_{-t}^{t} e^{-z^2/2}\,dz. \quad [\text{See (5.35)}]$$

This integral is often called the error function or the normal error integral. Its value as a function of t is tabulated in Appendix A. In particular,

$$Prob(\text{within } \sigma) = 68.27\%.$$

ESTIMATING X AND σ FROM N MEASURED VALUES

After N measurements of a normally distributed quantity x,

$$x_1, x_2, \ldots, x_N,$$

the best estimate for the true value X is the mean of our measurements,

$$(\text{best estimate for } X) = \bar{x} = \frac{\sum x_i}{N}, \qquad \text{[See (5.42)]}$$

and the best estimate for the width σ is the standard deviation of the measurements,

$$(\text{best estimate for } \sigma) = \sigma_x = \sqrt{\frac{\sum(x_i - \bar{x})^2}{(N-1)}}. \qquad \text{[See (5.45)]}$$

The uncertainties in these estimates are as follows: The uncertainty in $\bar{x}$ as an estimate of X is

$$(\text{uncertainty in } \bar{x}) = \text{SDOM} = \frac{\sigma_x}{\sqrt{N}} \qquad \text{[See (5.66)]}$$

and the uncertainty in σ_x as the estimate of the true width σ is given by

$$(\text{fractional uncertainty in } \sigma_x) = \frac{1}{\sqrt{2(N-1)}}. \qquad \text{[See (5.46)]}$$

ACCEPTABILITY OF A MEASURED ANSWER

Suppose we measure a quantity x in the standard form

$$(\text{value of } x) = x_{best} \pm \sigma,$$

where σ is the appropriate standard deviation. Suppose also that, based on some theory or on someone else's measurements, we expected the value x_{exp}. We say that x_{best} differs from x_{exp} by t standard deviations, where

$$t = \frac{|x_{best} - x_{exp}|}{\sigma}.$$

Assuming x is normally distributed about x_{exp} with width σ, we can find from Appendix A the probability $Prob(\text{outside } t\sigma)$ of a discrepancy as large as ours or larger. If this probability is less than some chosen level (1%, for example), we judge the agreement to be unacceptable at that level. [For example, if $Prob(\text{outside } t\sigma)$ is less than 1%, the agreement is unacceptable at the 1% level.]

Problems for Chapter 5

For Section 5.1: Histograms and Distributions

5.1. ★ Draw a bar histogram for Joe's grades, as given in Quick Check 5.1. The horizontal axis should show the five possible grades (or, better, the corresponding scores $s_k = 0, 1, \ldots, 4$) and the vertical axis, the fractions F_k. Compute his average grade as $\Sigma s_k F_k$.

5.2. ★ A health physicist places a weak radioactive sample in a liquid scintillation counter and records the number of decays in 10-minute intervals, starting at $t = 0$. His results are as follows (with t in minutes):

Interval for t:	0 to 10	10 to 20	20 to 30	30 to 40	40 to 50
Number:	9	6	3	1	1

and none beyond $t = 50$ min. Make a bin histogram of these events. Show t on the horizontal axis, and choose the vertical scale so that the area of each rectangle is the fraction of the decays that occurred in the corresponding bin.

5.3. ★★ A student measures the angular momenta L_i and L_f of a rotating system before and after adding an extra mass. To check the conservation of angular momentum, he calculates $L_i - L_f$ (expecting the answer zero). He repeats the measurement 50 times and collects his answers into bins as in Table 5.3, which shows his results

Table 5.3. Occurrences of values of $L_i - L_f$; for Problem 5.3.

	Bin								
After	$(-9, -7)$	$(-7, -5)$	$(-5, -3)$	$(-3, -1)$	$(-1, 1)$	$(1, 3)$	$(3, 5)$	$(5, 7)$	$(7, 9)$
5 trials	0	1	2	0	1	0	1	0	0
10 trials	0	1	2	2	3	1	1	0	0
50 trials	1	3	7	8	10	9	6	4	2

(in some unspecified units) after 5, 10, and 50 trials. Draw a bin histogram for each of these three cases. (Be careful to choose your scales so that the area of each rectangle is the fraction of events in the corresponding bin.)

5.4. ★★ A student makes 20 measurements of the time for a ball bearing to fall from the top to the bottom of a vertical cylinder of oil. She arranges her results in increasing order and counts how many times she got each different value, as follows (with the times in tenths of a second):

Time, t:	71	72	73	74	75	76	77	78	79	80
Occurrences:	2	0	3	5	4	1	3	1	0	1

(a) Draw a bin histogram of these results using bins of width 1, starting at 70.5. (Notice that, with bins of width 1, this bin histogram gives essentially the same result as a bar histogram.) **(b)** Redraw the histogram with a bin width of 2, again starting at 70.5. **(c)** Notice how the wider bins in part (b) give a smoother histogram. If the chosen bins are too wide, however, information starts to be lost. To illustrate this loss, redraw the histogram with a bin width of 10, again starting at 70.5.

For Section 5.2: Limiting Distributions

5.5. ★ The limiting distribution for the results in some hypothetical measurement is given by the triangular function shown in Figure 5.18, where the value of $f(0)$ is

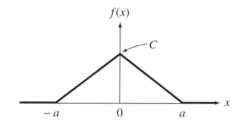

Figure 5.18. A triangular distribution; for Problem 5.5.

called C. **(a)** What is the probability of a measurement outside the range between $x = -a$ and $x = a$? **(b)** What is the probability of a measurement with $x > 0$? **(c)** Use the normalization condition (5.13) to find C in terms of a. **(d)** Sketch this function for the cases that $a = 1$ and $a = 2$.

5.6. ★ Consider an experiment similar to the one described in Problem 5.2, in which the experimenter counts the number of decays from a radioactive sample in a short time interval Δt from a radioactive source, starting at time $t = 0$. The limiting distribution for this kind of experiment is the *exponential distribution*,

$$f(t) = \frac{1}{\tau} e^{-t/\tau}, \tag{5.69}$$

where τ is a positive constant. **(a)** Sketch this function. (The distribution is zero for $t < 0$ because the experiment begins only at $t = 0$.) **(b)** Prove that this function satisfies the normalization condition (5.13). **(c)** Find the mean time $\bar{t}$ at which the decays occur, as given by Equation (5.15). (Your answer here shows the significance of the parameter τ: It is the mean time at which the atoms in a large sample decay.)

5.7. ★ For the exponential distribution of Problem 5.6, what is the probability for a result $t > \tau$? What for $t > 2\tau$? (Notice that these probabilities also give the fraction of the original atoms that live longer than τ and 2τ.)

5.8. ★★ The physicist of Problem 5.2 decides that the limiting distribution for his experiment is the exponential distribution (5.69) (from Problem 5.6) with the parameter $\tau = 14.4$ min. If you have not already done so, draw the histogram for the data of Problem 5.2, and then draw the supposed limiting distribution on the same plot. How well does it seem to match? (You have no quantitative way to measure the agreement, so you can decide only if it *looks* satisfactory.)

5.9. ★★   The limiting distribution for the results in some hypothetical measurement has the form

$$f(x) \;=\; \begin{cases} C & \text{for } |x| < a \\ 0 & \text{otherwise.} \end{cases}$$

(a) Use the normalization condition (5.13) to find C in terms of a. **(b)** Sketch this function. Describe in words the distribution of results governed by this distribution. **(c)** Use Equations (5.15) and (5.16) to calculate the mean and standard deviation that would be found after many measurements.

5.10. ★★   Explain clearly why the standard deviation for a limiting distribution $f(x)$ is given by (5.16). Your argument will parallel closely that leading from (5.14) to (5.15).

For Section 5.3: The Normal Distribution

5.11. ★   Using proper squared paper and clearly labeled axes, make good plots of the Gauss distribution

$$G_{X,\sigma}(x) \;=\; \frac{1}{\sigma\sqrt{2\pi}}\, e^{-(x-X)^2/2\sigma^2}$$

for $X = 2$, $\sigma = 1$, and for $X = 3$, $\sigma = 0.3$. Use your calculator to compute the values of $G_{X,\sigma}(x)$. (If it has two memories to store $\sigma\sqrt{2\pi}$ and $-2\sigma^2$, this feature will speed your calculations. If you remember that the function is symmetric about $x = X$, the number of calculations needed is halved.) Put both graphs on the same plot for comparison.

5.12. ★   The width of a Gauss distribution is usually characterized by the parameter σ. An alternative parameter with a simple geometric interpretation is the *full width at half maximum,* or FWHM. This parameter is the distance between the two points x where $G_{X,\sigma}(x)$ is half its maximum value, as in Figure 5.19. Prove that

$$\text{FWHM} \;=\; 2\sigma\sqrt{2\ln 2} \;=\; 2.35\,\sigma.$$

Some physicists use the *half width at half maximum,* shown in Figure 5.19 and defined as half the FWHM. Obviously HWHM = 1.17σ, or, very roughly, HWHM $\approx \sigma$.

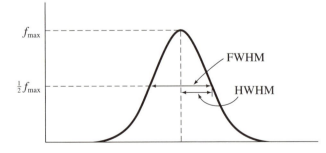

Figure 5.19. The full width at half maximum (FWHM) and the half width at half maximum (HWHM); for Problem 5.12.

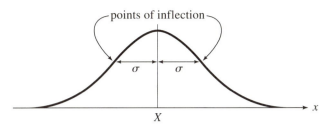

Figure 5.20. The points $X \pm \sigma$ are the points of inflection of the Gauss curve; for Problem 5.13.

5.13. ★   One way to define the width σ of the Gauss distribution is that the points $X \pm \sigma$ are the two points of inflection (Figure 5.20), where the curvature changes sign; that is, where the second derivative is zero. Prove this claim.

5.14. ★★   Make careful sketches of the normal distributions $G_{X,\sigma}(x)$ for the three cases:
   (a) $X = 0, \sigma = 1$
   (b) $X = 0, \sigma = 2$
   (c) $X = 5, \sigma = 1$
Describe in words the differences among the three curves.

5.15. ★★   If you have not yet done so, plot the third histogram of Problem 5.3. The student of that problem decides that the distribution of his results is consistent with the Gauss distribution $G_{X,\sigma}(x)$ centered on $X = 0$ with width $\sigma = 3.4$. Draw this distribution on the same graph and compare it with your histogram. (Read the hints to Problem 5.11. Note that you have no quantitative way to assess the fit; all you can do is see if the Gauss function *seems* to fit the histogram satisfactorily.)

5.16. ★★   Give in detail the steps leading from (5.30) to (5.31) to show that the standard deviation σ_x of many measurements normally distributed with width parameter σ is $\sigma_x = \sigma$.

For Section 5.4: The Standard Deviation as 68% Confidence Limit

5.17. ★   Some statisticians speak of a "68–95 rule." What do you suppose this rule is? What is the "68–95–99.7 rule"?

5.18. ★   We often say that the probability of a measurement in the range $X \pm 2\sigma$ is about 95%. What is this probability to four significant figures? What range $X \pm t\sigma$ actually has probability equal to 95.00%? (Give t to three significant figures. The needed probabilities can be found in Appendix A.)

5.19. ★★   A student measures a quantity y many times and calculates his mean as $\bar{y} = 23$ and his standard deviation as $\sigma_y = 1$. What fraction of his readings would you expect to find between
   (a) 22 and 24?   (d) 21 and 23?
   (b) 22.5 and 23.5?   (e) 24 and 25?
   (c) 21 and 25?

Finally, **(f)** within what limits (equidistant on either side of the mean) would you expect to find 50% of his readings? (The necessary information for all parts of this question is in Figure 5.13. More detailed information on these kinds of probabilities is in Appendixes A and B.)

5.20. ★★   An extensive survey reveals that the heights of men in a certain country are normally distributed, with a mean $\bar{h} = 69''$ and standard deviation $\sigma = 2''$. In a random sample of 1,000 men, how many would you expect to have a height

  (a) between $67''$ and $71''$?

  (b) more than $71''$?

  (c) more than $75''$?

  (d) between $65''$ and $67''$?

5.21. ★★   The Giraffe Club of Casterbridge is a club for young adults (18–24 years) who are unusually tall. There are 2,000 women between the ages of 18 and 24 in Casterbridge, and the heights of women in this age range are distributed normally with a mean of $5'5\frac{1}{2}''$ and standard deviation of $2\frac{1}{2}''$. **(a)** If the Giraffe Club initially sets its minimum height for women at $5'10''$, approximately how many women are eligible to join? **(b)** A year or so later, the club decides to double its female membership by lowering the minimum height requirement; what would you recommend that the club set as its new minimum height for women (to the nearest half inch)?

5.22. ★★   If the measurements of a quantity x are governed by the Gauss distribution $G_{X,\sigma}(x)$, the probability of obtaining a value between $X - t\sigma$ and $X + t\sigma$ is

$$Prob(\text{within } t\sigma) = \int_{X-t\sigma}^{X+t\sigma} G_{X,\sigma}(x)\, dx.$$

Prove carefully, showing all the necessary changes of variables, that

$$Prob(\text{within } t\sigma) = \frac{1}{\sqrt{2\pi}} \int_{-t}^{t} e^{-z^2/2}\, dz. \qquad (5.70)$$

With each change of variables, check carefully what happens to your limits of integration. The integral (5.70) is often called the *error function,* denoted erf(t), or the *normal error integral.*

5.23. ★★★   We have seen that for the normal distribution, the standard deviation gives the 68% confidence range. This result is not necessarily true for other distributions, as the following problem illustrates: Consider the exponential distribution of Problem 5.6, $f(t) = (1/\tau)e^{-t/\tau}$ (for $t \geqslant 0$; $f(t) = 0$ for $t < 0$). The parameter τ is the mean value of t (that is, after many measurements, $\bar{t} = \tau$). **(a)** Use the integral (5.16) to prove that τ is also the standard deviation, $\sigma_t = \tau$. (This result is a noteworthy property of this distribution—that the mean value is equal to the standard deviation.) **(b)** By doing the necessary integral, find the probability that any one value would fall in the range $\bar{t} \pm \sigma_t$.

For Section 5.5: Justification of the Mean as Best Estimate

5.24. ★   Suppose you have measured a quantity x six times, as follows:

$$51, 53, 54, 55, 52, 53.$$

(a) Assuming these measurements are normally distributed, what should be your best estimates for the true value X and the standard deviation σ? (b) Based on these estimates, what is the probability that a seventh measurement would fall outside the range of the first six? (Given that your results are rounded to the nearest integer, this probability is that for a result $x \leqslant 50.5$ or $x \geqslant 55.5$.)

5.25. ★   A student measures a time t eight times with the following results (in tenths of a second):

$$\begin{array}{lcccccc} \text{Value, } t_k: & 75 & 76 & 77 & 78 & 79 & 80 \\ \text{Occurrences, } n_k: & 2 & 3 & 0 & 0 & 2 & 1 \end{array}$$

(a) Assuming these measurements are normally distributed, what should be your best estimates for the true value and the standard deviation? (b) Based on these estimates, what is the probability that a ninth measurement would be 81 or more? (Because the measurements are rounded to the nearest integer, this probability is that for a value $t \geqslant 80.5$.)

5.26. ★★   Suppose we have N measurements $x_1, \ldots, x_N$ of the same quantity x, and we believe that their limiting distribution should be the Gauss function $G_{X,\sigma}(x)$, with X and σ unknown. The principle of maximum likelihood asserts that the best estimate for the width is the value of σ for which the probability $Prob_{X,\sigma}(x_1, \ldots, x_N)$ of the observed values $x_1, \ldots, x_N$ is largest. Differentiate $Prob_{X,\sigma}(x_1, \ldots, x_N)$ in (5.40) with respect to σ, and show that the maximum occurs when σ is given by (5.43). [As discussed after (5.43), this result means that the best estimate for the true width σ is the standard deviation of the N observed values $x_1, \ldots, x_N$. In practice, the true value X must be replaced with its best estimate $\bar{x}$, which requires replacement of N by $N - 1$, as proved in Appendix E.]

5.27. ★★   (a) Based on the data of Problem 5.24, find the mean, the standard deviation, and the uncertainty of your value for the SD [the last using Equation (5.46)]. (b) Recalculate the probability asked for in part (b) of Problem 5.24 assuming the true value of σ is $\sigma_x - \delta\sigma_x$, and, once again, assuming σ is really $\sigma_x + \delta\sigma_x$. Comment on the difference in your two answers here.

5.28. ★★   Based on several measurements of the same quantity x normally distributed about X with width σ, we can estimate X and σ. (a) Approximately how many measurements must we make to know σ within 30%? (b) Within 10%? (c) Within 3%?

Section 5.6: Justification of Addition in Quadrature

5.29. ★   The measurements of a certain quantity x are distributed according to the Gauss function $G_{X,\sigma}(x)$ with $X = 10$ and $\sigma = 2$. Sketch this distribution and the distributions for $q = x + 5$ and for $q' = x/2$, all on the same graph.

5.30. ★   Verify the identity (5.54) used in justifying addition in quadrature when propagating random errors.

For Section 5.7: Standard Deviation of the Mean

5.31. ★★   Listed here are 40 measurements $t_1, \ldots, t_{40}$ of the time for a stone to fall from a window to the ground (all in hundredths of a second).

```
63  58  74  78  70  74  75  82  68  69
76  62  72  88  65  81  79  77  66  76
86  72  79  77  60  70  65  69  73  77
72  79  65  66  70  74  84  76  80  69
```

(a) Compute the standard deviation σ_t for the 40 measurements. **(b)** Compute the means $\bar{t}_1, \ldots, \bar{t}_{10}$ of the four measurements in each of the 10 columns. You can think of the data as resulting from 10 experiments, in each of which you found the *mean of four timings.* Given the result of part (a), what would you expect for the standard deviation of the 10 averages $\bar{t}_1, \ldots, \bar{t}_{10}$? What is it? **(c)** Plot histograms for the 40 individual measurements $t_1, \ldots, t_{40}$ and for the 10 averages $\bar{t}_1, \ldots, \bar{t}_{10}$. [Use the same scales and bin sizes for both plots so they can be compared easily. Bin boundaries can be chosen in various ways; perhaps the simplest is to put one boundary at the mean of all 40 measurements (72.90) and to use bins whose width is the standard deviation of the 10 averages $\bar{t}_1, \ldots, \bar{t}_{10}$.]

5.32. ★★   In Problem 4.13 are listed 30 measurements of a certain time t. **(a)** If you haven't already done so, find the standard deviation of these 30 values. **(b)** Now think of the 30 data as 10 columns, each representing an experiment consisting of three measurements, and compute the 10 averages, $\bar{t}_1, \ldots, \bar{t}_{10}$, for these 10 experiments. **(c)** Given the result of part (a), what would you expect for the standard deviation of the 10 means of part (b)? What is it? **(d)** Draw histograms for the 30 separate measurements and for the 10 means, $\bar{t}_1, \ldots, \bar{t}_{10}$. (Use the same scale for both. A good way to choose your bin boundaries is to put one at the mean of all 30 measurements and to use the SD of the 10 means as the bin width.)

5.33. ★★   Based on the data of Problem 5.25, what would you give for the best estimate for the time t and its uncertainty? What is the uncertainty in your value for the uncertainty in t? [The uncertainty in t is the SDOM, $\sigma_{\bar{t}} = \sigma_t/\sqrt{N}$. You can find the uncertainty in the SD σ_t from (5.46), and from this result you can find the uncertainty in the SDOM by error propagation.]

For Section 5.8: Acceptability of a Measured Answer

5.34. ★   According to a proposed theory, the quantity x should have the value x_{th}. Having measured x in the usual form $x_{best} \pm \sigma$ (where σ is the appropriate SD), we would say that the discrepancy between x_{best} and x_{th} is t standard deviations, where $t = |x_{best} - x_{th}|/\sigma$. How large must t be for us to say the discrepancy is significant at the 5% level? At the 2% level? At the 1% level?

5.35. ★   A student measures g, the acceleration of gravity, repeatedly and carefully and gets a final answer of 9.5 m/s^2 with a standard deviation of 0.1. If his measurements were normally distributed with center at the accepted value 9.8 and with width 0.1, what would be the probability of his getting an answer that differs from

9.8 by as much as (or more than) his? Assuming he made no actual mistakes, do you think his experiment may have suffered from some undetected systematic errors?

5.36. ★★ Two students measure the same quantity x and get final answers $x_A = 13 \pm 1$ and $x_B = 15 \pm 1$, where the quoted uncertainties are the appropriate standard deviations. **(a)** Assuming all errors are independent and random, what is the discrepancy $x_A - x_B$, and what is its uncertainty? **(b)** Assuming all quantities were normally distributed as expected, what would be the probability of getting a discrepancy as large as they did? Do you consider their discrepancy significant (at the 5% level)?

5.37. ★★ A nuclear physicist wants to check the conservation of energy in a certain nuclear reaction and measures the initial and final energies as $E_i = 75 \pm 3$ MeV and $E_f = 60 \pm 9$ MeV, where both quoted uncertainties are the standard deviations of the answers. Is this discrepancy significant (at the 5% level)? Explain your reasoning clearly.

Part II

If you have read and understood Chapter 5, you are now ready, with surprisingly little difficulty, to study a number of more advanced topics. The chapters of Part II present seven such topics, some of which are applications of the statistical theory already developed, and others of which are further extensions of that theory. All are important, and you will probably study them sooner or later. Because you might not necessarily want to learn them all at once, these topics have been arranged into independent, short chapters that can be studied in any order, as your needs and interests dictate.

Chapter 6

Rejection of Data

This chapter discusses the awkward question of whether to discard a measurement that seems so unreasonable that it looks like a mistake. This topic is controversial; some scientists would argue that discarding a measurement just because it looks unreasonable is *never* justified. Nevertheless, there is a simple test you could at least consider applying if you find yourself confronted with this situation. The test is called *Chauvenet's criterion* and is a nice application of the statistical ideas developed in Chapters 4 and 5.

6.1 The Problem of Rejecting Data

Sometimes, one measurement in a series of measurements appears to disagree strikingly with all the others. When this happens, the experimenter must decide whether the anomalous measurement resulted from some mistake and should be rejected or was a *bona fide* measurement that should be used with all the others. For example, imagine we make six measurements of the period of a pendulum and get the results (all in seconds)

$$3.8, 3.5, 3.9. 3.9, 3.4, 1.8. \tag{6.1}$$

In this example, the value 1.8 is startlingly different from all the others, and we must decide what to do with it.

We know from Chapter 5 that a legitimate measurement *may* deviate significantly from other measurements of the same quantity. Nevertheless, a legitimate discrepancy as large as that of the last measurement in (6.1) is *very improbable,* so we are inclined to suspect that the time 1.8 s resulted from some undetected mistake or other external cause. Perhaps, for example, we simply misread the last time or our electric timer stopped briefly during the last measurement because of a momentary power failure.

If we have kept very careful records, we may sometimes be able to establish such a definite cause for the anomalous measurement. For example, our records might show that a different stopwatch was used for the last timing in (6.1), and a subsequent check might show that this watch runs slow. In this case, the anomalous measurement should definitely be rejected.

Unfortunately, establishing an external cause for an anomalous result is usually not possible. We must then decide whether or not to reject the anomaly simply by examining the results themselves, and here our knowledge of the Gauss distribution proves useful.

The rejection of data is a controversial question on which experts disagree. It is also an *important* question. In our example, the best estimate for the period of the pendulum is significantly affected if we reject the suspect 1.8 s. The average of all six measurements is 3.4 s, whereas that of the first five is 3.7 s, an appreciable difference.

Furthermore, the decision to reject data is ultimately a subjective one, and the scientist who makes this decision may reasonably be accused by other scientists of "fixing" the data. The situation is made worse by the possibility that the anomalous result may reflect some important effect. Indeed, many important scientific discoveries first appeared as anomalous measurements that looked like mistakes. In throwing out the time 1.8 s in the example (6.1), we just *might* be throwing out the most interesting part of the data.

In fact, faced with data like those in (6.1), our only really honest course is to repeat the measurement many, many times. If the anomaly shows up again, we will presumably be able to trace its cause, either as a mistake or a real physical effect. If it does not recur, then by the time we have made, say, 100 measurements, there will be no significant difference in our final answer whether we include the anomaly or not.

Nevertheless, repeating a measurement 100 times every time a result seems suspect is frequently impractical (especially in a teaching laboratory). We therefore need some criterion for rejecting a suspect result. There are various such criteria, some quite complicated. Chauvenet's criterion provides a simple and instructive application of the Gauss distribution.

6.2   Chauvenet's Criterion

Let us return to the six measurements of the example (6.1):

$$3.8, 3.5, 3.9, 3.9, 3.4, 1.8.$$

If we assume for the moment that these are six legitimate measurements of a quantity x, we can calculate the mean $\bar{x}$ and standard deviation σ_x,

$$\bar{x} = 3.4 \text{ s} \qquad\qquad (6.2)$$

and

$$\sigma_x = 0.8 \text{ s}. \qquad\qquad (6.3)$$

We can now quantify the extent to which the suspect measurement, 1.8, is anomalous. It differs from the mean 3.4 by 1.6, or two standard deviations. If we assume the measurements were governed by a Gauss distribution with center and width given by (6.2) and (6.3), we can calculate the probability of obtaining measurements that differ by at least this much from the mean. According to the probabilities shown in Appendix A, this probability is

$$Prob(\text{outside } 2\sigma) = 1 - Prob(\text{within } 2\sigma)$$
$$= 1 - 0.95$$
$$= 0.05.$$

In other words, assuming that the values (6.2) and (6.3) for $\bar{x}$ and σ_x are legitimate, we would expect one in every 20 measurements to differ from the mean by at least as much as the suspect 1.8 s does. If we had made 20 or more measurements, we should actually *expect* to get one or two measurements as deviant as the 1.8 s, and we would have no reason to reject it. But we have made only six measurements, so the expected number of measurements as deviant as 1.8 s was actually

(expected number as deviant as 1.8 s)

$$= (\text{number of measurements}) \times Prob(\text{outside } 2\sigma)$$
$$= 6 \times 0.05 = 0.3.$$

That is, in six measurements we would expect (on average) only one-third of a measurement as deviant as the suspect 1.8 s.

This result provides us with the needed quantitative measure of the "reasonableness" of our suspect measurement. If we choose to regard one-third of a measurement as "ridiculously improbable," then we will conclude that the value 1.8 s was not a legitimate measurement and should be rejected.

The decision of where to set the boundary of "ridiculous improbability" is up to the experimenter. Chauvenet's criterion, as normally given, states that if the expected number of measurements at least as deviant as the suspect measurement is less than one-half, then the suspect measurement should be rejected. Obviously, the choice of one-half is arbitrary, but it is also reasonable and can be defended.

The application of Chauvenet's criterion to a general problem can now be described easily. Suppose you have made N measurements

$$x_1, \ldots, x_N$$

of a single quantity x. From all N measurements, you calculate $\bar{x}$ and σ_x. If one of the measurements (call it x_{sus}) differs from $\bar{x}$ so much that it looks suspicious, then find

$$t_{sus} = \frac{|x_{sus} - \bar{x}|}{\sigma_x}, \tag{6.4}$$

the number of standard deviations by which x_{sus} differs from $\bar{x}$. Next, from Appendix A you can find the probability

$$Prob(\text{outside } t_{sus}\sigma)$$

that a legitimate measurement would differ from $\bar{x}$ by t_{sus} or more standard deviations. Finally, multiplying by N, the total number of measurements, gives

$$n = (\text{expected number as deviant as } x_{sus})$$
$$= N \times Prob(\text{outside } t_{sus}\sigma).$$

If this expected number n is less than one-half, then, according to Chauvenet's criterion, you can reject x_{sus}.

If you do decide to reject x_{sus}, you would naturally recalculate $\bar{x}$ and σ_x using just the remaining data; in particular, your final answer for x would be this new mean, with an uncertainty equal to the new SDOM.

Example: Ten Measurements of a Length

A student makes 10 measurements of one length x and gets the results (all in mm)

$$46, 48, 44, 38, 45, 47, 58, 44, 45, 43.$$

Noticing that the value 58 seems anomalously large, he checks his records but can find no evidence that the result was caused by a mistake. He therefore applies Chauvenet's criterion. What does he conclude?

Accepting provisionally all 10 measurements, he computes

$$\bar{x} = 45.8 \quad \text{and} \quad \sigma_x = 5.1.$$

The difference between the suspect value $x_{sus} = 58$ and the mean $\bar{x} = 45.8$ is 12.2, or 2.4 standard deviations; that is,

$$t_{sus} = \frac{x_{sus} - \bar{x}}{\sigma_x} = \frac{58 - 45.8}{5.1} = 2.4.$$

Referring to the table in Appendix A, he sees that the probability that a measurement will differ from $\bar{x}$ by $2.4\sigma_x$ or more is

$$Prob(\text{outside } 2.4\sigma) = 1 - Prob(\text{within } 2.4\sigma)$$
$$= 1 - 0.984$$
$$= 0.016.$$

In 10 measurements, he would therefore expect to find only 0.16 of one measurement as deviant as his suspect result. Because 0.16 is less than the number 0.5 set by Chauvenet's criterion, he should at least consider rejecting the result.

If he decides to reject the suspect 58, then he must recalculate $\bar{x}$ and σ_x as

$$\bar{x} = 44.4 \quad \text{and} \quad \sigma_x = 2.9.$$

As you would expect, his mean changes a bit, and his standard deviation drops appreciably.

Quick Check 6.1. A student makes 20 measurements of a certain voltage V and computes her mean and standard deviation as $\bar{V} = 51$ and $\sigma_V = 2$ (both in microvolts). The evening before the work is due, she starts to write her report and realizes that one of her measured values was $V_{sus} = 56$. What is the probability of her getting a measurement this deviant from $\bar{V}$? If she decides to use Chauvenet's criterion, should she reject the suspect value?

6.3   Discussion

You should be aware that some scientists believe that data should *never* be rejected without *external* evidence that the measurement in question is incorrect. A reasonable compromise is to use Chauvenet's criterion to identify data that could be *considered* for rejection; having made this identification, you could do all subsequent calculations twice, once including the suspect data and once excluding them, to see how much the questionable values affect your final conclusion.

One reason many scientists are uncomfortable with Chauvenet's criterion is that the choice of one-half as the boundary of rejection (in the condition that $n < \frac{1}{2}$) is arbitrary. Perhaps even more important, unless you have made a very large number of measurements ($N \approx 50$, say), the value of σ_x is extremely uncertain as an estimate for the true standard deviation of the measurements. (See Problem 6.7 for an example.) This means in turn that the number t_{sus} in (6.4) is very uncertain. Because the probability of a measurement outside t standard deviations is very sensitive to t, a large error in t_{sus} causes a very large error in this probability and casts serious doubt on the whole procedure. For both of these reasons, Chauvenet's criterion should be used only as a last resort, when you cannot check your measurements by repeating them.

So far, we have assumed that only one measurement is suspect. What should you do if you have several? Given that the use of Chauvenet's criterion to reject *one* measurement is open to doubt, clearly its use to reject *several* measurements is even more problematic. Nevertheless, if there is absolutely no way you can repeat your measurements (because you have rashly dismantled your equipment before writing your report, for example), you may have to confront this question.

Suppose first that you have two measurements that deviate from the mean by the same large amount. In this case, you could calculate the expected number of measurements this deviant, and if this number is less than one (that is, *two times* one-half), then both measurements could be considered candidates for rejection. If you have two suspect measurements, x_1 and x_2, with x_2 more deviant than x_1, you should first apply Chauvenet's criterion using the value x_1. If the expected number this deviant is less than one, you could reject *both* values. If this expected number is more than one, you certainly should not reject both but instead reapply Chauvenet's criterion using x_2 and, if the expected number this deviant is less than one-half, you could reject just x_2.

Having rejected any measurements that fail Chauvenet's criterion, you would naturally recalculate $\bar{x}$ and σ_x using just the remaining data. The resulting value of σ_x will be smaller than the original one, and with the new σ_x, some more measurements may fail Chauvenet's criterion. However, agreement seems widespread that Chauvenet's criterion should *not* be applied a second time using the recalculated values of $\bar{x}$ and σ_x.

Principal Definitions and Equations of Chapter 6

CHAUVENET'S CRITERION

If you make N measurements $x_1, \ldots, x_N$ of a single quantity x, and if one of the measurements (x_{sus}, say) is suspiciously different from all the others, Chauvenet's criterion gives a simple test for deciding whether to reject this suspect value. First, compute the mean and standard deviation of all N measurements and then find the number of standard deviations by which x_{sus} differs from $\bar{x}$,

$$t_{sus} = \frac{|x_{sus} - \bar{x}|}{\sigma_x}.$$

Next, find the probability (assuming the measurements are normally distributed about $\bar{x}$ with width σ_x) of getting a result as deviant as x_{sus}, and, hence, the number of measurements expected to deviate this much,

$$n = N \times Prob(\text{outside } t_{sus}\sigma).$$

If $n < \frac{1}{2}$, then according to Chauvenet's criterion, you can reject the value x_{sus}.

Because there are several objections to Chauvenet's criterion (especially if N is not very large), it should be used only as a last resort, when the measurements of x cannot be checked. The objections to Chauvenet's criterion are even greater if two or more measurements are suspect, but the test *can* be extended to this situation, as described in Section 6.3.

Problems for Chapter 6

For Section 6.2: Chauvenet's Criterion

6.1. ★   An enthusiastic student makes 50 measurements of the heat Q released in a certain reaction. Her average and standard deviation are

$$\overline{Q} = 4.8 \quad \text{and} \quad \sigma_Q = 0.4,$$

both in kilocalories. **(a)** Assuming her measurements are governed by the normal distribution, find the probability that any one measurement would differ from $\overline{Q}$ by 0.8 kcal or more. How many of her 50 measurements should she expect to differ from $\overline{Q}$ by 0.8 kcal or more? If one of her measurements is 4.0 kcal and she decides to use Chauvenet's criterion, would she reject this measurement? **(b)** Would she reject a measurement of 6.0 kcal?

6.2. ★   The Franck-Hertz experiment involves measuring the differences between a series of equally spaced voltages that cause the maximum current through a tube of mercury vapor. A student measures 10 such differences and obtains these results (all in volts):

0.48, 0.45, 0.49, 0.46, 0.44, 0.57, 0.45, 0.47, 0.51, 0.50.

(a) Calculate the mean and standard deviation of these results. (By all means, use the built-in functions on your calculator.)

(b) If he decides to use Chauvenet's criterion, should he reject the reading of 0.57 volts? Explain your reasoning clearly.

6.3. ★ A student makes 14 measurements of the period of a damped oscillator and obtains these results (in tenths of a second):

$$7, 3, 9, 3, 6, 9, 8, 7, 8, 12, 5, 9, 9, 3.$$

Believing that the result 12 is suspiciously high, she decides to apply Chauvenet's criterion. How many results should she expect to find as far from the mean as 12 is? Should she reject the suspect result?

6.4. ★★ A physicist makes 14 measurements of the density of tracks on an emulsion exposed to cosmic rays and obtains the following results (in tracks/cm²):

$$11, 9, 13, 15, 8, 10, 5, 11, 9, 12, 12, 13, 9, 14.$$

(a) What are his mean and standard deviation? (Use the built-in functions on your calculator.) (b) According to Chauvenet's criterion, would he be justified in rejecting the measurement of 5? Explain your reasoning clearly. (c) If he does reject this measurement, what does he get for his new mean and standard deviation? (Hint: You can almost certainly recalculate the mean and standard deviation by editing the contents of your calculator's statistical registers rather than re-entering all the data. If you don't know how to do this editing, take a moment to learn.)

6.5. ★★ In the course of a couple of hours, a nuclear engineer makes 12 measurements of the strength of a long-lived radioactive source with the following results, in millicuries:

$$12, 34, 22, 14, 22, 17, 24, 22, 18, 14, 18, 12.$$

(Because the source has a long life, its activity should not change appreciably during the time all the measurements are made.)

(a) What are his mean and standard deviation? (Use the built-in functions on your calculator.) (b) According to Chauvenet's criterion, would he be justified in rejecting the value 34 as a mistake? Explain your reasoning clearly. (c) If he does reject this measurement, what does he get for his new mean and standard deviation? (Read the hint to Problem 6.4.)

6.6. ★★ Chauvenet's criterion defines a boundary outside which a measurement is regarded as rejectable. If we make 10 measurements and one differs from the mean by more than about two standard deviations (in either direction), that measurement is considered rejectable. For 20 measurements, the corresponding boundary is approximately 2.2 standard deviations. Make a table showing the "boundary of rejectability" for 5, 10, 15, 20, 50, 100, 200, and 1,000 measurements of a quantity that is normally distributed.

6.7. ★★ Based on N measurements of a normally distributed quantity x, the best estimate of the width σ is the standard deviation σ_x of the N measured values. Unfortunately, if N is small, this estimate is fairly uncertain. Specifically, the fractional uncertainty in σ_x as an estimate of σ is given by (5.46) as $1/\sqrt{2(N-1)}$. If

$N = 6$, for example, our estimate for σ is about 30% uncertain. This large uncertainty in σ means you should regard Chauvenet's criterion with considerable skepticism when N is small, as the following example illustrates.

An experimenter measures a certain current six times and obtains the following results (in milliamps):

$$28, 25, 29, 18, 29, 24.$$

(a) Find the mean $\bar{x}$ and standard deviation σ_x of these measurements. (b) If he decided to apply Chauvenet's criterion, would the experimenter reject the value 18? (c) Find the uncertainty—call it $\delta\sigma_x$—in σ_x. (d) The true value of σ may well be as small as $\sigma_x - \delta\sigma_x$ or as large as $\sigma_x + \delta\sigma_x$. Re-apply Chauvenet's criterion, using each of these two values for σ. Comment on the results.

Chapter 7

Weighted Averages

This chapter addresses the problem of combining two or more separate and independent measurements of a single physical quantity. We will find that the best estimate of that quantity, based on the several measurements, is an appropriate *weighted average* of those measurements.

7.1 The Problem of Combining Separate Measurements

Often, a physical quantity is measured several times, perhaps in several separate laboratories, and the question arises how these measurements can be combined to give a single best estimate. Suppose, for example, that two students, A and B, measure a quantity x carefully and obtain these results:

$$\text{Student } A: \quad x = x_A \pm \sigma_A \tag{7.1}$$

and

$$\text{Student } B: \quad x = x_B \pm \sigma_B. \tag{7.2}$$

Each result will probably itself be the result of several measurements, in which case x_A will be the mean of all A's measurements and σ_A the standard deviation of that mean (and similarly for x_B and σ_B). The question is how best to combine x_A and x_B for a single best estimate of x.

Before examining this question, note that if the discrepancy $|x_A - x_B|$ between the two measurements is much greater than both uncertainties σ_A and σ_B, we should suspect that something has gone wrong in at least one of the measurements. In this situation, we would say that the two measurements are *inconsistent,* and we should examine both measurements carefully to see whether either (or both) was subject to unnoticed systematic errors.

Let us suppose, however, that the two measurements (7.1) and (7.2) are *consistent;* that is, the discrepancy $|x_A - x_B|$ is *not* significantly larger than both σ_A and σ_B. We can then sensibly ask what the best estimate x_{best} is of the true value X, based on the two measurements. Your first impulse might be to use the average $(x_A + x_B)/2$ of the two measurements. Some reflection should suggest, however, that this average is unsuitable if the two uncertainties σ_A and σ_B are unequal. The simple

173

average $(x_A + x_B)/2$ gives equal importance to both measurements, whereas the more precise reading should somehow be given more weight.

Throughout this chapter, I will assume all systematic errors have been identified and reduced to a negligible level. Thus, all remaining errors are random, and the measurements of x are distributed normally around the true value X.

7.2   The Weighted Average

We can solve our problem easily by using the principle of maximum likelihood, much as we did in Section 5.5. We are assuming that both measurements are governed by the Gauss distribution and denote the unknown true value of x by X. Therefore, the probability of Student A's obtaining his particular value x_A is

$$Prob_X(x_A) \; \propto \; \frac{1}{\sigma_A} e^{-(x_A - X)^2/2\sigma_A^2}, \tag{7.3}$$

and that of B's getting his observed x_B is

$$Prob_X(x_B) \; \propto \; \frac{1}{\sigma_B} e^{-(x_B - X)^2/2\sigma_B^2}. \tag{7.4}$$

The subscript X indicates explicitly that these probabilities depend on the unknown actual value.

The probability that A finds the value x_A *and* B the value x_B is just the product of the two probabilities (7.3) and (7.4). In a way that should now be familiar, this product will involve an exponential function whose exponent is the sum of the two exponents in (7.3) and (7.4). We write this as

$$Prob_X(x_A, x_B) \; = \; Prob_X(x_A) \, Prob_X(x_B)$$
$$\propto \; \frac{1}{\sigma_A \sigma_B} e^{-\chi^2/2}, \tag{7.5}$$

where I have introduced the convenient shorthand χ^2 (chi squared) for the exponent

$$\chi^2 \; = \; \left(\frac{x_A - X}{\sigma_A}\right)^2 + \left(\frac{x_B - X}{\sigma_B}\right)^2. \tag{7.6}$$

This important quantity is the sum of the squares of the deviations from X of the two measurements, each divided by its corresponding uncertainty.

The principle of maximum likelihood asserts, just as before, that our best estimate for the unknown true value X is that value for which the actual observations x_A, x_B are most likely. That is, the best estimate for X is the value for which the probability (7.5) is maximum or, equivalently, the exponent χ^2 is minimum. (Because maximizing the probability entails minimizing the "sum of squares" χ^2, this method for estimating X is sometimes called the "method of least squares.") Thus, to find the best estimate, we simply differentiate (7.6) with respect to X and set the derivative equal to zero,

$$2\frac{x_A - X}{\sigma_A^2} + 2\frac{x_B - X}{\sigma_B^2} \; = \; 0.$$

The solution of this equation for X is our best estimate and is easily seen to be

$$\text{(best estimate for } X) = \left(\frac{x_A}{\sigma_A{}^2} + \frac{x_B}{\sigma_B{}^2}\right) \bigg/ \left(\frac{1}{\sigma_A{}^2} + \frac{1}{\sigma_B{}^2}\right). \tag{7.7}$$

This rather ugly result can be made tidier if we define *weights*

$$w_A = \frac{1}{\sigma_A{}^2} \quad \text{and} \quad w_B = \frac{1}{\sigma_B{}^2}. \tag{7.8}$$

With this notation, we can rewrite (7.7) as the *weighted average* (denoted x_{wav})

$$\text{(best estimate for } X) = x_{\text{wav}} = \frac{w_A x_A + w_B x_B}{w_A + w_B}. \tag{7.9}$$

If the original two measurements are equally uncertain ($\sigma_A = \sigma_B$ and hence $w_A = w_B$), this answer reduces to the simple average $(x_A + x_B)/2$. In general, when $w_A \neq w_B$, the weighted average (7.9) is *not* the same as the ordinary average; it is similar to the formula for the center of gravity of two bodies, where w_A and w_B are the actual weights of the two bodies, and x_A and x_B their positions. In (7.9), the "weights" are the inverse squares of the uncertainties in the original measurements, as in (7.8). If A's measurement is more precise than B's, then $\sigma_A < \sigma_B$ and hence $w_A > w_B$, so the best estimate x_{best} is closer to x_A than to x_B, just as it should be.

Quick Check 7.1.   Workers from two laboratories report the lifetime of a certain particle as 10.0 ± 0.5 and 12 ± 1, both in nanoseconds. If they decide to combine the two results, what will be their respective weights as given by (7.8) and their weighted average as given by (7.9)?

Our analysis of two measurements can be generalized to cover any number of measurements. Suppose we have N separate measurements of a quantity x,

$$x_1 \pm \sigma_1, \quad x_2 \pm \sigma_2, \ldots, \quad x_N \pm \sigma_N,$$

with their corresponding uncertainties $\sigma_1, \sigma_2, \ldots, \sigma_N$. Arguing much as before, we find that the best estimate based on these measurements is the weighted average

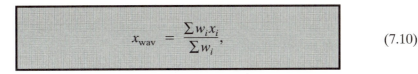

$$x_{\text{wav}} = \frac{\sum w_i x_i}{\sum w_i}, \tag{7.10}$$

where the sums are over all N measurements, $i = 1, \ldots, N$, and the *weight* w_i of each measurement is the reciprocal square of the corresponding uncertainty,

$$w_i = \frac{1}{\sigma_i{}^2} \tag{7.11}$$

for $i = 1, 2, \ldots, N$.

Because the weight $w_i = 1/\sigma_i^2$ associated with each measurement involves the *square* of the corresponding uncertainty σ_i, any measurement that is much less precise than the others contributes *very* much less to the final answer (7.10). For example, if one measurement is four times less precise than the rest, its weight is 16 times less than the other weights, and for many purposes this measurement could simply be ignored.

Because the weighted average x_{wav}, is a function of the original measured values $x_1, x_2, \ldots, x_N$, the uncertainty in x_{wav} can be calculated using error propagation. As you can easily check (Problem 7.8), the uncertainty in x_{wav} is

$$\sigma_{wav} = \frac{1}{\sqrt{\sum w_i}}. \tag{7.12}$$

This rather ugly result is perhaps a little easier to remember if we rewrite (7.11) as

$$\sigma_i = \frac{1}{\sqrt{w_i}}. \tag{7.13}$$

Paraphrasing Equation (7.13), we can say that the uncertainty in each measurement is the reciprocal square root of its weight. Returning to Equation (7.12), we can paraphrase it similarly to say that the uncertainty in the grand answer x_{wav} is the reciprocal square root of *the sum of all the individual weights;* in other words, the total weight of the final answer is the sum of the individual weights w_i.

Quick Check 7.2. What is the uncertainty in your final answer for Quick Check 7.1?

7.3 An Example

Here is an example involving three separate measurements of the same resistance.

Example: Three Measurements of a Resistance

Each of three students measures the same resistance several times, and their three final answers are (all in ohms):

$$\text{Student 1:} \quad R = 11 \pm 1$$
$$\text{Student 2:} \quad R = 12 \pm 1$$
$$\text{Student 3:} \quad R = 10 \pm 3$$

Given these three results, what is the best estimate for the resistance R?

The three uncertainties $\sigma_1, \sigma_2, \sigma_3$ are 1, 1, and 3. Therefore, the corresponding weights $w_i = 1/\sigma_i^2$ are

$$w_1 = 1, \quad w_2 = 1, \quad w_3 = \tfrac{1}{9}.$$

The best estimate for R is the weighted average, which according to (7.10) is

$$R_{\text{wav}} = \frac{\sum w_i R_i}{\sum w_i}$$

$$= \frac{(1 \times 11) + (1 \times 12) + (\frac{1}{9} \times 10)}{1 + 1 + \frac{1}{9}} = 11.42 \text{ ohms.}$$

The uncertainty in this answer is given by (7.12) as

$$\sigma_{\text{wav}} = \frac{1}{\sqrt{\sum w_i}} = \frac{1}{\sqrt{1 + 1 + \frac{1}{9}}} = 0.69.$$

Thus, our final conclusion (properly rounded) is

$$R = 11.4 \pm 0.7 \text{ ohms.}$$

For interest, let us see what answer we would get if we were to ignore completely the third student's measurement, which is three times less accurate and hence nine times less important. Here, a simple calculation gives $R_{\text{best}} = 11.50$ (compared with 11.42) with an uncertainty of 0.71 (compared with 0.69). Obviously, the third measurement does not have a big effect.

Principal Definitions and Equations of Chapter 7

If $x_1, x_2, \ldots, x_N$ are measurements of a single quantity x, with known uncertainties $\sigma_1, \sigma_2, \ldots, \sigma_N$, then the best estimate for the true value of x is the *weighted average*

$$x_{\text{wav}} = \frac{\sum w_i x_i}{\sum w_i}, \qquad \text{[See (7.10)]}$$

where the sums are over all N measurements, $i = 1, \ldots, N$, and the weights w_i are the reciprocal squares of the corresponding uncertainties,

$$w_i = \frac{1}{\sigma_i^2}.$$

The uncertainty in x_{wav} is

$$\sigma_{\text{wav}} = \frac{1}{\sqrt{\sum w_i}}, \qquad \text{[See (7.12)]}$$

where, again, the sum runs over all of the measurements $i = 1, 2, \ldots, N$.

Problems for Chapter 7

For Section 7.2: The Weighted Average

7.1. ★   Find the best estimate and its uncertainty based on the following four measurements of a certain voltage:

$$1.4 \pm 0.5, \quad 1.2 \pm 0.2, \quad 1.0 \pm 0.25, \quad 1.3 \pm 0.2.$$

7.2. ★   Three groups of particle physicists measure the mass of a certain elementary particle with the results (in units of MeV/c^2):

$$1{,}967.0 \pm 1.0, \quad 1{,}969 \pm 1.4, \quad 1{,}972.1 \pm 2.5.$$

Find the weighted average and its uncertainty.

7.3. ★   **(a)** Two measurements of the speed of sound u give the answers 334 ± 1 and 336 ± 2 (both in m/s). Would you consider them consistent? If so, calculate the best estimate for u and its uncertainty. **(b)** Repeat part (a) for the results 334 ± 1 and 336 ± 5. Is the second measurement worth including in this case?

7.4. ★★   Four measurements are made of the wavelength of light emitted by a certain atom. The results, in nanometers, are:

$$503 \pm 10, \quad 491 \pm 8, \quad 525 \pm 20, \quad 570 \pm 40.$$

Find the weighted average and its uncertainty. Is the last measurement worth including?

7.5. ★★   Two students measure a resistance by different methods. Each makes 10 measurements and computes the mean and its standard deviation, and their final results are as follows:

Student A: $R = 72 \pm 8$ ohms
Student B: $R = 78 \pm 5$ ohms.

(a) Including both measurements, what are the best estimate of R and its uncertainty? **(b)** Approximately how many measurements (using his same technique) would student A need to make to give his result the same weight as B's? (Remember that each student's final uncertainty is the SDOM, which is equal to the SD/$\sqrt{N}$.)

7.6. ★★   Two physicists measure the rate of decay of a long-lived radioactive source. Physicist A monitors the sample for 4 hours and observes 412 decays; physicist B monitors it for 6 hours and observes 576 decays. **(a)** Find the uncertainties in these two counts using the square-root rule (3.2). (Remember that the square-root rule gives the uncertainty in the actual counted number.) **(b)** What should each physicist report for the decay rate in decays per hour, with its uncertainty? **(c)** What is the proper weighted average of these two rates, with its uncertainty?

7.7. ★★   Suppose that N separate measurements of a quantity x all have the same uncertainty. Show clearly that in this situation the weighted average (7.10) reduces to the ordinary average, or mean, $\bar{x} = \Sigma x_i/N$, and that the uncertainty (7.12) reduces to the familiar standard deviation of the mean.

7.8. ★★   The weighted average (7.10) of N separate measurements is a simple function of $x_1, x_2, \ldots, x_N$. Therefore, the uncertainty in x_{wav} can be found by error propagation. Prove in this way that the uncertainty in x_{wav} is as claimed in (7.12).

7.9. ★★★   **(a)** If you have access to a spreadsheet program such as Lotus 123 or Excel, create a spreadsheet to calculate the weighted average of three measurements x_i with given uncertainties σ_i. In the first column, give the trial number i, and use columns 2 and 3 to enter the data x_i and the uncertainties σ_i. In columns 4 and 5, put functions to calculate the weights w_i and the products $w_i x_i$; at the bottoms of these columns, you can calculate the sums Σw_i and $\Sigma w_i x_i$. Finally, in some convenient position, place functions to calculate x_{wav} and its uncertainty (7.12). Test your spreadsheet using the data of Section 7.3. **(b)** Try to modify your spreadsheet so that it can handle *any number* of measurements up to some maximum (20 say). (The main difficulty is that you will probably need to use some logical functions to make sure that empty cells in column 3 don't get counted as zeros and cause trouble with the function that calculates $w_i = 1/\sigma_i^2$.) Test your new spreadsheet using the data in Section 7.3 and in Problem 7.1.

Chapter 8

Least-Squares Fitting

Our discussion of the statistical analysis of data has so far focused exclusively on the repeated measurement of one single quantity, not because the analysis of many measurements of one quantity is the most interesting problem in statistics, but because this simple problem must be well understood before more general ones can be discussed. Now we are ready to discuss our first, and very important, more general problem.

8.1 Data That Should Fit a Straight Line

One of the most common and interesting types of experiment involves the measurement of several values of two different physical variables to investigate the mathematical relationship between the two variables. For instance, an experimenter might drop a stone from various different heights $h_1, \ldots, h_N$ and measure the corresponding times of fall $t_1, \ldots, t_N$ to see if the heights and times are connected by the expected relation $h = \frac{1}{2}gt^2$.

Probably the most important experiments of this type are those for which the expected relation is *linear*. For instance, if we believe that a body is falling with constant acceleration g, then its velocity v should be a linear function of the time t,

$$v = v_o + gt.$$

More generally, we will consider any two physical variables x and y that we suspect are connected by a linear relation of the form

$$y = A + Bx, \tag{8.1}$$

where A and B are constants. Unfortunately, many different notations are used for a linear relation; beware of confusing the form (8.1) with the equally popular $y = ax + b$.

If the two variables y and x are linearly related as in (8.1), then a graph of y against x should be a straight line that has slope B and intersects the y axis at $y = A$. If we were to measure N different values $x_1, \ldots, x_N$ and the corresponding values $y_1, \ldots, y_N$ and if our measurements were subject to no uncertainties, then each of the points (x_i, y_i) would lie exactly on the line $y = A + Bx$, as in Figure 8.1(a). In

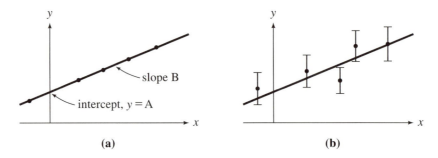

Figure 8.1. (a) If the two variables y and x are linearly related as in Equation (8.1), and if there were no experimental uncertainties, then the measured points (x_i, y_i) would all lie exactly on the line $y = A + Bx$. (b) In practice, there always are uncertainties, which can be shown by error bars, and the points (x_i, y_i) can be expected only to lie reasonably close to the line. Here, only y is shown as subject to appreciable uncertainties.

practice, there *are* uncertainties, and the most we can expect is that the distance of each point (x_i, y_i) from the line will be reasonable compared with the uncertainties, as in Figure 8.1(b).

When we make a series of measurements of the kind just described, we can ask two questions. First, if we take for granted that y and x *are* linearly related, then the interesting problem is to find the straight line $y = A + Bx$ that best fits the measurements, that is, to find the best estimates for the constants A and B based on the data $(x_1, y_1), \ldots, (x_N, y_N)$. This problem can be approached graphically, as discussed briefly in Section 2.6. It can also be treated analytically, by means of the principle of maximum likelihood. This analytical method of finding the best straight line to fit a series of experimental points is called *linear regression,* or the *least-squares fit for a line,* and is the main subject of this chapter.

The second question that can be asked is whether the measured values (x_1, y_1), $\ldots, (x_N, y_N)$ do really bear out our expectation that y is linear in x. To answer this question, we would first find the line that best fits the data, but we must then devise some measure of *how well* this line fits the data. If we already know the uncertainties in our measurements, we can draw a graph, like that in Figure 8.1(b), that shows the best-fit straight line and the experimental data with their error bars. We can then judge visually whether or not the best-fit line passes sufficiently close to all of the error bars. If we do not know the uncertainties reliably, we must judge how well the points fit a straight line by examining the distribution of the points themselves. We take up this question in Chapter 9.

8.2   Calculation of the Constants A and B

Let us now return to the question of finding the best straight line $y = A + Bx$ to fit a set of measured points $(x_1, y_1), \ldots, (x_N, y_N)$. To simplify our discussion, we will suppose that, although our measurements of y suffer appreciable uncertainty, the uncertainty in our measurements of x is negligible. This assumption is often reasonable, because the uncertainties in one variable often are much larger than those in the other, which we can therefore safely ignore. We will further assume that the uncertainties in y all have the same magnitude. (This assumption is also reasonable in many experiments, but if the uncertainties are different, then our analysis can be generalized to weight the measurements appropriately; see Problem 8.9.) More specifically, we assume that the measurement of each y_i is governed by the Gauss distribution, with the same width parameter σ_y for all measurements.

If we knew the constants A and B, then, for any given value x_i (which we are assuming has no uncertainty), we could compute the true value of the corresponding y_i,

$$(\text{true value for } y_i) \ = \ A + Bx_i. \tag{8.2}$$

The measurement of y_i is governed by a normal distribution centered on this true value, with width parameter σ_y. Therefore, the probability of obtaining the observed value y_i is

$$Prob_{A,B}(y_i) \ \propto \ \frac{1}{\sigma_y} e^{-(y_i - A - Bx_i)^2/2\sigma_y^2}, \tag{8.3}$$

where the subscripts A and B indicate that this probability depends on the (unknown) values of A and B. The probability of obtaining our complete set of measurements $y_1, \ldots, y_N$ is the product

$$Prob_{A,B}(y_1, \ldots, y_N) \ = \ Prob_{A,B}(y_1) \cdots Prob_{A,B}(y_N)$$

$$\propto \ \frac{1}{\sigma_y^N} e^{-\chi^2/2}, \tag{8.4}$$

where the exponent is given by

$$\chi^2 \ = \ \sum_{i=1}^{N} \frac{(y_i - A - Bx_i)^2}{\sigma_y^2}. \tag{8.5}$$

In the now-familiar way, we will assume that the best estimates for the unknown constants A and B, based on the given measurements, are those values of A and B for which the probability $Prob_{A,B}(y_1, \ldots, y_N)$ is maximum, or for which the sum of squares χ^2 in (8.5) is a minimum. (This is why the method is known as least-squares fitting.) To find these values, we differentiate χ^2 with respect to A and B and set the derivatives equal to zero:

$$\frac{\partial \chi^2}{\partial A} \ = \ \frac{-2}{\sigma_y^2} \sum_{i=1}^{N} (y_i - A - Bx_i) \ = \ 0 \tag{8.6}$$

and

$$\frac{\partial \chi^2}{\partial B} \ = \ \frac{-2}{\sigma_y^2} \sum_{i=1}^{N} x_i(y_i - A - Bx_i) \ = \ 0. \tag{8.7}$$

These two equations can be rewritten as simultaneous equations for A and B:

$$AN + B\Sigma x_i \ = \ \Sigma y_i \tag{8.8}$$

and

$$A\Sigma x_i + B\Sigma x_i^2 \ = \ \Sigma x_i y_i. \tag{8.9}$$

Here, I have omitted the limits $i = 1$ to N from the summation signs Σ. In the following discussion, I also omit the subscripts i when there is no serious danger of confusion; thus, $\Sigma x_i y_i$ is abbreviated to Σxy and so on.

The two equations (8.8) and (8.9), sometimes called *normal equations,* are easily solved for the least-squares estimates for the constants A and B,

$$A = \frac{\Sigma x^2 \Sigma y - \Sigma x \Sigma xy}{\Delta} \qquad (8.10)$$

and

$$B = \frac{N\Sigma xy - \Sigma x \Sigma y}{\Delta}, \qquad (8.11)$$

where I have introduced the convenient abbreviation for the denominator,

$$\Delta = N\Sigma x^2 - (\Sigma x)^2. \qquad (8.12)$$

The results (8.10) and (8.11) give the best estimates for the constants A and B of the straight line $y = A + Bx$, based on the N measured points $(x_1, y_1), \ldots, (x_N, y_N)$. The resulting line is called the *least-squares fit* to the data, or the *line of regression* of y on x.

Example: Length versus Mass for a Spring Balance

A student makes a scale to measure masses with a spring. She attaches its top end to a rigid support, hangs a pan from its bottom, and places a meter stick behind the arrangement to read the length of the spring. Before she can use the scale, she must calibrate it; that is, she must find the relationship between the mass in the pan and the length of the spring. To do this calibration, she gets five accurate 2-kg masses, which she adds to the pan one by one, recording the corresponding lengths l_i as shown in the first three columns of Table 8.1. Assuming the spring obeys Hooke's law, she anticipates that l should be a linear function of m,

$$l = A + Bm. \qquad (8.13)$$

(Here, the constant A is the unloaded length of the spring, and B is g/k, where k is the usual spring constant.) The calibration equation (8.13) will let her find any unknown mass m from the corresponding length l, once she knows the constants A and B. To find these constants, she uses the method of least squares. What are her answers for A and B? Plot her calibration data and the line given by her best fit (8.13). If she puts an unknown mass m in the pan and observes the spring's length to be $l = 53.2$ cm, what is m?

Table 8.1. Masses m_i (in kg) and lengths l_i (in cm) for a spring balance. The "x" and "y" in quotes indicate which variables play the roles of x and y in this example.

Trial number i	"x" Load, m_i	"y" Length, l_i	m_i^2	$m_i l_i$
1	2	42.0	4	84
2	4	48.4	16	194
3	6	51.3	36	308
4	8	56.3	64	450
5	10	58.6	100	586
$N = 5$	$\Sigma m_i = 30$	$\Sigma l_i = 256.6$	$\Sigma m_i^2 = 220$	$\Sigma m_i l_i = 1{,}622$

As often happens in such problems, the two variables are not called x and y, and one must be careful to identify which is which. Comparing (8.13) with the standard form, $y = A + Bx$, we see that the length l plays the role of the dependent variable y, while the mass m plays the role of the independent variable x. The constants A and B are given by (8.10) through (8.12), with the replacements

$$x_i \leftrightarrow m_i \qquad \text{and} \qquad y_i \leftrightarrow l_i.$$

(This correspondence is indicated by the headings "x" and "y" in Table 8.1.) To find A and B, we need to find the sums Σm_i, Σl_i, Σm_i^2, and $\Sigma m_i l_i$; therefore, the last two columns of Table 8.1 show the quantities m_i^2 and $m_i l_i$, and the corresponding sum is shown at the bottom of each column.

Computing the constants A and B is now straightforward. According to (8.12), the denominator Δ is

$$\Delta = N\Sigma m^2 - (\Sigma m)^2$$
$$= 5 \times 220 - 30^2 = 200.$$

Next, from (8.10) we find the intercept (the unstretched length)

$$A = \frac{\Sigma m^2 \Sigma l - \Sigma m \Sigma ml}{\Delta}$$

$$= \frac{220 \times 256.6 - 30 \times 1622}{200} = 39.0 \text{ cm.}$$

Finally, from (8.11) we find the slope

$$B = \frac{N\Sigma ml - \Sigma m \Sigma l}{\Delta}$$

$$= \frac{5 \times 1622 - 30 \times 256.6}{200} = 2.06 \text{ cm/kg.}$$

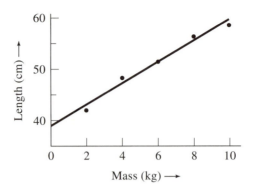

Figure 8.2. A plot of the data from Table 8.1 and the best-fit line (8.13).

A plot of the data and the line (8.13) using these values of A and B is shown in Figure 8.2. If the mass m stretches the spring to 53.2 cm, then according to (8.13) the mass is

$$m = \frac{l - A}{B} = \frac{(53.2 - 39.0) \text{ cm}}{2.06 \text{ cm/kg}} = 6.9 \text{ kg}.$$

Quick Check 8.1.   Find the least-squares best-fit line $y = A + Bx$ through the three points $(x, y) = (-1, 0)$, $(0, 6)$, and $(1, 6)$. Using squared paper, plot the points and your line. [Note that because the three values of x $(-1, 0,$ and $1)$ are symmetric about zero, $\Sigma x = 0$, which simplifies the calculation of A and B. In some experiments, the values of x can be arranged to be symmetrically spaced in this way, which saves some trouble.]

Now that we know how to find the best estimates for the constants A and B, we naturally ask for the uncertainties in these estimates. Before we can find these uncertainties, however, we must discuss the uncertainty σ_y in the original measurements of $y_1, y_2, \ldots, y_N$.

8.3   Uncertainty in the Measurements of y

In the course of measuring the values $y_1, \ldots, y_N$, we have presumably formed some idea of their uncertainty. Nonetheless, knowing how to calculate the uncertainty by analyzing the data themselves is important. Remember that the numbers $y_1, \ldots, y_N$ are *not* N measurements of the same quantity. (They might, for instance, be the times for a stone to fall from N different heights.) Thus, we certainly do not get an idea of their reliability by examining the spread in their values.

Nevertheless, we can easily estimate the uncertainty σ_y in the numbers $y_1, \ldots, y_N$. The measurement of each y_i is (we are assuming) normally distributed about its true value $A + Bx_i$, with width parameter σ_y. Thus the *deviations* $y_i - A - Bx_i$ are normally distributed, all with the same central value zero and the same width σ_y. This situation immediately suggests that a good estimate for σ_y would be given by a sum of squares with the familiar form

$$\sigma_y = \sqrt{\frac{1}{N}\Sigma(y_i - A - Bx_i)^2}. \tag{8.14}$$

In fact, this answer can be confirmed by means of the principle of maximum likelihood. As usual, the best estimate for the parameter in question (σ_y here) is that value for which the probability (8.4) of obtaining the observed values $y_1, \ldots, y_N$ is maximum. As you can easily check by differentiating (8.4) with respect to σ_y and setting the derivative equal to zero, this best estimate is precisely the answer (8.14). (See Problem 8.12.)

Unfortunately, as you may have suspected, the estimate (8.14) for σ_y is not quite the end of the story. The numbers A and B in (8.14) are the unknown true values of the constants A and B. In practice, these numbers must be replaced by our *best estimates* for A and B, namely, (8.10) and (8.11), and this replacement slightly reduces the value of (8.14). It can be shown that this reduction is compensated for if we replace the factor N in the denominator by $(N - 2)$. Thus, our final answer for the uncertainty in the measurements $y_1, \ldots, y_N$ is

$$\sigma_y = \sqrt{\frac{1}{N-2}\sum_{i=1}^{N}(y_i - A - Bx_i)^2}, \tag{8.15}$$

with A and B given by (8.10) and (8.11). If we already have an independent estimate of our uncertainty in $y_1, \ldots, y_N$, we would expect this estimate to compare with σ_y as computed from (8.15).

I will not attempt to justify the factor of $(N - 2)$ in (8.15) but can make some comments. First, as long as N is moderately large, the difference between N and $(N - 2)$ is unimportant anyway. Second, that the factor $(N - 2)$ is *reasonable* becomes clear if we consider measuring just two pairs of data (x_1, y_1) and (x_2, y_2). With only two points, we can always find a line that passes *exactly* through both points, and the least-squares fit will give this line. That is, with just two pairs of data, we cannot possibly deduce anything about the reliability of our measurements. Now, since both points lie exactly on the best line, the two terms of the sum in (8.14) and (8.15) are zero. Thus, the formula (8.14) (with $N = 2$ in the denominator) would give the absurd answer $\sigma_y = 0$; whereas (8.15), with $N - 2 = 0$ in the denominator, gives $\sigma_y = 0/0$, indicating correctly that σ_y is undetermined after only two measurements.

The presence of the factor $(N - 2)$ in (8.15) is reminiscent of the $(N - 1)$ that appeared in our estimate of the standard deviation of N measurements of one quantity x, in Equation (5.45). There, we made N measurements $x_1, \ldots, x_N$ of the one quantity x. Before we could calculate σ_x, we had to use our data to find the mean $\bar{x}$. In a certain sense, this computation left only $(N - 1)$ independent measured values, so we say that, having computed $\bar{x}$, we have only $(N - 1)$ *degrees of freedom* left. Here, we made N measurements, but before calculating σ_y we had to compute the *two* quantities A and B. Having done this, we had only $(N - 2)$ degrees of freedom left. In general, we define the *number of degrees of freedom* at any stage in a statistical calculation as the number of independent measurements *minus* the number of parameters calculated from these measurements. We can show (but will not do so here) that the number of degrees of freedom, *not* the number of measurements, is what should appear in the denominator of formulas such as (8.15) and (5.45). This fact explains why (8.15) contains the factor $(N - 2)$ and (5.45) the factor $(N - 1)$.

8.4   Uncertainty in the Constants A and B

Having found the uncertainty σ_y in the measured numbers $y_1, \ldots, y_N$, we can easily return to our estimates for the constants A and B and calculate their uncertainties. The point is that the estimates (8.10) and (8.11) for A and B are well-defined functions of the measured numbers $y_1, \ldots, y_N$. Therefore, the uncertainties in A and B are given by simple error propagation in terms of those in $y_1, \ldots, y_N$. I leave it as an exercise for you to check (Problem 8.16) that

$$\sigma_A = \sigma_y \sqrt{\frac{\sum x^2}{\Delta}} \tag{8.16}$$

and

$$\sigma_B = \sigma_y \sqrt{\frac{N}{\Delta}} \tag{8.17}$$

where Δ is given by (8.12) as usual.

The results of this and the previous two sections were based on the assumptions that the measurements of y were all equally uncertain and that any uncertainties in x were negligible. Although these assumptions often are justified, we need to discuss briefly what happens when they are not. First, if the uncertainties in y are not all equal, we can use the method of *weighted least squares,* as described in Problem 8.9. Second, if there are uncertainties in x but not in y, we can simply interchange the roles of x and y in our analysis. The remaining case is that in which both x and y have uncertainties—a case that certainly *can* occur. The least-squares fitting of a general curve when both x and y have uncertainties is rather complicated and even controversial. In the important special case of a *straight line* (which is all we have discussed so far), uncertainties in both x and y make surprisingly little difference, as we now discuss.

Suppose, first, that our measurements of x are subject to uncertainty but those of y are not, and we consider a particular measured point (x, y). This point and the true line $y = A + Bx$ are shown in Figure 8.3. The point (x, y) does not lie on the line because of the error—call it Δx—in our measurement of x. Now, we can see

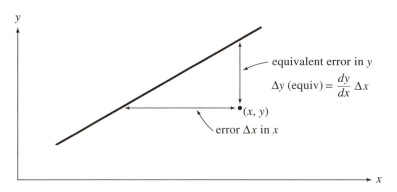

Figure 8.3. A measured point (x, y) and the line $y = A + Bx$ on which the point is supposed to lie. The error Δx in x, with y exact, produces the same effect as an error Δy(equiv) $= (dy/dx)\Delta x$ in y, with x exact. (Here, dy/dx denotes the slope of the expected line.)

easily from the picture that the error Δx in x, with no error in y, produces exactly the same effect as if there had been no error in x but an error in y given by

$$\Delta y(\text{equiv}) = \frac{dy}{dx} \Delta x \qquad (8.18)$$

(where "equiv" stands for "equivalent"). The standard deviation σ_x is just the root-mean-square value of Δx that would result from repeating this measurement many times. Thus, according to (8.18), the problem with uncertainties σ_x in x can be replaced with an equivalent problem with uncertainties in y, given by

$$\sigma_y(\text{equiv}) = \frac{dy}{dx} \sigma_x. \qquad (8.19)$$

The result (8.19) is true whatever the curve of y vs x, but (8.19) is especially simple if the curve is a straight line, because the slope dy/dx is just the constant B. Therefore, for a straight line

$$\sigma_y(\text{equiv}) = B\sigma_x. \qquad (8.20)$$

In particular, if all the uncertainties σ_x are equal, the same is true of the equivalent uncertainties σ_y(equiv). Therefore, the problem of fitting a line to points (x_i, y_i) with equal uncertainties in x but no uncertainties in y is equivalent to the problem of equal uncertainties in y but none in x. This equivalence means we can safely use the method already described for either problem. [In practice, the points do not lie *exactly* on the line, and the two "equivalent" problems will not give *exactly* the same line. Nevertheless, the two lines should usually agree within the uncertainties given by (8.16) and (8.17). See Problem 8.17.]

We can now extend this argument to the case that *both* x and y have uncertainties. The uncertainty in x is equivalent to an uncertainty in y as given by (8.20). In addition, y is already subject to its own uncertainty σ_y. These two uncertainties are independent and must be combined in quadrature. Thus, the original problem, with uncertainties in both x and y, can be replaced with an equivalent problem in which only y has uncertainty, given by

$$\sigma_y(\text{equiv}) = \sqrt{\sigma_y^2 + (B\sigma_x)^2}. \tag{8.21}$$

Provided all the uncertainties σ_x are the same, and likewise all the uncertainties σ_y, the equivalent uncertainties (8.21) are all the same, and we can safely use the formulas (8.10) through (8.17).

If the uncertainties in x (or in y) are not all the same, we can still use (8.21), but the resulting uncertainties will not all be the same, and we will need to use the method of weighted least squares. If the curve to which we are fitting our points is not a straight line, a further complication arises because the slope dy/dx is not a constant and we cannot replace (8.19) with (8.20). Nevertheless, we can still use (8.21) (with dy/dx in place of B) to replace the original problem (with uncertainties in both x and y) by an equivalent problem in which only y has uncertainties as given by (8.21).[1]

8.5   An Example

Here is a simple example of least-squares fitting to a straight line; it involves the constant-volume gas thermometer.

Example: Measurement of Absolute Zero with a Constant-Volume Gas Thermometer

If the volume of a sample of an ideal gas is kept constant, its temperature T is a linear function of its pressure P,

$$T = A + BP. \tag{8.22}$$

Here, the constant A is the temperature at which the pressure P would drop to zero (if the gas did not condense into a liquid first); it is called the *absolute zero of temperature,* and has the accepted value

$$A = -273.15°C$$

The constant B depends on the nature of the gas, its mass, and its volume.[2] By measuring a series of values for T and P, we can find the best estimates for the constants A and B. In particular, the value of A gives the absolute zero of temperature.

One set of five measurements of P and T obtained by a student was as shown in the first three columns of Table 8.2. The student judged that his measurements of

---

[1] This procedure is quite complicated in practice. Before we can use (8.21) to find the uncertainty $\sigma_y(\text{equiv})$, we need to know the slope B (or, more generally, dy/dx), which is not known until we have solved the problem! Nevertheless, we can get a reasonable first approximation for the slope using the method of unweighted least squares, ignoring all of the complications discussed here. This method gives an approximate value for the slope B, which can then be used in (8.21) to give a reasonable approximation for $\sigma_y(\text{equiv})$.

[2] The difference $T - A$ is called the *absolute temperature.* Thus (8.22) can be rewritten to say that the absolute temperature is proportional to the pressure (at constant volume).

Table 8.2. Pressure (in mm of mercury) and temperature (°C) of a gas at constant volume.

Trial number i	"x" Pressure P_i	"y" Temperature T_i	$A + BP_i$
1	65	-20	-22.2
2	75	17	14.9
3	85	42	52.0
4	95	94	89.1
5	105	127	126.2

P had negligible uncertainty, and those of T were all equally uncertain with an uncertainty of "a few degrees." Assuming his points should fit a straight line of the form (8.22), he calculated his best estimate for the constant A (the absolute zero) and its uncertainty. What should have been his conclusions?

All we have to do here is use formulas (8.10) and (8.16), with x_i replaced by P_i and y_i by T_i, to calculate all the quantities of interest. This requires us to compute the sums ΣP, ΣP^2, ΣT, ΣPT. Many pocket calculators can evaluate all these sums automatically, but even without such a machine, we can easily handle these calculations if the data are properly organized. From Table 8.2, we can calculate

$$\Sigma P = 425,$$
$$\Sigma P^2 = 37,125,$$
$$\Sigma T = 260,$$
$$\Sigma PT = 25,810,$$
$$\Delta = N\Sigma P^2 - (\Sigma P)^2 = 5,000.$$

In this kind of calculation, it is important to keep plenty of significant figures because we have to take differences of these large numbers. Armed with these sums, we can immediately calculate the best estimates for the constants A and B:

$$A = \frac{\Sigma P^2 \Sigma T - \Sigma P \Sigma PT}{\Delta} = -263.35$$

and

$$B = \frac{N \Sigma PT - \Sigma P \Sigma T}{\Delta} = 3.71.$$

This calculation already gives the student's best estimate for absolute zero, $A = -263°C$.

Knowing the constants A and B, we can next calculate the numbers $A + BP_i$, the temperatures "expected" on the basis of our best fit to the relation $T = A + BP$. These numbers are shown in the far right column of the table, and as we would hope, all agree reasonably well with the observed temperatures. We can now take the difference between the figures in the last two columns and calculate

$$\sigma_T = \sqrt{\frac{1}{N-2} \Sigma(T_i - A - BP_i)^2} = 6.7.$$

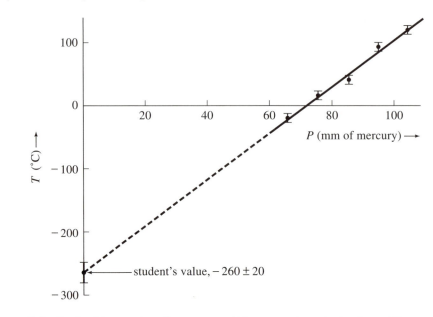

Figure 8.4. Graph of temperature T vs pressure P for a gas at constant volume. The error bars extend one standard deviation, σ_T, on each side of the five experimental points, and the line is the least-squares best fit. The absolute zero of temperature was found by extrapolating the line back to its intersection with the T axis.

This result agrees reasonably with the student's estimate that his temperature measurements were uncertain by "a few degrees."

Finally, we can calculate the uncertainty in A using (8.16):

$$\sigma_A = \sigma_T \sqrt{\Sigma P^2 / \Delta} = 18.$$

Thus, our student's final conclusion, suitably rounded, should be

$$\text{absolute zero, } A = -260 \pm 20°C,$$

which agrees satisfactorily with the accepted value, $-273°C$.

As is often true, these results become much clearer if we graph them, as in Figure 8.4. The five data points, with their uncertainties of $\pm 7°$ in T, are shown on the upper right. The best straight line passes through four of the error bars and close to the fifth.

To find a value for absolute zero, the line was extended beyond all the data points to its intersection with the T axis. This process of *extrapolation* (extending a curve beyond the data points that determine it) can introduce large uncertainties, as is clear from the picture. A very small change in the line's slope will cause a large change in its intercept on the distant T axis. Thus, any uncertainty in the data is greatly magnified if we have to extrapolate any distance. This magnification explains why the uncertainty in the value of absolute zero ($\pm 18°$) is so much larger than that in the original temperature measurements ($\pm 7°$).

8.6   Least-Squares Fits to Other Curves

So far in this chapter, we have considered the observation of two variables satisfying a linear relation, $y = A + Bx$, and we have discussed the calculation of the constants A and B. This important problem is a special case of a wide class of curve-fitting problems, many of which can be solved in a similar way. In this section, I mention briefly a few more of these problems.

FITTING A POLYNOMIAL

Often, one variable, y, is expected to be expressible as a polynomial in a second variable, x,

$$y = A + Bx + Cx^2 + \cdots + Hx^n. \tag{8.23}$$

For example, the height y of a falling body is expected to be quadratic in the time t,

$$y = y_0 + v_0 t - \tfrac{1}{2} g t^2,$$

where y_0 and v_0 are the initial height and velocity, and g is the acceleration of gravity. Given a set of observations of the two variables, we can find best estimates for the constants $A, B, \ldots, H$ in (8.23) by an argument that exactly parallels that of Section 8.2, as I now outline.

To simplify matters, we suppose that the polynomial (8.23) is actually a quadratic,

$$y = A + Bx + Cx^2. \tag{8.24}$$

(You can easily extend the analysis to the general case if you wish.) We suppose, as before, that we have a series of measurements (x_i, y_i), $i = 1, \ldots, N$, with the y_i all equally uncertain and the x_i all exact. For each x_i, the corresponding true value of y_i is given by (8.24), with A, B, and C as yet unknown. We assume that the measurements of the y_i are governed by normal distributions, each centered on the appropriate true value and all with the same width σ_y. This assumption lets us compute the probability of obtaining our observed values $y_1, \ldots, y_N$ in the familiar form

$$Prob(y_1, \ldots, y_N) \propto e^{-\chi^2/2}, \tag{8.25}$$

where now

$$\chi^2 = \sum_{i=1}^{N} \frac{(y_i - A - Bx_i - Cx_i^2)^2}{\sigma_y^2}. \tag{8.26}$$

[This equation corresponds to Equation (8.5) for the linear case.] The best estimates for A, B, and C are those values for which $Prob(y_1, \ldots, y_N)$ is largest, or χ^2 is smallest. Differentiating χ^2 with respect to A, B, and C and setting these derivatives equal to zero, we obtain the three equations (as you should check; see Problem 8.21):

$$\begin{aligned}
AN + B\textstyle\sum x + C\textstyle\sum x^2 &= \textstyle\sum y, \\
A\textstyle\sum x + B\textstyle\sum x^2 + C\textstyle\sum x^3 &= \textstyle\sum xy, \\
A\textstyle\sum x^2 + B\textstyle\sum x^3 + C\textstyle\sum x^4 &= \textstyle\sum x^2 y.
\end{aligned} \tag{8.27}$$

For any given set of measurements (x_i, y_i), these simultaneous equations for A, B, and C (known as the *normal equations*) can be solved to find the best estimates for A, B, and C. With A, B, and C calculated in this way, the equation $y = A + Bx + Cx^2$ is called the least-squares polynomial fit, or the polynomial regression, for the given measurements. (For an example, see Problem 8.22.)

The method of polynomial regression generalizes easily to a polynomial of any degree, although the resulting normal equations become cumbersome for polynomials of high degree. In principle, a similar method can be applied to *any* function $y = f(x)$ that depends on various unknown parameters A, B, ... Unfortunately, the resulting normal equations that determine the best estimates for A, B, . . . can be difficult or impossible to solve. However, one large class of problems *can* always be solved, namely, those problems in which the function $y = f(x)$ depends linearly on the parameters A, B, ... These include all polynomials (obviously the polynomial (8.23) is linear in its coefficients A, B, . . .) but they also include many other functions. For example, in some problems y is expected to be a sum of trigonometric functions, such as

$$y = A \sin x + B \cos x. \tag{8.28}$$

For this function, and in fact for any function that is linear in the parameters A, B, . . ., the normal equations that determine the best estimates for A, B, . . . are simultaneous linear equations, which can always be solved (see Problems 8.23 and 8.24).

EXPONENTIAL FUNCTIONS

One of the most important functions in physics is the exponential function

$$y = Ae^{Bx}, \tag{8.29}$$

where A and B are constants. The intensity I of radiation, after passing a distance x through a shield, falls off exponentially:

$$I = I_0 e^{-\mu x},$$

where I_0 is the original intensity and μ characterizes the absorption by the shield. The charge on a short-circuited capacitor drains away exponentially:

$$Q = Q_0 e^{-\lambda t}$$

where Q_0 is the original charge and $\lambda = 1/(RC)$, where R and C are the resistance and capacitance.

If the constants A and B in (8.29) are unknown, we naturally seek estimates of them based on measurements of x and y. Unfortunately, direct application of our previous arguments leads to equations for A and B that cannot be conveniently solved. We can, however, transform the nonlinear relation (8.29) between y and x into a linear relation, to which we can apply our least-squares fit.

To effect the desired "linearization," we simply take the natural logarithm of (8.29) to give

$$\ln y = \ln A + Bx. \tag{8.30}$$

We see that, even though y is not linear in x, $\ln y$ *is*. This conversion of the nonlinear (8.29) into the linear (8.30) is useful in many contexts besides that of least-squares fitting. If we want to check the relation (8.29) graphically, then a direct plot of y against x will produce a curve that is hard to identify visually. On the other hand, a plot of $\ln y$ against x (or of $\log y$ against x) should produce a straight line, which can be identified easily. (Such a plot is especially easy if you use "semilog" graph paper, on which the graduations on one axis are spaced logarithmically. Such paper lets you plot $\log y$ directly without even calculating it.)

The usefulness of the linear equation (8.30) in least-squares fitting is readily apparent. If we believe that y and x should satisfy $y = Ae^{Bx}$, then the variables $z = \ln y$ and x should satisfy (8.30), or

$$z = \ln A + Bx. \tag{8.31}$$

If we have a series of measurements (x_i, y_i), then for each y_i we can calculate $z_i = \ln y_i$. Then the pairs (x_i, z_i) should lie on the line (8.31). This line can be fitted by the method of least squares to give best estimates for the constants $\ln A$ (from which we can find A) and B.

Example: A Population of Bacteria

Many populations (of people, bacteria, radioactive nuclei, etc.) tend to vary exponentially over time. If a population N is decreasing exponentially, we write

$$N = N_0 e^{-t/\tau}, \tag{8.32}$$

where τ is called the population's *mean life* [closely related to the *half-life*, $t_{1/2}$; in fact, $t_{1/2} = (\ln 2)\tau$]. A biologist suspects that a population of bacteria is decreasing exponentially as in (8.32) and measures the population on three successive days; he obtains the results shown in the first two columns of Table 8.3. Given these data, what is his best estimate for the mean life τ?

Table 8.3. Population of bacteria.

Time t_i (days)	Population N_i	$z_i = \ln N_i$
0	153,000	11.94
1	137,000	11.83
2	128,000	11.76

If N varies as in (8.32), then the variable $z = \ln N$ should be linear in t:

$$z = \ln N = \ln N_0 - \frac{t}{\tau}. \tag{8.33}$$

Our biologist therefore calculates the three numbers $z_i = \ln N_i$ ($i = 0, 1, 2$) shown in the third column of Table 8.3. Using these numbers, he makes a least-squares fit to the straight line (8.33) and finds as best estimates for the coefficients $\ln N_0$ and $(-1/\tau)$,

$$\ln N_0 = 11.93 \quad \text{and} \quad (-1/\tau) = -0.089 \text{ day}^{-1}.$$

The second of these coefficients implies that his best estimate for the mean life is

$$\tau = 11.2 \text{ days.}$$

The method just described is attractively simple (especially with a calculator that performs linear regression automatically) and is frequently used. Nevertheless, the method is not quite logically sound. Our derivation of the least-squares fit to a straight line $y = A + Bx$ was based on the assumption that the measured values $y_1, \ldots, y_N$ were all equally uncertain. Here, we are performing our least-squares fit using the variable $z = \ln y$. Now, if the measured values y_i are all equally uncertain, then the values $z_i = \ln y_i$ are *not*. In fact, from simple error propagation we know that

$$\sigma_z = \left| \frac{dz}{dy} \right| \sigma_y = \frac{\sigma_y}{y}. \tag{8.34}$$

Thus, if σ_y is the same for all measurements, then σ_z varies (with σ_z larger when y is smaller). Evidently, the variable $z = \ln y$ does not satisfy the requirement of equal uncertainties for all measurements, if y itself does.

The remedy for this difficulty is straightforward. The least-squares procedure can be modified to allow for different uncertainties in the measurements, provided the various uncertainties are known. (This method of *weighted least squares* is outlined in Problem 8.9). If we know that the measurements of $y_1, \ldots, y_N$ really are equally uncertain, then Equation (8.34) tells us how the uncertainties in $z_1, \ldots, z_N$ vary, and we can therefore apply the method of weighted least squares to the equation $z = \ln A + Bx$.

In practice, we often cannot be sure that the uncertainties in $y_1, \ldots, y_N$ really are constant; so we can perhaps argue that we could just as well assume the uncertainties in $z_1, \ldots, z_N$ to be constant and use the simple unweighted least squares. Often the variation in the uncertainties is small, and which method is used makes little difference, as in the preceding example. In any event, when the uncertainties are unknown, straightforward application of the ordinary (unweighted) least-squares fit is an unambiguous and simple way to get *reasonable* (if not *best*) estimates for the constants A and B in the equation $y = Ae^{Bx}$, so it is frequently used in this way.

MULTIPLE REGRESSION

Finally, we have so far discussed only observations of *two* variables, x and y, and their relationship. Many real problems, however, have more than two variables to be considered. For example, in studying the pressure P of a gas, we find that it depends on the volume V and temperature T, and we must analyze P as a function of V and T. The simplest example of such a problem is when one variable, z, depends linearly on two others, x and y:

$$z = A + Bx + Cy. \tag{8.35}$$

This problem can be analyzed by a very straightforward generalization of our two-variable method. If we have a series of measurements (x_i, y_i, z_i), $i = 1, \ldots, N$ (with the z_i all equally uncertain, and the x_i and y_i exact), then we can use the principle of maximum likelihood exactly as in Section 8.2 to show that the best estimates for the constants A, B, and C are determined by normal equations of the form

$$
\begin{aligned}
AN + B\Sigma x + C\Sigma y &= \Sigma z, \\
A\Sigma x + B\Sigma x^2 + C\Sigma xy &= \Sigma xz, \\
A\Sigma y + B\Sigma xy + C\Sigma y^2 &= \Sigma yz.
\end{aligned}
\qquad (8.36)
$$

The equations can be solved for A, B, and C to give the best fit for the relation (8.35). This method is called *multiple regression* ("multiple" because there are more than two variables), but we will not discuss it further here.

Principal Definitions and Equations of Chapter 8

Throughout this chapter, we have considered N pairs of measurements $(x_1, y_1), \ldots, (x_N, y_N)$ of two variables x and y. The problem addressed was finding the best values of the parameters of the curve that a graph of y vs x is expected to fit. We assume that only the measurements of y suffered appreciable uncertainties, whereas those for x were negligible. [For the case in which both x and y have significant uncertainties, see the discussion following Equation (8.17).] Various possible curves can be analyzed, and there are two different assumptions about the uncertainties in y. Some of the more important cases are as follows:

A STRAIGHT LINE, $y = A + Bx$; EQUAL WEIGHTS

If y is expected to lie on a straight line $y = A + Bx$, and if the measurements of y all have the same uncertainties, then the best estimates for the constants A and B are:

$$
A = \frac{\Sigma x^2 \, \Sigma y - \Sigma x \, \Sigma xy}{\Delta}
$$

and

$$
B = \frac{N \Sigma xy - \Sigma x \, \Sigma y}{\Delta},
$$

where the denominator, Δ, is

$$
\Delta = N \Sigma x^2 - (\Sigma x)^2. \qquad \text{[See (8.10) to (8.12)]}
$$

Based on the observed points, the best estimate for the uncertainty in the measurements of y is

$$
\sigma_y = \sqrt{\frac{1}{N-2} \sum_{i=1}^{N} (y_i - A - Bx_i)^2}.
$$

[See (8.15)]

The uncertainties in A and B are:

$$\sigma_A = \sigma_y \sqrt{\frac{\sum x^2}{\Delta}}$$

and

$$\sigma_B = \sigma_y \sqrt{\frac{N}{\Delta}}.$$ [See (8.16) & (8.17)]

STRAIGHT LINE THROUGH THE ORIGIN ($y = Bx$); EQUAL WEIGHTS

If y is expected to lie on a straight line through the origin, $y = Bx$, and if the measurements of y all have the same uncertainties, then the best estimate for the constant B is:

$$B = \frac{\sum xy}{\sum x^2}.$$ [See Problem 8.5]

Based on the measured points, the best estimate for the uncertainty in the measurements of y is:

$$\sigma_y = \sqrt{\frac{1}{N-1}\sum(y_i - Bx_i)^2}$$

and the uncertainty in B is:

$$\sigma_B = \frac{\sigma_y}{\sqrt{\sum x^2}}.$$ [See Problem 8.18]

WEIGHTED FIT FOR A STRAIGHT LINE, $y = A + Bx$

If y is expected to lie on a straight line $y = A + Bx$, and if the measured values y_i have different, known uncertainties σ_i, then we introduce the *weights* $w_i = 1/\sigma_i^2$, and the best estimates for the constants A and B are:

$$A = \frac{\sum wx^2 \sum wy - \sum wx \sum wxy}{\Delta}$$

and

$$B = \frac{\sum w \sum wxy - \sum wx \sum wy}{\Delta},$$

where

$$\Delta = \sum w \sum wx^2 - (\sum wx)^2.$$ [See Problem 8.9]

The uncertainties in A and B are:

$$\sigma_A = \sqrt{\frac{\sum wx^2}{\Delta}}$$

and

$$\sigma_B = \sqrt{\frac{\Sigma w}{\Delta}}.$$ [See Problem 8.19]

OTHER CURVES

If y is expected to be a polynomial in x, that is,

$$y = A + Bx + \ldots + Hx^n,$$

then an exactly analogous method of least-squares fitting can be used, although the equations are quite cumbersome if n is large. (For examples, see Problems 8.21 and 8.22.) Curves of the form

$$y = Af(x) + Bg(x) + \ldots + Hk(x),$$

where $f(x), \ldots, k(x)$ are known functions, can also be handled in the same way. (For examples, see Problems 8.23 and 8.24.)

If y is expected to be given by the exponential function

$$y = Ae^{Bx},$$

then we can "linearize" the problem by using the variable $z = \ln(y)$, which should satisfy the linear relation

$$z = \ln(y) = \ln(A) + Bx.$$ [See (8.31)]

We can then apply the linear least-squares fit to z as a function of x. Note, however, that if the uncertainties in the measured values of y are all equal, the same is certainly *not* true of the values of z. Then, strictly speaking, the method of weighted least squares should be used. (See Problem 8.26 for an example.)

Problems for Chapter 8

For Section 8.2: Calculation of the Constants A and B

8.1. ★   Use the method of least squares to find the line $y = A + Bx$ that best fits the three points $(1, 6)$, $(3, 5)$, and $(5, 1)$. Using squared paper, plot the three points and your line. Your calculator probably has a built-in function to calculate A and B; if you don't know how to use it, take a moment to learn and then check your own answers to this problem.

8.2. ★   Use the method of least squares to find the line $y = A + Bx$ that best fits the four points $(-3, 3)$, $(-1, 4)$, $(1, 8)$, and $(3, 9)$. Using squared paper, plot the four points and your line. Your calculator probably has a built-in function to calculate A and B; if you don't know how to use it, take a moment to learn and then check your own answers to this problem.

8.3. ★   The best estimates for the constants A and B are determined by Equations (8.8) and (8.9). The solutions to these equations were given in Equations (8.10) through (8.12). Verify that these are indeed the solutions of (8.8) and (8.9).

8.4. ★   Prove the following useful fact: The least-squares fit for a line through any set of points $(x_1, y_1), \ldots, (x_N, y_N)$ always passes through the "center of gravity" $(\bar{x}, \bar{y})$ of the points, where the bar denotes the average of the N values concerned. [Hint: You know that A and B satisfy Equation (8.8).]

8.5. ★★ **Line Through the Origin.**   Suppose two variables x and y are known to satisfy a relation $y = Bx$. That is, $y \propto x$, and a graph of y vs x is a line *through the origin.* (For example, Ohm's law, $V = RI$, tells us that a graph of voltage V vs current I should be a straight line through the origin.) Suppose further that you have N measurements (x_i, y_i) and that the uncertainties in x are negligible and those in y are all equal. Using arguments similar to those of Section 8.2, prove that the best estimate for B is

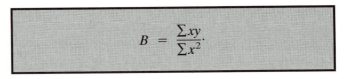

$$B = \frac{\Sigma xy}{\Sigma x^2}.$$

8.6. ★★   For measurements of just two points (x_1, y_1) and (x_2, y_2), the line that best fits the points is obviously the line *through* the points. Prove that the least-squares line for two points does indeed pass through both points. [One way to do this is to use Equations (8.8) and (8.9) to show that either of the points does satisfy the equation $y = A + Bx$.]

8.7. ★★   To find the spring constant of a spring, a student loads it with various masses M and measures the corresponding lengths l. Her results are shown in Table 8.4. Because the force $mg = k(l - l_o)$, where l_o is the unstretched length of the

Table 8.4. Length versus load for a spring; for Problem 8.7.

"x": Load m (grams)	200	300	400	500	600	700	800	900
"y": Length l (cm)	5.1	5.5	5.9	6.8	7.4	7.5	8.6	9.4

spring, these data should fit a straight line, $l = l_o + (g/k)m$. Make a least-squares fit to this line for the given data, and find the best estimates for the unstretched length l_o and the spring constant k. Do the calculations yourself, and then check your answers using the built-in functions on your calculator.

8.8. ★★   A student measures the velocity of a glider on a horizontal air track. She uses a multiflash photograph to find the glider's position s at five equally spaced times as in Table 8.5. **(a)** One way to find v would be to calculate $v = \Delta s/\Delta t$ for each of the four successive two-second intervals and then average them. Show that this procedure gives $v = (s_5 - s_1)/(t_5 - t_1)$, which means that the middle three measurements are completely ignored by this method. Prove this without putting in numbers; then find a numerical answer. **(b)** A better procedure is to make a least-

Table 8.5. Position versus time data; for Problem 8.8.

"x": Time, t (s)	−4	−2	0	2	4
"y": Position, s (cm)	13	25	34	42	56

squares fit to the equation $s = s_0 - vt$ using all five data points. Follow this procedure to find the best estimate for v and compare your result with that from part (a).
Do the calculations yourself and check your answers using the built-in functions on
your calculator. (The five times have negligible uncertainty; the positions are all
equally uncertain.)

8.9. ★★ Weighted Least Squares. Suppose we measure N pairs of values (x_i, y_i)
of two variables x and y that are supposed to satisfy a linear relation $y = A + Bx$.
Suppose the x_i have negligible uncertainty and the y_i have *different* uncertainties σ_i.
(That is, y_1 has uncertainty σ_1, while y_2 has uncertainty σ_2, and so on.) As in
Chapter 7, we can define the *weight* of the ith measurement as $w_i = 1/\sigma_i^2$. Review
the derivation of the least-squares fit in Section 8.2 and generalize it to cover this
situation, where the measurements of the y_i have different weights. Show that the
best estimates of A and B are

$$A = \frac{\sum wx^2 \sum wy - \sum wx \sum wxy}{\Delta} \qquad (8.37)$$

and

$$B = \frac{\sum w \sum wxy - \sum wx \sum wy}{\Delta} \qquad (8.38)$$

where

$$\Delta = \sum w \sum wx^2 - (\sum wx)^2. \qquad (8.39)$$

Obviously, this method of *weighted least squares* can be applied only when the
uncertainties σ_i (or at least their relative sizes) are known.

8.10. ★★ Suppose y is known to be linear in x, so that $y = A + Bx$, and we
have three measurements of (x, y):

$$(1, 2 \pm 0.5), (2, 3 \pm 0.5), \text{ and } (3, 2 \pm 1).$$

(The uncertainties in x are negligible.) Use the method of weighted least squares,
Equations (8.37) to (8.39), to calculate the best estimates for A and B. Compare your
results with what you would get if you ignored the variation in the uncertainties, that
is, used the unweighted fit of Equations (8.10) to (8.12). Plot the data and both
lines, and try to understand the differences.

8.11. ★★ **(a)** If you have access to a spreadsheet program such as Lotus 123 or Excel, create a spreadsheet that will calculate the coefficients A and B for the least-squares fit for up to 10 points $(x_1, y_1), \ldots$. Use the layout of Table 8.1. **(b)** Test your spreadsheet with the data of Problems 8.1 and 8.7.

For Section 8.3: Uncertainty in the Measurements of y

8.12. ★★ Use the principle of maximum likelihood, as outlined in the discussion of Equation (8.14), to show that (8.14) gives the best estimate for the uncertainty σ_y in y in a series of measurements $(x_1, y_1), \ldots, (x_N, y_N)$ that are supposed to lie on a straight line. [Note that in (8.14), A and B are the true values of these two constants. When we replace these true values by our best estimates, as given by (8.10) and (8.11), the expression is decreased, and the N in the denominator must be replaced by $N - 2$ to get the best estimate, as in (8.15).]

8.13. ★★★ If you have a reliable estimate of the uncertainty δy in the measurements of y, then by comparing this estimate with σ_y as given by (8.15), you can assess whether your data confirm the expected linear relation $y = A + Bx$. The quantity σ_y is roughly the average distance by which the points (x_i, y_i) fail to lie on the best-fit line. If σ_y is about the same as the expected uncertainty δy, the data are consistent with the expected linear relation; if σ_y is much larger than δy, there is good reason to doubt the linear relation. The following problem illustrates these ideas.

A student measures the velocity of a glider coasting along a horizontal air track using a multiflash photograph to find its position s at four equally spaced times, as shown in Table 8.6. Assuming the glider moves with constant velocity, he fits these

Table 8.6. Positions and times of a coasting glider; for Problem 8.13.

"x": Time, t (s)	-3	-1	1	3
"y": Position, s (cm)	4.0	7.5	10.3	12.0

data to a line $s = s_0 + vt$. **(a)** Use the method of least squares to find his best estimates for a s_0 and v and for the standard deviation σ_s in the measurements of s. **(b)** Suppose the quality of his photograph was poor and he believed his measurements of s were uncertain by $\delta s \approx 1$ cm. By comparing this value of δs with σ_s, decide if his data are consistent with his assumption that the velocity was constant. Draw a graph of s vs t, with error bars showing his uncertainties $\delta s \approx 1$ cm, to confirm your conclusion. **(c)** Suppose, instead, that he was confident his measurements were good within 0.1 cm. In this case, are his measurements consistent with a constant velocity? After examining your graph of the data and best line, can you suggest an explanation?

For Section 8.4: Uncertainty in the Constants A and B

8.14. ★ Use the method of least squares to find the student's best estimate for the velocity v of the glider in Problem 8.8 and the uncertainty in v. (By all means, use the built-in functions on your calculator to find the best-fit line; unfortunately, most calculators do not have built-in functions to find the uncertainty in A and B, so you will probably have to do this part of the calculation yourself.)

8.15. ★★ Kundt's tube is a device for measuring the wavelength λ of sound. The experimenter sets up a standing wave inside a glass tube in which he or she has sprinkled a light powder. The vibration of the air causes the powder to move and eventually to collect in small piles at the displacement nodes of the standing wave, as shown in Figure 8.5. Because the distance between the nodes is $\lambda/2$, this lets the

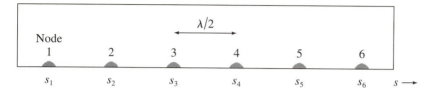

Figure 8.5. Kundt's tube, with small piles of powder at the nodes of a standing wave; for Problem 8.15.

experimenter find λ. A student finds six nodes (numbered $n = 1, \ldots, 6$) as shown in Table 8.7. Because the nodes should be equal distances $\lambda/2$ apart, their positions should satisfy $s_n = A + Bn$, where $B = \lambda/2$. Use the method of least squares to fit the data to this line and find the wavelength and its uncertainty.

Table 8.7. Positions of nodes in Kundt's tube; for Problem 8.15.

"x": Node number n	1	2	3	4	5	6
"y": Position s_n (cm)	5.0	14.4	23.1	32.3	41.0	50.4

8.16. ★★ Use error propagation to verify that the uncertainties in the parameters of a straight line $y = A + Bx$ are given by (8.16) and (8.17). [Hint: Use the general error propagation formula (3.47) and remember that, by assumption, only the y_i are subject to uncertainty. When you differentiate a sum like Σy with respect to y_i, writing the sum out as $y_1 + \ldots + y_N$ may help. Any quantity, such as the denominator Δ, that involves only the x_i has no uncertainty.]

8.17. ★★ The least-squares fit to a set of points $(x_1, y_1), \ldots, (x_N, y_N)$ treats the variables x and y unsymmetrically. Specifically, the best fit for a line $y = A + Bx$ is found on the assumption that the numbers $y_1, \ldots, y_N$ are all equally uncertain, whereas $x_1, \ldots, x_N$ have negligible uncertainty. If the situation were reversed, the roles of x and y would have to be exchanged and x and y fitted to a line $x = A' + B'y$. The resulting two lines, $y = A + Bx$ and $x = A' + B'y$, would be the same if the N points lay *exactly* on a line, but in general the two lines will be slightly different. The following problem illustrates this small difference.

(a) Find the best fit to the line $y = A + Bx$ for the data of Problem 8.1 and the uncertainties in A and B. (b) Now reverse the roles of x and y and fit the same data to the line $x = A' + B'y$. If you solve this equation for y in terms of x, you can find values for A and B based on the values of A' and B'. Comment on the difference between the two approaches.

8.18. ★★ **Uncertainties for a Line Through the Origin.** Consider the situation described in Problem 8.5, in which it is known that $y = Bx$ [that is, the points (x, y) lie on a straight line known to pass through the origin]. The best value for the slope B is given in Problem 8.5. Arguing as in Section 8.3, we can show that the uncertainty σ_y in the measurements of y is given by

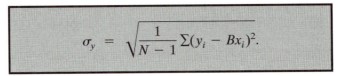

$$\sigma_y = \sqrt{\frac{1}{N-1}\Sigma(y_i - Bx_i)^2}.$$

Following the argument sketched in Problem 8.16, prove that the uncertainty in the constant B is given by

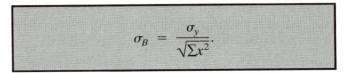

$$\sigma_B = \frac{\sigma_y}{\sqrt{\Sigma x^2}}.$$

8.19. ★★★ **Uncertainties in Weighted Least-Squares Fits.** Consider the method of weighted least squares outlined in Problem 8.9. Use error propagation to prove that the uncertainties in the constants A and B are given by

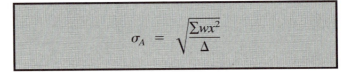

$$\sigma_A = \sqrt{\frac{\Sigma wx^2}{\Delta}}$$

and

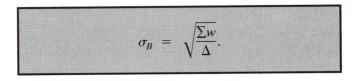

$$\sigma_B = \sqrt{\frac{\Sigma w}{\Delta}}.$$

For Section 8.5: An Example

8.20. ★ A student measures the pressure P of a gas at five different temperatures T, keeping the volume constant. His results are shown in Table 8.8. His data should

Table 8.8. Temperature (in °C) versus Pressure (in mm of mercury); for Problem 8.20.

"x": Pressure P	79	82	85	88	90
"y": Temperature T	8	17	30	37	52

fit a straight line of the form $T = A + BP$, where A is the absolute zero of temperature (whose accepted value is -273°C). Find the best fit to the student's data and hence his best estimate for absolute zero and its uncertainty.

For Section 8.6: Least-Squares Fits to Other Curves

8.21. ★★ Consider the problem of fitting a set of measurements (x_i, y_i), with $i = 1, \ldots, N$, to the polynomial $y = A + Bx + Cx^2$. Use the principle of maximum likelihood to show that the best estimates for A, B, and C based on the data are given by Equations (8.27). Follow the arguments outlined between Equations (8.24) and (8.27).

8.22. ★★ One way to measure the acceleration of a freely falling body is to measure its heights y_i at a succession of equally spaced times t_i (with a multiflash photograph, for example) and to find the best fit to the expected polynomial

$$y = y_o + v_o t - \tfrac{1}{2}gt^2. \tag{8.40}$$

Use the equations (8.27) to find the best estimates for the three coefficients in (8.40) and hence the best estimate for g, based on the five measurements in Table. 8.9.

Table 8.9. Height (in cm) versus time (in tenths of a second) for a falling body; for Problem 8.22.

"x": Time t	−2	−1	0	1	2
"y": Height y	131	113	89	51	7

(Note that we can name the times however we like. A more natural choice might seem to be $t = 0, 1, \ldots, 4$. When you solve the problem, however, you will see that defining the times to be symmetrically spaced about $t = 0$ causes approximately half of the sums involved to be zero and greatly simplifies the algebra. This trick can be used whenever the values of the independent variable are equally spaced.)

8.23. ★★ Suppose y is expected to have the form $y = Af(x) + Bg(x)$, where A and B are unknown coefficients and f and g are fixed, known functions (such as $f = x$ and $g = x^2$, or $f = \cos x$ and $g = \sin x$). Use the principle of maximum likelihood to show that the best estimates for A and B, based on data (x_i, y_i), $i = 1, \ldots, N$, must satisfy

$$\begin{aligned}
A\Sigma[f(x_i)]^2 + B\Sigma f(x_i)g(x_i) &= \Sigma y_i f(x_i), \\
A\Sigma f(x_i)g(x_i) + B\Sigma[g(x_i)]^2 &= \Sigma y_i g(x_i).
\end{aligned} \tag{8.41}$$

8.24. ★★ A weight oscillating on a vertical spring should have height given by

$$y = A \cos \omega t + B \sin \omega t.$$

A student measures ω to be 10 rad/s with negligible uncertainty. Using a multiflash photograph, she then finds y for five equally spaced times, as shown in Table 8.10.

Table 8.10. Positions (in cm) and times (in tenths of a second) for an oscillating mass; for Problem 8.24.

"x": Time t	−4	−2	0	2	4
"y": Position y	3	−16	6	9	−8

Use Equations (8.41) to find best estimates for A and B. Plot the data and your best fit. (If you plot the data first, you will have the opportunity to consider how hard it would be to choose a best fit without the least-squares method.) If the student judges that her measured values of y were uncertain by "a couple of centimeters," would you say the data are an acceptable fit to the expected curve?

8.25. ★★   The rate at which a sample of radioactive material emits radiation decreases exponentially as the material is depleted. To study this exponential decay, a student places a counter near a sample and records the number of decays in 15 seconds. He repeats this five times at 10-minute intervals and obtains the results shown in Table 8.11. (Notice that, because it takes nearly 10 minutes to prepare the equipment, his first measurement is made at $t = 10$ min.)

Table 8.11. Number $\nu(t)$ of emissions in a 15-second interval versus total time elapsed t (in minutes); for Problem 8.25.

"x": Elapsed time t	10	20	30	40	50
"y": Number $\nu(t)$	409	304	260	192	170

If the sample does decay exponentially, the number $\nu(t)$ should satisfy

$$\nu(t) = \nu_o e^{-t/\tau}, \tag{8.42}$$

where τ is the (unknown) mean life of the material in question and ν_o is another unknown constant. To find the mean life τ, the student takes the natural log of Equation (8.42) to give the linear relation

$$z = \ln(\nu) = \ln(\nu_o) - t/\tau \tag{8.43}$$

and makes a least-squares fit to this line. What is his answer for the mean life τ? How many decays would he have counted in 15 seconds at $t = 0$?

8.26. ★★★   The student of Problem 8.25 decides to investigate the exponential character of radioactive decay further and repeats his experiment, monitoring the decays for a longer total time. He decides to count the decays in 15-second periods as before but makes six measurements at 30-minute intervals, as shown in Table 8.12.

Table 8.12. Number $\nu(t)$ of emissions in a 15-second interval versus total time elapsed t (in minutes); for Problem 8.26.

"x": Time t	10	40	70	100	130	160
"y": Number $\nu(t)$	188	102	60	18	16	5

According to the square-root rule of Section 3.2, the uncertainty in each of the counts ν is $\sqrt{\nu}$, which obviously varies widely during this experiment. Therefore, he decides to make a *weighted* least-squares fit to the straight line (8.43). **(a)** To this end, he must first calculate the logs, $z_i = \ln(\nu_i)$, and their uncertainties. Use error propagation to show that the uncertainty in z_i is $1/\sqrt{\nu_i}$, which means that the weight of z_i is just ν_i. **(b)** Now make the weighted least-squares fit to the line (8.43). What is his answer for the material's mean life τ and its uncertainty? [The best values for the coefficients A and B are given in Equations (8.37) and (8.38) of Problem 8.9, and their uncertainties are given in Problem 8.19. Make a table showing t, ν, $z = \ln(\nu)$, and the various sums in (8.37) to (8.39).] **(c)** Draw a graph of $z = \ln(\nu)$ vs t, showing the best-fit straight line and the data with error bars. Is your graph consistent with the expectation that the decay rate should decrease exponentially?

Chapter 9

Covariance and Correlation

This chapter introduces the important concept of covariance. Because this concept arises naturally in the propagation of errors, Section 9.1 starts with a quick review of error propagation. This review sets the stage for Section 9.2, which defines covariance and discusses its role in the propagation of errors. Then, Section 9.3 uses the covariance to define the coefficient of linear correlation for a set of measured points $(x_1, y_1), \ldots, (x_N, y_N)$. This coefficient, denoted r, provides a measure of how well the points fit a straight line of the form $y = A + Bx$; its use is described in Sections 9.4 and 9.5.

9.1 Review of Error Propagation

This and the next section provide a final look at the important question of error propagation. We first discussed error propagation in Chapter 3, where we reached several conclusions. We imagined measuring two quantities x and y to calculate some function $q(x, y)$, such as $q = x + y$ or $q = x^2 \sin y$. [In fact, we discussed a function $q(x, \ldots, z)$ of an arbitrary number of variables $x, \ldots, z$; for simplicity, we will now consider just two variables.] A simple argument suggested that the uncertainty in our answer for q is just

$$\delta q \approx \left| \frac{\partial q}{\partial x} \right| \delta x + \left| \frac{\partial q}{\partial y} \right| \delta y. \tag{9.1}$$

We first derived this approximation for the simple special cases of sums, differences, products, and quotients. For instance, if q is the sum $q = x + y$, then (9.1) reduces to the familiar $\delta q \approx \delta x + \delta y$. The general result (9.1) was derived in Equation (3.43).

We next recognized that (9.1) is often probably an overstatement of δq, because there may be partial cancellation of the errors in x and y. We stated, without proof, that when the errors in x and y are independent and random, a better value for the uncertainty in the calculated value of $q(x, y)$ is the quadratic sum

$$\delta q = \sqrt{\left(\frac{\partial q}{\partial x}\delta x\right)^2 + \left(\frac{\partial q}{\partial y}\delta y\right)^2}.$$ (9.2)

We also stated, without proof, that whether or not the errors are independent and random, the simpler formula (9.1) always gives an upper bound on δq; that is, the uncertainty δq is never any worse than is given by (9.1).

Chapter 5 gave a proper definition and proof of (9.2). First, we saw that a good measure of the uncertainty δx in a measurement is given by the standard deviation σ_x; in particular, we saw that if the measurements of x are normally distributed, we can be 68% confident that the measured value lies within σ_x of the true value. Second, we saw that if the measurements of x and y are governed by independent normal distributions, with standard deviations σ_x and σ_y, the values of $q(x, y)$ are also normally distributed, with standard deviation

$$\sigma_q = \sqrt{\left(\frac{\partial q}{\partial x}\sigma_x\right)^2 + \left(\frac{\partial q}{\partial y}\sigma_y\right)^2}.$$ (9.3)

This result provides the justification for the claim (9.2).

In Section 9.2, I will derive a precise formula for the uncertainty in q that applies whether or not the errors in x and y are independent and normally distributed. In particular, I will prove that (9.1) always provides an upper bound on the uncertainty in q.

Before I derive these results, let us first review the definition of the standard deviation. The standard deviation σ_x of N measurements $x_1, \ldots, x_N$ was originally defined by the equation

$$\sigma_x^2 = \frac{1}{N}\sum_{i=1}^{N}(x_i - \bar{x})^2.$$ (9.4)

If the measurements of x are normally distributed, then in the limit that N is large, the definition (9.4) is equivalent to defining σ_x as the width parameter that appears in the Gauss function

$$\frac{1}{\sigma_x\sqrt{2\pi}}e^{-(x-X)^2/2\sigma_x^2}$$

that governs the measurements of x. Because we will now consider the possibility that the errors in x may not be normally distributed, this second definition is no longer available to us. We can, and will, still define σ_x by (9.4), however. Whether or not the distribution of errors is normal, this definition of σ_x gives a reasonable measure of the random uncertainties in our measurement of x. (As in Chapter 5, I will suppose all systematic errors have been identified and reduced to a negligible level, so that all remaining errors *are* random.)

The usual ambiguity remains as to whether to use the definition (9.4) of σ_x or the "improved" definition with the factor N in the denominator replaced by $(N - 1)$. Fortunately, the discussion that follows applies to either definition, as long as we are consistent in our use of one or the other. For convenience, I will use the definition (9.4), with N in the denominator throughout this chapter.

9.2   Covariance in Error Propagation

Suppose that to find a value for the function $q(x, y)$, we measure the two quantities x and y several times, obtaining N pairs of data, $(x_1, y_1), \ldots, (x_N, y_N)$. From the N measurements $x_1, \ldots, x_N$, we can compute the mean $\bar{x}$ and standard deviation σ_x in the usual way; similarly, from $y_1, \ldots, y_N$, we can compute $\bar{y}$ and σ_y. Next, using the N pairs of measurements, we can compute N values of the quantity of interest

$$q_i = q(x_i, y_i), \qquad (i = 1, \ldots, N).$$

Given $q_1, \ldots, q_N$, we can now calculate their mean $\bar{q}$, which we assume gives our best estimate for q, and their standard deviation σ_q, which is our measure of the random uncertainty in the values q_i.

I will assume, as usual, that all our uncertainties are small and hence that all the numbers $x_1, \ldots, x_N$ are close to $\bar{x}$ and that all the $y_1, \ldots, y_N$ are close to $\bar{y}$. We can then make the approximation

$$
\begin{aligned}
q_i &= q(x_i, y_i) \\
&\approx q(\bar{x}, \bar{y}) + \frac{\partial q}{\partial x}(x_i - \bar{x}) + \frac{\partial q}{\partial y}(y_i - \bar{y}).
\end{aligned}
\tag{9.5}
$$

In this expression, the partial derivatives $\partial q/\partial x$ and $\partial q/\partial y$ are taken at the point $x = \bar{x}$, $y = \bar{y}$, and are therefore the same for all $i = 1, \ldots, N$. With this approximation, the mean becomes

$$
\begin{aligned}
\bar{q} &= \frac{1}{N}\sum_{i=1}^{N} q_i \\
&= \frac{1}{N}\sum_{i=1}^{N}\left[q(\bar{x}, \bar{y}) + \frac{\partial q}{\partial x}(x_i - \bar{x}) + \frac{\partial q}{\partial y}(y_i - \bar{y}) \right].
\end{aligned}
$$

This equation gives $\bar{q}$ as the sum of three terms. The first term is just $q(\bar{x}, \bar{y})$, and the other two are exactly zero. [For example, it follows from the definition of $\bar{x}$ that $\sum(x_i - \bar{x}) = 0$.] Thus, we have the remarkably simple result

$$\bar{q} = q(\bar{x}, \bar{y}); \tag{9.6}$$

that is, to find the mean $\bar{q}$ we have only to calculate the function $q(x, y)$ at the point $x = \bar{x}$ and $y = \bar{y}$.

The standard deviation in the N values $q_1, \ldots, q_N$ is given by

$$\sigma_q{}^2 = \frac{1}{N}\Sigma(q_i - \bar{q})^2.$$

Substituting (9.5) and (9.6), we find that

$$
\begin{aligned}
\sigma_q{}^2 &= \frac{1}{N}\Sigma\left[\frac{\partial q}{\partial x}(x_i - \bar{x}) + \frac{\partial q}{\partial y}(y_i - \bar{y})\right]^2 \\
&= \left(\frac{\partial q}{\partial x}\right)^2 \frac{1}{N}\Sigma(x_i - \bar{x})^2 + \left(\frac{\partial q}{\partial y}\right)^2 \frac{1}{N}\Sigma(y_i - \bar{y})^2 \\
&\quad + 2\frac{\partial q}{\partial x}\frac{\partial q}{\partial y}\frac{1}{N}\Sigma(x_i - \bar{x})(y_i - \bar{y}).
\end{aligned}
\tag{9.7}
$$

The sums in the first two terms are those that appear in the definition of the standard deviations σ_x and σ_y. The final sum is one we have not encountered before. It is called the *covariance*[1] of x and y and is denoted

$$\sigma_{xy} = \frac{1}{N} \sum_{i=1}^{N} (x_i - \bar{x})(y_i - \bar{y}). \tag{9.8}$$

With this definition, Equation (9.7) for the standard deviation σ_q becomes

$$\sigma_q^2 = \left(\frac{\partial q}{\partial x}\right)^2 \sigma_x^2 + \left(\frac{\partial q}{\partial y}\right)^2 \sigma_y^2 + 2\frac{\partial q}{\partial x}\frac{\partial q}{\partial y}\sigma_{xy}. \tag{9.9}$$

This equation gives the standard deviation σ_q, whether or not the measurements of x and y are independent or normally distributed.

If the measurements of x and y *are* independent, we can easily see that, after many measurements, the covariance σ_{xy} should approach zero: Whatever the value of y_i, the quantity $x_i - \bar{x}$ is just as likely to be negative as it is to be positive. Thus, after many measurements, the positive and negative terms in (9.8) should nearly balance; in the limit of infinitely many measurements, the factor $1/N$ in (9.8) guarantees that σ_{xy} is zero. (After a finite number of measurements, σ_{xy} will not be exactly zero, but it should be *small* if the errors in x and y really are independent and random.) With σ_{xy} zero, Equation (9.9) for σ_q reduces to

$$\sigma_q^2 = \left(\frac{\partial q}{\partial x}\right)^2 \sigma_x^2 + \left(\frac{\partial q}{\partial y}\right)^2 \sigma_y^2, \tag{9.10}$$

the familiar result for independent and random uncertainties.

If the measurements of x and y are *not* independent, the covariance σ_{xy} need not be zero. For instance, it is easy to imagine a situation in which an overestimate of x will always be accompanied by an overestimate of y, and vice versa. The numbers $(x_i - \bar{x})$ and $(y_i - \bar{y})$ will then always have the same sign (both positive or both negative), and their product will always be positive. Because all terms in the sum (9.8) are positive, σ_{xy} will be positive (and nonzero), even in the limit that we make infinitely many measurements. Conversely, you can imagine situations in which an overestimate of x is always accompanied by an underestimate of y, and vice versa; in this case $(x_i - \bar{x})$ and $(y_i - \bar{y})$ will always have opposite signs, and σ_{xy} will be negative. This case is illustrated in the example below.

When the covariance σ_{xy} is not zero (even in the limit of infinitely many measurements), we say that the errors in x and y are *correlated*. In this case, the uncertainty σ_q in $q(x, y)$ as given by (9.9) is *not* the same as we would get from the formula (9.10) for independent, random errors.

---

[1] The name *covariance* for σ_{xy} (for two variables x, y) parallels the name *variance* for σ_x^2 (for one variable x). To emphasize this parallel, the covariance (9.8) is sometimes denoted σ_{xy}^2, not an especially apt notation, because the covariance can be negative. A convenient feature of the definition (9.8) is that σ_{xy} has the dimensions of xy, just as σ_x has the dimensions of x.

Example: Two Angles with a Negative Covariance

Each of five students measures the same two angles α and β and obtains the results shown in the first three columns of Table 9.1.

Table 9.1. Five measurements of two angles α and β (in degrees).

Student	α	β	$(\alpha - \overline{\alpha})$	$(\beta - \overline{\beta})$	$(\alpha - \overline{\alpha})(\beta - \overline{\beta})$
A	35	50	2	−2	−4
B	31	55	−2	3	−6
C	33	51	0	−1	0
D	32	53	−1	1	−1
E	34	51	1	−1	−1

Find the average and standard deviation for each of the two angles, and then find the covariance $\sigma_{\alpha\beta}$ as defined by (9.8). The students now calculate the sum $q = \alpha + \beta$. Find their best estimate for q as given by (9.6) and the standard deviation σ_q as given by (9.9). Compare the standard deviation with what you would get if you assumed (incorrectly) that the errors in α and β were independent and that σ_q was given by (9.10).

The averages are immediately seen to be $\overline{\alpha} = 33$ and $\overline{\beta} = 52$. With these values, we can find the deviations $(\alpha - \overline{\alpha})$ and $(\beta - \overline{\beta})$, as shown in Table 9.1, and from these deviations we easily find

$$\sigma_\alpha^2 = 2.0 \quad \text{and} \quad \sigma_\beta^2 = 3.2.$$

[Here I have used the definition (9.4), with the N in the denominator.]

You can see from Table 9.1 that high values of α seem to be correlated with low values of β and vice versa, because $(\alpha - \overline{\alpha})$ and $(\beta - \overline{\beta})$ always have opposite signs. (For an experiment in which this kind of correlation arises, see Problem 9.6.) This correlation means that the products $(\alpha - \overline{\alpha})(\beta - \overline{\beta})$ shown in the last column of the table are all negative (or zero). Thus, the covariance $\sigma_{\alpha\beta}$ as defined by (9.8) is negative,

$$\sigma_{\alpha\beta} = \frac{1}{N}\Sigma(\alpha - \overline{\alpha})(\beta - \overline{\beta}) = \frac{1}{5} \times (-12) = -2.4.$$

The best estimate for the sum $q = \alpha + \beta$ is given by (9.6) as

$$q_{\text{best}} = \overline{q} = \overline{\alpha} + \overline{\beta} = 33 + 52 = 85.$$

To find the standard deviation using (9.9), we need the two partial derivatives, which are easily seen to be $\partial q/\partial \alpha = \partial q/\partial \beta = 1$. Therefore, according to (9.9),

$$\sigma_q = \sqrt{\sigma_\alpha^2 + \sigma_\beta^2 + 2\sigma_{\alpha\beta}}$$
$$= \sqrt{2.0 + 3.2 - 2 \times 2.4} = 0.6.$$

If we overlooked the correlation between the measurements of α and β and treated them as independent, then according to (9.10) we would get the incorrect answer

$$\sigma_q = \sqrt{\sigma_\alpha^2 + \sigma_\beta^2}$$
$$= \sqrt{2.0 + 3.2} = 2.3.$$

We see from this example that a correlation of the right sign can cause a dramatic difference in a propagated error. In this case we can see why there is this difference: The errors in each of the angles α and β are a degree or so, suggesting that $q = \alpha + \beta$ would be uncertain by a couple of degrees. But, as we have noted, the positive errors in α are accompanied by negative errors in β, and vice versa. Thus, when we add α and β, the errors tend to cancel, leaving an uncertainty of only a fraction of a degree.

Quick Check 9.1. Each of three students measures the two sides, x and y, of a rectangle and obtains the results shown in Table 9.2. Find the means $\bar{x}$ and $\bar{y}$,

Table 9.2. Three measurements of x and y (in mm); for Quick Check 9.1.

Student	x	y
A	25	33
B	27	34
C	29	38

and then make a table like Table 9.1 to find the covariance σ_{xy}. If the students calculate the sum $q = x + y$, find the standard deviation σ_q using the correct formula (9.9), and compare it with the value you would get if you ignored the covariance and used (9.10). (Notice that in this example, high values of x seem to correlate with high values of y and vice versa. Specifically, student C appears consistently to overestimate and student A to underestimate. Remember also that with just three measurements, the results of any statistical calculation are only a rough guide to the uncertainties concerned.)

Using the formula (9.9), we can derive an upper limit on σ_q that is always valid. It is a simple algebraic exercise (Problem 9.7) to prove that the covariance σ_{xy} satisfies the so-called *Schwarz inequality*

$$|\sigma_{xy}| \leq \sigma_x \sigma_y. \tag{9.11}$$

If we substitute (9.11) into the expression (9.9) for the uncertainty σ_q, we find that

$$\sigma_q^2 \leq \left(\frac{\partial q}{\partial x}\right)^2 \sigma_x^2 + \left(\frac{\partial q}{\partial y}\right)^2 \sigma_y^2 + 2\left|\frac{\partial q}{\partial x}\frac{\partial q}{\partial y}\right| \sigma_x \sigma_y$$

$$= \left[\left| \frac{\partial q}{\partial x} \right| \sigma_x + \left| \frac{\partial q}{\partial y} \right| \sigma_y \right]^2 ;$$

that is,

$$\sigma_q \leqslant \left| \frac{\partial q}{\partial x} \right| \sigma_x + \left| \frac{\partial q}{\partial y} \right| \sigma_y . \tag{9.12}$$

With this result, we have finally established the precise significance of our original, simple expression

$$\delta q \approx \left| \frac{\partial q}{\partial x} \right| \delta x + \left| \frac{\partial q}{\partial y} \right| \delta y \tag{9.13}$$

for the uncertainty δq. If we adopt the standard deviation σ_q as our measure of the uncertainty in q, then (9.12) shows that the old expression (9.13) is really the *upper limit* on the uncertainty. Whether or not the errors in x and y are independent and normally distributed, the uncertainty in q will never exceed the right side of (9.13). If the measurements of x and y are correlated in just such a way that $|\sigma_{xy}| = \sigma_x \sigma_y$, its largest possible value according to (9.11), then the uncertainty in q can actually be as large as given by (9.13), but it can never be any larger.

In an introductory physics laboratory, students usually do not make measurements for which the covariance σ_{xy} can be estimated reliably. Thus, you will probably not have occasion to use the result (9.9) explicitly. If, however, you suspect that two variables x and y may be correlated, you should probably consider using the bound (9.12) instead of the quadratic sum (9.10). Our next topic is an application of covariance that you will almost certainly be able to use.

9.3   Coefficient of Linear Correlation

The notion of covariance σ_{xy} introduced in Section 9.2 enables us to answer the question raised in Chapter 8 of how well a set of measurements $(x_1, y_1), \ldots, (x_N, y_N)$ of two variables supports the hypothesis that x and y are linearly related.

Let us suppose we have measured N pairs of values $(x_1, y_1), \ldots, (x_N, y_N)$ of two variables that we suspect should satisfy a linear relation of the form

$$y = A + Bx.$$

Note that $x_1, \ldots, x_N$ are no longer measurements of one single number, as they were in the past two sections; rather, they are measurements of N different values of some variable (for example, N different heights from which we have dropped a stone). The same applies to $y_1, \ldots, y_N$.

Using the method of least squares, we can find the values of A and B for the line that best fits the points $(x_1, y_1), \ldots, (x_N, y_N)$. If we already have a reliable estimate of the uncertainties in the measurements, we can see whether the measured points do lie reasonably close to the line (compared with the known uncertainties). If they do, the measurements support our suspicion that x and y are linearly related.

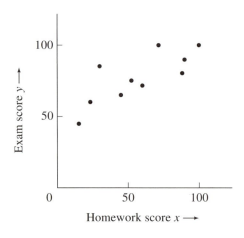

Figure 9.1. A "scatter plot" showing students' scores on exams and homework. Each of the 10 points (x_i, y_i) shows a student's homework score, x_i, and exam score, y_i.

Unfortunately, in many experiments, getting a reliable estimate of the uncertainties in advance is hard, and we must use the data themselves to decide whether the two variables appear to be linearly related. In particular, there is a type of experiment for which knowing the size of uncertainties in advance is *impossible*. This type of experiment, which is more common in the social than the physical sciences, is best explained by an example.

Suppose a professor, anxious to convince his students that doing homework will help them do well in exams, keeps records of their scores on homework and exams and plots the scores on a "scatter plot" as in Figure 9.1. In this figure, homework scores are plotted horizontally and exam scores vertically. Each point (x_i, y_i) shows one student's homework score, x_i, and exam score, y_i. The professor hopes to show that high exam scores tend to be *correlated* with high homework scores, and vice versa (and his scatter plot certainly suggests this is approximately so). This kind of experiment has no uncertainties in the points; each student's two scores are known exactly. The uncertainty lies rather in the extent to which the scores *are correlated;* and this has to be decided from the data.

The two variables x and y (in either a typical physics experiment or one like that just described) may, of course, be related by a more complicated relation than the simple linear one, $y = A + Bx$. For example, plenty of physical laws lead to quadratic relations of the form $y = A + Bx + Cx^2$. Nevertheless, I restrict my discussion here to the simpler problem of deciding whether a given set of points supports the hypothesis of a *linear* relation $y = A + Bx$.

The extent to which a set of points $(x_1, y_1), \ldots, (x_N, y_N)$ supports a linear relation between x and y is measured by the *linear correlation coefficient*, or just *correlation coefficient*,

$$r = \frac{\sigma_{xy}}{\sigma_x \sigma_y}, \qquad (9.14)$$

where the covariance σ_{xy} and standard deviations σ_x and σ_y are defined exactly as before, in Equations (9.8) and (9.4).[2] Substituting these definitions into (9.14), we can rewrite the correlation coefficient as

$$r = \frac{\sum (x_i - \bar{x})(y_i - \bar{y})}{\sqrt{\sum (x_i - \bar{x})^2 \sum (y_i - \bar{y})^2}}. \tag{9.15}$$

As I will show directly, the number r is an indicator of how well the points (x_i, y_i) fit a straight line. It is a number between -1 and 1. If r is close to ± 1, the points lie close to some straight line; if r is close to 0, the points are uncorrelated and have little or no tendency to lie on a straight line.

To prove these assertions, we first observe that the Schwarz inequality (9.11), $|\sigma_{xy}| \leq \sigma_x \sigma_y$, implies immediately that $|r| \leq 1$ or

$$-1 \leq r \leq 1$$

as claimed. Next, let us suppose that the points (x_i, y_i) all lie *exactly* on the line $y = A + Bx$. In this case $y_i = A + Bx_i$ for all i, and hence $\bar{y} = A + B\bar{x}$. Subtracting these two equations, we see that

$$y_i - \bar{y} = B(x_i - \bar{x})$$

for each i. Inserting this result into (9.15), we find that

$$r = \frac{B \sum (x_i - \bar{x})^2}{\sqrt{\sum (x_i - \bar{x})^2 B^2 \sum (x_i - \bar{x})^2}} = \frac{B}{|B|} = \pm 1. \tag{9.16}$$

That is, if the points $(x_1, y_1), \ldots, (x_N, y_N)$ lie perfectly on a line, then $r = \pm 1$, and its sign is determined by the slope of the line ($r = 1$ for B positive, and $r = -1$ for B negative).[3] Even when the variables x and y really are linearly related, we do not expect our experimental points to lie *exactly* on a line. Thus, we do not expect r to be exactly ± 1. On the other hand, we do expect a value of r that is *close to* ± 1, if we believe that x and y are linearly related.

Suppose, on the other hand, there is no relationship between the variables x and y. Whatever the value of y_i, each x_i would then be just as likely to be above $\bar{x}$ as below $\bar{x}$. Thus, the terms in the sum

$$\sum (x_i - \bar{x})(y_i - \bar{y})$$

in the numerator of r in (9.15) are just as likely to be positive as negative. Meanwhile, the terms in the denominator of r are all positive. Thus, in the limit that N, the number of measurements, approaches infinity, the correlation coefficient r will

---

[2] Notice, however, that their significance is slightly different. For example, in Section 9.2 $x_1, \ldots, x_N$ were measurements of *one number,* and if these measurements were precise, σ should be small. In the present case $x_1, \ldots, x_N$ are measurements of *different* values of a variable, and even if the measurements are precise, there is no reason to think σ_x will be small. Note also that some authors use the number r^2, called the *coefficient of determination.*

[3] If the line is exactly horizontal, then $B = 0$, and (9.16) gives $r = 0/0$; that is, r is undefined. Fortunately, this special case is not important in practice, because it corresponds to y being a constant, independent of x.

be zero. With a finite number of data points, we do not expect r to be exactly zero, but we do expect it to be *small* (if the two variables really are unrelated).

If two variables x and y are such that, in the limit of infinitely many measurements, their covariance σ_{xy} is zero (and hence $r = 0$), we say that the variables are *uncorrelated*. If, after a finite number of measurements, the correlation coefficient $r = \sigma_{xy}/\sigma_x\sigma_y$ is small, the hypothesis that x and y are uncorrelated is supported.

As an example, consider the exam and homework scores shown in Figure 9.1. These scores are given in Table 9.3. A simple calculation (Problem 9.12) shows that

Table 9.3. Students' scores.

Student i	1	2	3	4	5	6	7	8	9	10
Homework x_i	90	60	45	100	15	23	52	30	71	88
Exam y_i	90	71	65	100	45	60	75	85	100	80

the correlation coefficient for these 10 pairs of scores is $r = 0.8$. The professor concludes that this value is "reasonably close" to 1 and so can announce to next year's class that, because homework and exam scores show good correlation, it is important to do the homework.

If our professor had found a correlation coefficient r close to zero, he would have been in the embarrassing position of having shown that homework scores have no bearing on exam scores. If r had turned out to be close to -1, then he would have made the even more embarrassing discovery that homework and exam scores show a *negative* correlation; that is, that students who do a good job on homework tend to do poorly on the exam.

Quick Check 9.2.   Find the correlation coefficient for the data of Quick Check 9.1. Note that these measurements show a positive correlation; that is, high values of x correlate with high values of y, and vice versa.

9.4   Quantitative Significance of r

The example of the homework and exam scores clearly shows that we do not yet have a complete answer to our original question about how well data points support a linear relation between x and y. Our professor found a correlation coefficient $r = 0.8$, and judged this value "reasonably close" to 1. But how can we decide objectively what is "reasonably close" to 1? Would $r = 0.6$ have been reasonably close? Or $r = 0.4$? These questions are answered by the following argument.

Suppose the two variables x and y are in reality *uncorrelated;* that is, in the limit of infinitely many measurements, the correlation coefficient r would be zero.

After a finite number of measurements, r is very unlikely to be exactly zero. One can, in fact, calculate the probability that r will exceed any specific value. We will denote by

$$Prob_N\left(|r| \geq r_o\right)$$

the probability that N measurements of two uncorrelated variables x and y will give a coefficient r larger[4] than any particular r_o. For instance, we could calculate the probability

$$Prob_N\left(|r| \geq 0.8\right)$$

that, after N measurements of the uncorrelated variables x and y, the correlation coefficient would be at least as large as our professor's 0.8. The calculation of these probabilities is quite complicated and will not be given here. The results for a few representative values of the parameters are shown in Table 9.4, however, and a more complete tabulation is given in Appendix C.

Table 9.4. The probability $Prob_N(|r| \geq r_o)$ that N measurements of two uncorrelated variables x and y would produce a correlation coefficient with $|r| \geq r_o$. Values given are percentage probabilities, and blanks indicate values less than 0.05%.

N	0	0.1	0.2	0.3	0.4	r_o 0.5	0.6	0.7	0.8	0.9	1
3	100	94	87	81	74	67	59	51	41	29	0
6	100	85	70	56	43	31	21	12	6	1	0
10	100	78	58	40	25	14	7	2	0.5		0
20	100	67	40	20	8	2	0.5	0.1			0
50	100	49	16	3	0.4						0

Although we have not shown how the probabilities in Table 9.4 are calculated, we can understand their general behavior and put them to use. The first column shows the number of data points N. (In our example, the professor recorded 10 students' scores, so $N = 10$.) The numbers in each succeeding column show the percentage probability that N measurements of two *uncorrelated* variables would yield a coefficient r at least as big as the number at the top of the column. For example, we see that the probability that 10 uncorrelated data points would give $|r| \geq 0.8$ is only 0.5%, not a large probability. Our professor can therefore say it is *very unlikely* that uncorrelated scores would have produced a coefficient with $|r|$ greater than or equal to the 0.8 that he obtained. In other words, it is very *likely* that the scores on homework and examinations really are correlated.

Several features of Table 9.4 deserve comment. All entries in the first column are 100%, because $|r|$ is always greater than or equal to zero; thus, the probability

---

[4] Because a correlation is indicated if r is close to $+1$ *or* to -1, we consider the probability of getting the *absolute value* $|r| \geq r_o$.

of finding $|r| \geq 0$ is always 100%. Similarly, the entries in the last column are all zero, because the probability of finding $|r| \geq 1$ is zero.[5] The numbers in the intermediate columns vary with the number of data points N. This variation also is easily understood. If we make just three measurements, the chance of their having a correlation coefficient with $|r| \geq 0.5$, say, is obviously quite good (67%, in fact); but if we make 20 measurements and the two variables really are uncorrelated, the chance of finding $|r| \geq 0.5$ is obviously very small (2%).

Armed with the probabilities in Table 9.4 (or in the more complete table in Appendix C), we now have the most complete possible answer to the question of how well N pairs of values (x_i, y_i) support a linear relation between x and y. From the measured points, we can first calculate the observed correlation coefficient r_o (the subscript o stands for "observed"). Next, using one of these tables, we can find the probability $Prob_N(|r| \geq |r_o|)$ that N uncorrelated points would have given a coefficient at least as large as the observed coefficient r_o. If this probability is "sufficiently small," we conclude that it is very *improbable* that x and y are uncorrelated and hence very *probable* that they really are correlated.

We still have to choose the value of the probability we regard as "sufficiently small." One fairly common choice is to regard an observed correlation r_o as "significant" if the probability of obtaining a coefficient r with $|r| \geq |r_o|$ from uncorrelated variables is less than 5%. A correlation is sometimes called "highly significant" if the corresponding probability is less than 1%. Whatever choice we make, we do *not* get a definite answer that the data are, or are not, correlated; instead, we have a quantitative measure of how improbable it is that they are uncorrelated.

Quick Check 9.3. The professor of Section 9.3 teaches the same course the following year and this time has 20 students. Once again, he records homework and exam scores and this time finds a correlation coefficient $r = 0.6$. Would you describe this correlation as significant? Highly significant?

9.5   Examples

Suppose we measure three pairs of values (x_i, y_i) and find that they have a correlation coefficient of 0.7 (or -0.7). Does this value support the hypothesis that x and y are linearly related?

Referring to Table 9.4, we see that even if the variables x and y were completely uncorrelated, the probability is 51% for getting $|r| \geq 0.7$ when $N = 3$. In other words, it is entirely possible that x and y are uncorrelated, so we have no worthwhile evidence of correlation. In fact, with only three measurements, getting convincing evidence of a correlation would be very difficult. Even an observed coefficient as large as 0.9 is quite insufficient, because the probability is 29% for getting $|r| \geq 0.9$ from three measurements of uncorrelated variables.

---

[5]Although it is *impossible* that $|r| > 1$, it is, in principle, possible that $|r| = 1$. However, r is a continuous variable, and the probability of getting $|r|$ exactly equal to 1 is zero. Thus $Prob_N(|r| \geq 1) = 0$.

If we found a correlation of 0.7 from six measurements, the situation would be a little better but still not good enough. With $N = 6$, the probability of getting $|r| \geq 0.7$ from uncorrelated variables is 12%. This probability is not small enough to rule out the possibility that x and y are uncorrelated.

On the other hand, if we found $r = 0.7$ after 20 measurements, we would have strong evidence for a correlation, because when $N = 20$, the probability of getting $|r| \geq 0.7$ from two uncorrelated variables is only 0.1%. By any standards this is very improbable, and we could confidently argue that a correlation is indicated. In particular, the correlation could be called "highly significant," because the probability concerned is less than 1%.

Principal Definitions and Equations of Chapter 9

COVARIANCE

Given N pairs of measurements $(x_1, y_1), \ldots, (x_N, y_N)$ of two quantities x and y, we define their covariance to be

$$\sigma_{xy} = \frac{1}{N} \Sigma (x_i - \bar{x})(y_i - \bar{y}).$$

[See (9.8)]

If we now use the measured values to calculate a function $q(x, y)$, the standard deviation of q is given by

$$\sigma_q^2 = \left(\frac{\partial q}{\partial x}\right)^2 \sigma_x^2 + \left(\frac{\partial q}{\partial y}\right)^2 \sigma_y^2 + 2 \frac{\partial q}{\partial x} \frac{\partial q}{\partial y} \sigma_{xy}.$$

[See (9.9)]

If the errors in x and y are independent, then $\sigma_{xy} = 0$, and this equation reduces to the usual formula for error propagation. Whether or not the errors are independent, the Schwarz inequality (9.11) implies the upper bound

$$\sigma_q \leq \left|\frac{\partial q}{\partial x}\right| \sigma_x + \left|\frac{\partial q}{\partial y}\right| \sigma_y.$$

[See (9.12)]

CORRELATION COEFFICIENT

Given N measurements $(x_1, y_1), \ldots, (x_N, y_N)$ of two variables x and y, we define the correlation coefficient r as

$$r = \frac{\sigma_{xy}}{\sigma_x \sigma_y} = \frac{\Sigma (x_i - \bar{x})(y_i - \bar{y})}{\sqrt{\Sigma (x_i - \bar{x})^2 \Sigma (y_i - \bar{y})^2}}.$$

[See (9.15)]

An equivalent form, which is sometimes more convenient, is

$$r = \frac{\Sigma x_i y_i - N\bar{x}\bar{y}}{\sqrt{(\Sigma x_i^2 - N\bar{x}^2)(\Sigma y_i^2 - N\bar{y}^2)}}.$$

[See Problem 9.10]

Values of r near 1 or -1 indicate strong linear correlation; values near 0 indicate little or no correlation. The probability $Prob_N(|r| > r_o)$ that N measurements of two *uncorrelated* variables would give a value of r larger than any observed value r_o is tabulated in Appendix C. The smaller this probability, the better the evidence that the variables x and y really are correlated. If the probability is less than 5%, we say the correlation is *significant*; if it is less than 1%, we say the correlation is *highly significant*.

Problems for Chapter 9

For Section 9.2: Covariance in Error Propagation

9.1. ★   Calculate the covariance for the following four measurements of two quantities x and y.

$$x: \quad 20 \quad 23 \quad 23 \quad 22$$
$$y: \quad 30 \quad 32 \quad 35 \quad 31$$

9.2. ★   Each of five students measures the two times (t and T) for a stone to fall from the third and sixth floors of a tall building. Their results are shown in Table 9.5. Calculate the two averages $\bar{t}$ and $\bar{T}$, and find the covariance σ_{tT} using the layout of Table 9.1.

Table 9.5. Five measurements of two times, t and T (in tenths of a second); for Problem 9.2.

Student	t	T
A	14	20
B	12	18
C	13	18
D	15	22
E	16	22

[As you examine the data, note that students B and C get lower-than-average answers for both times, whereas D's and E's answers are both higher than average. Although this difference could be just a chance fluctuation, it suggests B and C may have a systematic tendency to underestimate their times and D and E to overestimate. (For instance, B and C could tend to anticipate the landing, whereas D and E could tend to anticipate the launch.) Under these conditions, we would *expect* to get a correlation of the type observed.]

9.3 ★★   **(a)** For the data of Problem 9.1, calculate the variances σ_x^2 and σ_y^2 and the covariance σ_{xy}. **(b)** If you now decide to calculate the sum $q = x + y$, what will be its standard deviation according to (9.9)? **(c)** What would you have found for the standard deviation if you had ignored the covariance and used Equation (9.10)?

(d) In a simple situation like this, an easier way to find the standard deviation of q is just to calculate four values of q [one for each pair (x, y)] and then find σ_q from these four values. Show that this procedure gives the same answer as you got in part (b).

9.4. ★★ **(a)** For the data of Problem 9.2, calculate the variances σ_t^2 and σ_T^2 and the covariance σ_{tT}. **(b)** If the students decide to calculate the difference $T - t$, what will be its standard deviation according to (9.9)? **(c)** What would they have found for the standard deviation if they had ignored the covariance and used Equation (9.10)? **(d)** In a simple situation like this, an easier way to find the standard deviation of $T - t$ is just to calculate five values of $T - t$ [one for each pair (t, T)] and then find the standard deviation of these five values. Show that this procedure gives the same answer as you got in part (b).

9.5. ★★ Imagine a series of N measurements of two fixed lengths x and y that were made to find the value of some function $q(x, y)$. Suppose each pair is measured with a different tape; that is, the pair (x_1, y_1) is measured with one tape, (x_2, y_2) is measured with a second tape, and so on. **(a)** Assuming the main source of errors is that some of the tapes have shrunk and some stretched (uniformly, in either case), show that the covariance σ_{xy} is bound to be positive. **(b)** Show further, under the same conditions, that $\sigma_{xy} = \sigma_x \sigma_y$; that is, σ_{xy} is as large as permitted by the Schwarz inequality (9.11).

[Hint: Assume that the ith tape has shrunk by a factor λ_i, that is, present length = (design length)/λ_i, so that a length that is really X will be measured as $x_i = \lambda_i X$. The moral of this problem is that there are situations in which the covariance is certainly not negligible.]

9.6. ★★ Here is an example of an experiment in which we would expect a negative correlation between two measured quantities (high values of one correlated with low values of the other). Figure 9.2 represents a photograph taken in a bubble chamber, where charged subatomic particles leave clearly visible tracks. A positive particle called the K^+ has entered the chamber at the bottom of the picture. At point A, it has collided with an invisible neutron (n) and has undergone the reaction

$$K^+ + n \rightarrow K^o + p.$$

The proton's track (p) is clearly visible, going off to the right, but the path of the K^o (shown dotted) is really invisible because the K^o is uncharged. At point B, the K^o decays into two charged pions,

$$K^o \rightarrow \pi^+ + \pi^-,$$

whose tracks are again clearly visible. To investigate the conservation of momentum in the second process, the experimenter needs to measure the angles α and β between the paths of the two pions and the invisible path of the K^o, and this measurement requires drawing in the dotted line that joins A and B. The main source of error in finding α and β is in deciding on the direction of this line, because A and B are often close together (less than a cm), and the tracks that define A and B are rather wide. For the purpose of this problem, let us suppose that this is the *only* source of error.

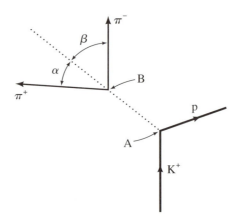

Figure 9.2. Tracks of charged particles in a bubble chamber. The dotted line shows the direction of an invisible K°, which was formed at A and decayed at B; for Problem 9.6.

Suppose several students are given copies of Figure 9.2, and each draws in his best estimate for the line AB and then measures the two angles α and β. The students then combine their results to find the means $\bar\alpha$ and $\bar\beta$, the standard deviations σ_α and σ_β, and the covariance $\sigma_{\alpha\beta}$. Assuming that the only source of error is in deciding the direction of the line AB, explain why an overestimate of α is inevitably accompanied by an underestimate of β. Prove that $\sigma_\alpha = \sigma_\beta$ and that the covariance $\sigma_{\alpha\beta}$ is negative and equal to the largest value allowed by the Schwartz inequality, $\sigma_{\alpha\beta} = -\sigma_\alpha\sigma_\beta$.

(Hint: Suppose that the ith student draws his line AB to the right of the true direction by an amount Δ_i. Then his value for α will be $\alpha_i = \alpha_{\text{true}} + \Delta_i$. Write the corresponding expression for his value β_i and compute the various quantities of interest in terms of the Δ_i and $\bar\Delta$.)

9.7. ★★ Prove that the covariance σ_{xy} defined in (9.8) satisfies the Schwarz inequality (9.11),

$$|\sigma_{xy}| \leqslant \sigma_x\sigma_y. \qquad (9.17)$$

[Hint: Let t be an arbitrary number and consider the function

$$A(t) = \frac{1}{N}\sum[(x_i - \bar x) + t(y_i - \bar y)]^2 \geqslant 0. \qquad (9.18)$$

Because $A(t) \geqslant 0$ whatever the value of t, even its minimum $A_{\min} \geqslant 0$. Find the minimum $A_{\min}$, and set $A_{\min} \geqslant 0$.]

For Section 9.3: Coefficient of Linear Correlation

9.8. ★ Calculate the correlation coefficient r for the following five pairs of measurements:

$$x = 1\ \ 2\ \ 3\ \ 4\ \ 5$$
$$y = 8\ \ 8\ \ 5\ \ 6\ \ 3$$

Do the calculations yourself, but if your calculator has a built-in function to compute r, make sure you know how it works, and use it to check your value.

9.9. ★   Calculate the correlation coefficient r for the following six pairs of measurements:

$$x = 1 \quad 2 \quad 3 \quad 5 \quad 6 \quad 7$$
$$y = 5 \quad 6 \quad 6 \quad 8 \quad 8 \quad 9$$

Do the calculations yourself, but if your calculator has a built-in function to compute r, make sure you know how it works, and use it to check your value.

9.10. ★★   **(a)** Prove the identity

$$\sum (x_i - \bar{x})(y_i - \bar{y}) = \sum x_i y_i - N\bar{x}\bar{y}.$$

(b) Hence, prove the correlation coefficient r defined in (9.15) can be written as

$$r = \frac{\sum x_i y_i - N\bar{x}\bar{y}}{\sqrt{(\sum x_i^2 - N\bar{x}^2)(\sum y_i^2 - N\bar{y}^2)}}. \tag{9.19}$$

Many calculators use this result to find r because it avoids the need to store all the data before calculating the means and deviations.

For Section 9.4: Quantitative Significance of r

9.11. ★   In the photoelectric effect, the kinetic energy K of electrons ejected from a metal by light is supposed to be a linear function of the light's frequency f,

$$K = hf - \phi, \tag{9.20}$$

where h and ϕ are constants. To check this linearity, a student measures K for N different values of f and calculates the correlation coefficient r for her results. **(a)** If she makes five measurements ($N = 5$) and finds $r = 0.7$, does she have significant support for the linear relation (9.20)? **(b)** What if $N = 20$ and $r = 0.5$?

9.12. ★   **(a)**   Check that the correlation coefficient r for the 10 pairs of test scores in Table 9.3 is approximately 0.8. (By all means, use the built-in function on your calculator, if it has one.) **(b)** Using the table of probabilities in Appendix C, find the probability that 10 *uncorrelated* scores would have given $|r| \geq 0.8$. Is the correlation of the test scores significant? Highly significant?

9.13. ★   A psychologist, investigating the relation between the intelligence of fathers and sons, measures the Intelligence Quotients of 10 fathers and sons and obtains the following results:

Father:	74	83	85	96	98	100	106	107	120	124
Son:	76	103	99	109	111	107	91	101	120	119

Do these data support a correlation between the intelligence of fathers and sons?

9.14. ★   Eight aspiring football players are timed in the 100-meter dash and the 1,500-meter run. Their times (in seconds) are as follows:

Dash:	12	11	13	14	12	15	12	16
Run:	280	290	220	260	270	240	250	230

Calculate the correlation coefficient r. What kind of correlation does your result suggest? Is there, in fact, significant evidence for a correlation?

9.15. ★★   Draw a scatter plot for the six data pairs of Problem 9.9 and the least-squares line that best fits these points. Find their correlation coefficient r. Based on the probabilities listed in Appendix C, would you say these data show a significant linear correlation? Highly significant?

9.16. ★★   **(a)** Draw a scatter plot for the five data pairs of Problem 9.8 and the least-squares line that best fits these points. Find their correlation coefficient r. Based on the probabilities listed in Appendix C, would you say these data show a significant linear correlation? Highly significant? **(b)** Repeat for the following data:

$$x = 1 \quad 2 \quad 3 \quad 4 \quad 5$$
$$y = 4 \quad 6 \quad 3 \quad 0 \quad 2$$

Chapter 10

The Binomial Distribution

The Gauss, or normal, distribution is the only example of a distribution we have studied so far. We will now discuss two other important examples, the binomial distribution (in this chapter) and the Poisson distribution (in Chapter 11).

10.1 Distributions

Chapter 5 introduced the idea of a *distribution,* the function that describes the proportion of times a repeated measurement yields each of its various possible answers. For example, we could make N measurements of the period T of a pendulum and find the distribution of our various measured values of T, or we could measure the heights h of N Americans and find the distribution of the various measured heights h.

I next introduced the notion of the *limiting distribution,* the distribution that would be obtained in the limit that the number of measurements N becomes very large. The limiting distribution can be viewed as telling us the *probability* that one measurement will yield any of the possible values: the probability that one measurement of the period will yield any particular value T; the probability that one American (chosen at random) will have any particular height h. For this reason, the limiting distribution is also sometimes called the *probability distribution.*

Of the many possible limiting distributions, the only one we have discussed is the Gauss, or normal, distribution, which describes the distribution of answers for any measurement subject to many sources of error that are all random and small. As such, the Gauss distribution is the most important of all limiting distributions for the physical scientist and amply deserves the prominence given it here. Nevertheless, several other distributions have great theoretical or practical importance, and this and the next chapter present two examples.

This chapter describes the binomial distribution, which is not of great practical importance to the experimental physicist. Its simplicity, however, makes it an excellent introduction to many properties of distributions, and it is theoretically important, because we can derive the all-important Gauss distribution from it.

10.2 Probabilities in Dice Throwing

The binomial distribution can best be described by an example. Suppose we under-take as our "experiment" to throw three dice and record the number of aces show-ing. The possible results of the experiment are the answers 0, 1, 2, or 3 aces. If we repeat the experiment an enormous number of times, we will find the limiting distri-bution, which will tell us the probability that in any one throw (of all three dice) we get ν aces, where $\nu = $ 0, 1, 2, or 3.

This experiment is sufficiently simple that we can easily calculate the probabil-ity of the four possible outcomes. We observe first that, assuming the dice are true, the probability of getting an ace when throwing *one* die is $\frac{1}{6}$. Let us now throw all three dice and ask first for the probability of getting three aces ($\nu = 3$). Because each separate die has probability $\frac{1}{6}$ of showing an ace, and because the three dice roll independently, the probability for three aces is

$$Prob(\text{3 aces in 3 throws}) = \left(\frac{1}{6}\right)^3 \approx 0.5\%.$$

Calculating the probability for two aces ($\nu = 2$) is a little harder because we can throw two aces in several ways. The first and second dice could show aces and the third not ($A, A, \text{not } A$), or the first and third could show aces and the second not ($A, \text{not } A, A$), and so on. Here, we argue in two steps. First, we consider the proba-bility of throwing two aces in any definite order, such as ($A, A, \text{not } A$). The probabil-ity that the first die will show an ace is $\frac{1}{6}$, and likewise for the second. On the other hand, the probability that the last die will *not* show an ace is $\frac{5}{6}$. Thus, the probability for two aces in this particular order is

$$Prob(A, A, \text{not } A) = \left(\frac{1}{6}\right)^2 \times \left(\frac{5}{6}\right).$$

The probability for two aces in any other definite order is the same. Finally, there are three different orders in which we could get our two aces: ($A, A, \text{not } A$), or ($A, \text{not } A, A$), or ($\text{not } A, A, A$). Thus, the total probability for getting two aces (in any order) is

$$Prob(\text{2 aces in 3 throws}) = 3 \times \left(\frac{1}{6}\right)^2 \times \left(\frac{5}{6}\right) \approx 6.9\%. \tag{10.1}$$

Similar calculations give the probabilities for one ace in three throws (34.7%) and for no aces in three throws (57.9%). Our numerical conclusions can be summa-rized by drawing the probability distribution for the number of aces obtained when throwing three dice, as in Figure 10.1. This distribution is an example of the bino-mial distribution, the general form of which we now describe.

10.3 Definition of the Binomial Distribution

To describe the general binomial distribution, I need to introduce some terminology. First, imagine making n independent *trials,* such as throwing n dice, tossing n coins, or testing n firecrackers. Each trial can have various outcomes: A die can show any

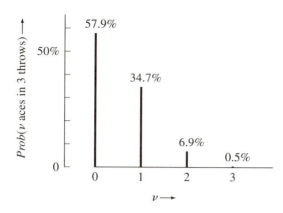

Figure 10.1. Probability of getting v aces when throwing three dice. This function is the binomial distribution $B_{n,p}(v)$, with $n = 3$ and $p = 1/6$.

face from 1 to 6, a coin can show heads or tails, a firecracker can explode or fizzle. We refer to the outcome in which we happen to be interested as a *success*. Thus "success" could be throwing an ace on a die, or a head on a coin, or having a firecracker explode. We denote by p the probability of success in any one trial, and by $q = 1 - p$ that of "failure" (that is, of getting any outcome other than the one of interest). Thus, $p = \frac{1}{6}$ for getting an ace on a die, $p = \frac{1}{2}$ for heads on a coin, and p might be 95% for a given brand of firecracker to explode properly.

Armed with these definitions, we can now ask for the probability of getting v successes in n trials. A calculation I will sketch in a moment shows that this probability is given by the so-called *binomial distribution:*

$$Prob(v \text{ successes in } n \text{ trials}) = B_{n,p}(v)$$

$$= \frac{n(n-1)\cdots(n-v+1)}{1 \times 2 \times \cdots \times v} p^v q^{n-v}. \quad (10.2)$$

Here the letter B stands for "binomial"; the subscripts n and p on $B_{n,p}(v)$ indicate that the distribution depends on n, the number of trials made, and p, the probability of success in one trial.

The distribution (10.2) is called the binomial distribution because of its close connection with the well-known binomial expansion. Specifically, the fraction in (10.2) is the *binomial coefficient,* often denoted

$$\binom{n}{v} = \frac{n(n-1)\cdots(n-v+1)}{1 \times 2 \times \cdots \times v} \quad (10.3)$$

$$= \frac{n!}{v!(n-v)!}, \quad (10.4)$$

where we have introduced the useful *factorial* notation,

$$n! = 1 \times 2 \times \cdots \times n.$$

[By convention, $0! = 1$, and so $\binom{n}{0} = 1$.] The binomial coefficient appears in the binomial expansion

$$(p + q)^n = p^n + np^{n-1}q + \cdots + q^n$$

$$= \sum_{v=0}^{n} \binom{n}{v} p^v q^{n-v}, \tag{10.5}$$

which holds for any two numbers p and q and any positive integer n (see Problems 10.5 and 10.6).

With the notation (10.3), we can rewrite the binomial distribution in the more compact form

---

The Binomial Distribution

$$Prob(v \text{ successes in } n \text{ trials}) = B_{n,p}(v)$$

$$= \binom{n}{v} p^v q^{n-v}, \tag{10.6}$$

---

where, as usual, p denotes the probability of success in one trial and $q = 1 - p$.

The derivation of the result (10.6) is similar to that of the example of the dice in (10.1),

$$Prob(2 \text{ aces in } 3 \text{ throws}) = 3 \times \left(\frac{1}{6}\right)^2 \times \left(\frac{5}{6}\right). \tag{10.7}$$

In fact, if we set $v = 2$, $n = 3$, $p = \frac{1}{6}$, and $q = \frac{5}{6}$ in (10.6), we obtain precisely (10.7), as you should check. Furthermore, the significance of each factor in (10.6) is the same as that of the corresponding factor in (10.7). The factor p^v is the probability of getting all successes in any definite v trials, and q^{n-v} is the probability of failure in the remaining $n - v$ trials. The binomial coefficient $\binom{n}{v}$ is easily shown to be the number of different orders in which there can be v successes in n trials. This establishes that the binomial distribution (10.6) is indeed the probability claimed.

Example: Tossing Four Coins

Suppose we toss four coins ($n = 4$) and count the number of heads obtained, v. What is the probability of obtaining the various possible values $v = 0, 1, 2, 3, 4$?

Because the probability of getting a head on one toss is $p = \frac{1}{2}$, the required probability is simply the binomial distribution $B_{n,p}(v)$, with $n = 4$ and $p = q = \frac{1}{2}$,

$$Prob(v \text{ heads in } 4 \text{ tosses}) = \binom{4}{v}\left(\frac{1}{2}\right)^4.$$

These probabilities are easily evaluated. For example,

$$Prob(0 \text{ heads in } 4 \text{ tosses}) = 1 \times \left(\frac{1}{2}\right)^4 = 0.0625.$$

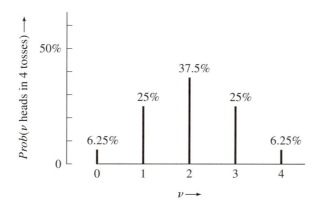

Figure 10.2. The binomial distribution $B_{n,p}(\nu)$ with $n = 4$, $p = \frac{1}{2}$. This gives the probability of getting ν heads when throwing four coins.

All five probabilities are shown in Figure 10.2.

We see that the most probable number of heads is $\nu = 2$, as we would expect. Here, the probabilities are symmetric about this most probable value. That is, the probability for three heads is the same as that for one, and the probability for four heads is the same as that for none. As we will see, this symmetry occurs only when $p = \frac{1}{2}$.

Quick Check 10.1. If you draw one card at random from a full deck of 52 playing cards, the probability of drawing any particular suit is $\frac{1}{4}$. If you draw three times, replacing your card after each draw, what is the probability of drawing three hearts? Of drawing exactly two hearts? Of drawing two or more hearts?

10.4   Properties of the Binomial Distribution

The binomial distribution $B_{n,p}(\nu)$ gives the probability of having ν "successes" in n trials, when p is the probability of success in a single trial. If we repeat our whole experiment, consisting of n trials, many times, then it is natural to ask what our average number of successes $\bar{\nu}$ would be. To find this average, we just sum over all possible values of ν, each multiplied by its probability. That is,

$$\bar{\nu} \;=\; \sum_{\nu=0}^{n} \nu B_{n,p}(\nu) \tag{10.8}$$

and is easily evaluated (Problem 10.14) as

$$\bar{\nu} \;=\; np. \tag{10.9}$$

That is, if we repeat our series of n trials many times, the average number of successes will be just the probability of success in one trial (p) times n, as you would expect. We can similarly calculate the standard deviation σ_ν in our number of successes (Problem 10.15). The result is

$$\sigma_\nu = \sqrt{np(1-p)}. \tag{10.10}$$

When $p = \frac{1}{2}$ (as in a coin-tossing experiment), the average number of successes is just $n/2$. Furthermore, it is easy to prove for $p = \frac{1}{2}$ that

$$B_{n,1/2}(\nu) = B_{n,1/2}(n - \nu) \tag{10.11}$$

(see Problem 10.13). That is, the binomial distribution with $p = \frac{1}{2}$ is symmetric about the average value $n/2$, as we noticed in Figure 10.2.

In general, when $p \neq \frac{1}{2}$, the binomial distribution $B_{n,p}(\nu)$ is not symmetric. For example, Figure 10.1 is clearly not symmetric; the most probable number of successes is $\nu = 0$, and the probability diminishes steadily for $\nu = 1, 2$, and 3. Also, the average number of successes ($\bar{\nu} = 0.5$) here is not the same as the most probable number of successes ($\nu = 0$).

It is interesting to compare the binomial distribution $B_{n,p}(\nu)$ with the more familiar Gauss distribution $G_{X,\sigma}(x)$. Perhaps the biggest difference is that the experiment described by the binomial distribution has outcomes given by the *discrete*[1] values $\nu = 0, 1, 2, \ldots, n$, whereas those of the Gauss distribution are given by the *continuous* values of the measured quantity x. The Gauss distribution is a symmetric peak centered on the average value $x = X$, which means that the average value X is also the most probable value [that for which $G_{X,\sigma}(x)$ is maximum]. As we have seen, the binomial distribution is symmetric only when $p = \frac{1}{2}$, and in general the average value does not coincide with the most probable value.

GAUSSIAN APPROXIMATION TO THE BINOMIAL DISTRIBUTION

For all their differences, the binomial and Gauss distributions have an important connection. If we consider the binomial distribution $B_{n,p}(\nu)$ for any fixed value of p, then when n is large $B_{n,p}(\nu)$ is closely approximated by the Gauss distribution $G_{X,\sigma}(\nu)$ with the same mean and same standard deviation; that is,

$$B_{n,p}(\nu) \approx G_{X,\sigma}(\nu) \qquad (n \text{ large}) \tag{10.12}$$

with

$$X = np \text{ and } \sigma = \sqrt{np(1-p)}. \tag{10.13}$$

We refer to (10.12) as the Gaussian approximation to the binomial distribution.

---

[1]The word *discrete* (not to be confused with discreet) means "detached from one another" and is the opposite of continuous.

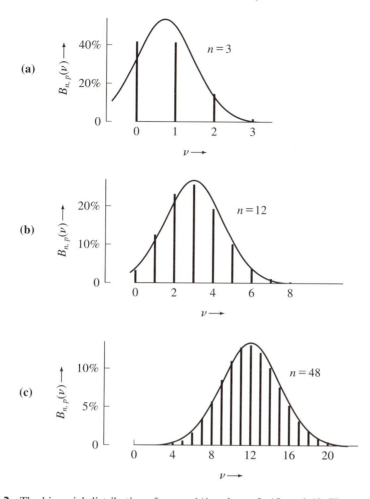

Figure 10.3. The binomial distributions for $p = 1/4$ and $n = 3$, 12, and 48. The continuous curve superimposed on each picture is the Gauss function with the same mean and same standard deviation.

I will not prove the result here,[2] but its truth is clearly illustrated in Figure 10.3, which shows the binomial distribution for $p = \frac{1}{4}$ and for three successively larger values of n ($n = 3$, 12, 48). Superimposed on each binomial distribution is the Gaussian distribution with the same mean and standard deviation. With just three trials ($n = 3$), the binomial distribution is quite different from the corresponding Gaussian. In particular, the binomial distribution is distinctly asymmetric, whereas the Gaussian is, of course, perfectly symmetric about its mean. By the time $n = 12$, the asymmetry of the binomial distribution is much less pronounced, and the two distributions are quite close to one another. When $n = 48$, the difference between the binomial and the corresponding Gauss distribution is so slight that the two are almost indistinguishable on the scale of Figure 10.3(c).

---

[2] For proofs, see S. L. Meyer, *Data Analysis for Scientists and Engineers* (John Wiley, 1975), p. 226, or H. D. Young, *Statistical Treatment of Experimental Data* (McGraw-Hill, 1962), Appendix C.

That the binomial distribution can be approximated by the Gauss function when n is large is very useful in practice. Calculation of the binomial function with n greater than 20 or so is extremely tedious, whereas calculation of the Gauss function is always simple whatever the values of X and σ. To illustrate this, suppose we wanted to know the probability of getting 23 heads in 36 tosses of a coin. This probability is given by the binomial distribution $B_{36,1/2}(\nu)$ because the probability of a head in one toss is $p = \frac{1}{2}$. Thus

$$Prob(23 \text{ heads in } 36 \text{ tosses}) = B_{36,\,1/2}(23) \qquad (10.14)$$

$$= \frac{36!}{23!\,13!} \left(\frac{1}{2}\right)^{36}, \qquad (10.15)$$

which a fairly tedious calculation[3] shows to be

$$Prob(23 \text{ heads}) = 3.36\%.$$

On the other hand, because the mean of the distribution is $np = 18$ and the standard deviation is $\sigma = \sqrt{np(1 - p)} = 3$, we can approximate (10.14) by the Gauss function $G_{18,3}(23)$, and a simple calculation gives

$$Prob(23 \text{ heads}) \approx G_{18,3}(23) = 3.32\%.$$

For almost all purposes, this approximation is excellent.

The usefulness of the Gaussian approximation is even more obvious if we want the probability of several outcomes. For example, the probability of getting 23 *or more* heads in 36 tosses is

$$Prob(23 \text{ or more heads}) = Prob(23 \text{ heads}) + Prob(24 \text{ heads}) + \cdots$$
$$+ Prob\ (36 \text{ heads}),$$

a tedious sum to calculate directly. If we approximate the binomial distribution by the Gaussian, however, then the probability is easily found. Because the calculation of Gaussian probabilities treats ν as a continuous variable, the probability for $\nu = 23, 24, \ldots$ is best calculated as $Prob_{Gauss}(\nu \geq 22.5)$, the probability for any $\nu \geq 22.5$. Now, $\nu = 22.5$ is 1.5 standard deviations above the mean value, 18. (Remember, $\sigma = 3$, so $4.5 = 1.5\sigma$.) The probability of a result more than 1.5σ above the mean equals the area under the Gauss function shown in Figure 10.4. It is easily calculated with the help of the table in Appendix B, and we find

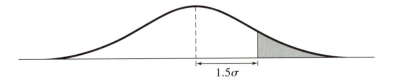

Figure 10.4. The probability of a result more than 1.5σ above the mean is the shaded area under the Gauss curve.

---

[3] Some hand calculators are preprogrammed to compute $n!$ automatically, and with such a calculator computation of (10.15) is easy. In most such calculators, however, $n!$ overflows when $n \geq 70$, so for $n \geq 70$ this preprogrammed function is no help.

$$Prob(23 \text{ or more heads}) \approx Prob_{\text{Gauss}}(\nu \geqslant X + 1.5\sigma) = 6.7\%.$$

This value compares well with the exact result (to two significant figures) 6.6%.

10.5   The Gauss Distribution for Random Errors

In Chapter 5, I claimed that a measurement subject to many small random errors will be distributed normally. We are now in a position to prove this claim, using a simple model for the kind of measurement concerned.

Let us suppose we measure a quantity x whose true value is X. We assume our measurements are subject to negligible systematic error but there are n independent sources of random error (effects of parallax, reaction times, and so on). To simplify our discussion, suppose further that all these sources produce random errors of the same fixed size ε. That is, each source of error pushes our result upward or downward by ε, and these two possibilities occur with equal probability, $p = \frac{1}{2}$. For exam-

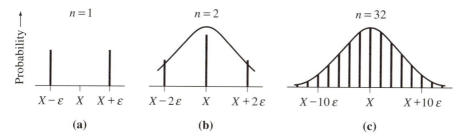

Figure 10.5. Distribution of measurements subject to n random errors of magnitude ε, for $n = 1$, 2, and 32. The continuous curves superimposed on (b) and (c) are Gaussians with the same center and width. (The vertical scales differ in the three graphs.)

ple, if the true value is X and there is just one source of error, our possible answers are $x = X - \varepsilon$ and $x = X + \varepsilon$, both equally likely. If there are two sources of error, a measurement could yield $x = X - 2\varepsilon$ (if both errors happened to be negative), or $x = X$ (if one was negative and one positive), or $x = X + 2\varepsilon$ (if both happened to be positive). These possibilities are shown in Figures 10.5(a) and (b).

In general, if there are n sources of error, our answer could range from $x = X - n\varepsilon$ to $x = X + n\varepsilon$. In a given measurement, if ν sources happen to give positive errors and $(n - \nu)$ negative errors, our answer will be

$$\begin{aligned} x &= X + \nu\varepsilon - (n - \nu)\varepsilon \\ &= X + (2\nu - n)\varepsilon. \end{aligned} \tag{10.16}$$

The probability of this result occurring is just the binomial probability

$$Prob(\nu \text{ positive errors}) = B_{n,1/2}(\nu). \tag{10.17}$$

Thus, the possible results of our measurement are symmetrically distributed around the true value X, and the probabilities are given by the binomial function (10.17).

This distribution is illustrated in Figure 10.5 for $n = 1$, 2, and 32.

I now claim that if the number of sources of error, n, is large and the size of the individual errors, ε, is small, then our measurements are normally distributed. To be more precise, we note that the standard deviation of the binomial distribution is $\sigma_\nu = \sqrt{np(1-p)} = \sqrt{n}/4$. Therefore, according to (10.16), the standard deviation of our measurements of x is $\sigma_x = 2\varepsilon\sigma_\nu = \varepsilon\sqrt{n}$. Accordingly, we let $n \to \infty$ and $\varepsilon \to 0$ in such a way that $\sigma_x = \varepsilon\sqrt{n}$ remains fixed. Two things happen as a result. First, as discussed in the previous section, the binomial distribution approaches the Gauss distribution with center X and width σ_x. This approach is clearly visible in Figures 10.5(b) and (c), on which the appropriate Gauss functions have been superimposed. Second, as $\varepsilon \to 0$, the possible results of our measurement move closer together (as is also clear in Figure 10.5), so that the discrete distribution approaches a continuous distribution that is precisely the expected Gauss distribution.

10.6   Applications; Testing of Hypotheses

Once we know how the results of an experiment should be distributed, we can ask whether the actual results of the experiment *were* distributed as expected. This kind of test of a distribution is an important technique in the physical sciences and perhaps even more so in the biological and social sciences. One important, general test, the chi-squared test, is the subject of Chapter 12. Here, I give two examples of a simpler test that can be applied to certain problems involving the binomial distribution.

TESTING A NEW SKI WAX

Suppose a manufacturer of ski waxes claims to have developed a new wax that greatly reduces the friction between skis and snow. To test this claim, we might take 10 pairs of skis and treat one ski from each pair with the wax. We could then hold races between the treated and untreated members of each pair by letting them slide down a suitable snow-covered incline.

If the treated skis won all 10 races, we would obviously have strong evidence that the wax works. Unfortunately, we seldom get such a clear-cut result, and even when we do, we would like some quantitative measure of the strength of the evidence. Thus, we have to address two questions. First, how can we quantify the evidence that the wax works (or doesn't work)? Second, where would we draw the line? If the treated skis won nine of the races, would this result be conclusive? What if they won eight races? Or seven?

Precisely these same questions arise in a host of similar statistical tests. If we wanted to test the efficacy of a fertilizer, we would organize "races" between treated and untreated plants. To predict which candidate is going to win an election, we would choose a random sample of voters and hold "races" between the candidates among the members of our sample.

To answer our questions, we need to decide more precisely what we should expect from our tests. In the accepted terminology, we must formulate a *statistical hypothesis.* In the example of the ski wax, the simplest hypothesis is the *null hypothesis,* that the new wax actually makes no difference. Subject to this hypothesis, we can calculate the probability of the various possible results of our test and then judge the significance of our particular result.

Suppose we take as our hypothesis that the ski wax makes no difference. In any one race, the treated and untreated skis would then be equally likely to win; that is, the probability for a treated ski to win is $p = \frac{1}{2}$. The probability that the treated skis will win ν of the 10 races is then the binomial probability:

$$Prob(\nu \text{ wins in 10 races}) = B_{10,1/2}(\nu)$$

$$= \frac{10!}{\nu!(10 - \nu)!} \left(\frac{1}{2}\right)^{10}. \qquad (10.18)$$

According to (10.18), the probability that the treated skis would win all 10 races is

$$Prob(10 \text{ wins in 10 races}) = (\tfrac{1}{2})^{10} \approx 0.1\%. \qquad (10.19)$$

That is, if our null hypothesis is correct, the treated skis would be *very* unlikely to win all 10 races. Conversely, if the treated skis *did* win all 10 races, the null hypothesis is very unlikely to be correct. In fact, the probability (10.19) is so small, we could say the evidence in favor of the wax is "highly significant," as we will discuss shortly.

Suppose instead that the treated skis had won eight of the 10 races. Here, we would calculate the probability of eight *or more* wins:

$$Prob(8 \text{ or more wins in 10 races})$$
$$= Prob(8 \text{ wins}) + Prob(9 \text{ wins}) + Prob(10 \text{ wins}) \approx 5.5\%. \qquad (10.20)$$

For the treated skis to win eight or more races is still quite unlikely but not nearly as unlikely as their winning all 10.

To decide what conclusion to draw from the eight wins, we must recognize that there are really just two alternatives. Either

(a) our null hypothesis is correct (the wax makes no difference), but by chance, an unlikely event has occurred (the treated skis have won eight races)

or

(b) our null hypothesis is false, and the wax does help.

In statistical testing, by tradition we pick some definite probability (5%, for example) to define the boundary below which an event is considered unacceptably improbable. If the probability of the actual outcome (eight or more wins, in our case) is below this boundary, we choose alternative (b), reject the hypothesis, and say the result of our experiment was *significant.*

By common practice, we call a result significant if its probability is less than 5% and call it highly significant if its probability is less than 1%. Because the probability (10.20) is 5.5%, we see that eight wins for the waxed skis are just *not* enough to give significant evidence that the wax works. On the other hand, we saw that the probability of 10 wins is 0.1%. Because this value is less than 1%, we can say that 10 wins would constitute highly significant evidence that the wax helps.[4]

GENERAL PROCEDURE

The methods of the example just described can be applied to any set of n similar but independent tests (or "races"), each of which has the same two possible outcomes, "success" or "failure." First, a hypothesis is formulated, here simply an assumed value for the probability p of success in any one test. This assumed value of p determines the expected mean number of successes, $\bar{\nu} = np$, in n trials.[5] If the actual number of successes, ν, in our n trials is close to np, there is no evidence against the hypothesis. (If the waxed skis win five out of 10 races, there is no evidence the wax makes any difference.) If ν is appreciably larger than np, we calculate the probability (given our hypothesis) of getting ν or more successes. If this probability is less than our chosen significance level (for example, 5% or 1%), we argue that our observed number is unacceptably improbable (if our hypothesis is correct) and hence that our hypothesis should be rejected. In the same way, if our number of successes ν is appreciably less than np, we can argue similarly, except that we would calculate the probability of getting ν or less successes.[6]

As you should have expected, this procedure does not provide a simple answer that our hypothesis is certainly true or certainly false. But it does give a quantitative measure of the reasonableness of our results in light of the hypothesis, so we can choose an objective, if arbitrary, criterion for rejection of the hypothesis. When experimenters state conclusions based on this kind of reasoning, they must state clearly the criterion used and the calculated probability so that readers can judge the reasonableness of the conclusion for themselves.

AN OPINION POLL

As a second example, consider an election between two candidates, A and B. Suppose candidate A claims that extensive research has established that he is favored by 60% of the electorate, and suppose candidate B asks us to check this claim (in the hope, of course, of showing that the number favoring A is significantly less than 60%).

Here, our statistical hypothesis would be that 60% of voters favor A, so the probability that a randomly selected voter will favor A would be $p = 0.6$. Recognizing that we cannot poll every single voter, we carefully select a random sample of

---

[4] Note the great simplicity of the test just described. We could have measured various additional parameters, such as the time taken by each ski, the maximum speed of each ski, and so on. Instead, we simply recorded which ski won each race. Tests that do not use such additional parameters are called *nonparametric* tests. They have the great advantages of simplicity and wide applicability.

[5] As usual, $\bar{\nu} = np$ is the mean number of successes expected if we were to repeat our whole set of n trials many times.

[6] As we discuss below, in some experiments the relevant probability is the "two-tailed" probability of getting a value of ν that deviates *in either direction* from np by as much as, or more than, the value actually obtained.

600 and ask their preferences. If 60% really favor A, the expected number in our sample who favor A is $np = 600 \times 0.6 = 360$. If in fact 330 state a preference for A, can we claim to have cast significant doubt on the hypothesis that 60% favor A?

To answer this question, we note that (according to the hypothesis) the probability that ν voters will favor A is the binomial probability

$$Prob(\nu \text{ voters favor } A) = B_{n,p}(\nu) \qquad (10.21)$$

with $n = 600$ and $p = 0.6$. Because n is so large, it is an excellent approximation to replace the binomial function by the appropriate Gauss function, with center at $np = 360$ and standard deviation $\sigma_\nu = \sqrt{np(1-p)} = 12$.

$$Prob(\nu \text{ voters favor } A) \approx G_{360,12}(\nu). \qquad (10.22)$$

The mean number expected to favor A is 360. Thus, the number who actually favored A in our sample (namely 330) is 30 less than expected. Because the standard deviation is 12, our result is 2.5 standard deviations below the supposed mean. The probability of a result this low or lower (according to the table in Appendix B) is 0.6%.[7] Thus, our result is highly significant, and at the 1% level we can confidently reject the hypothesis that A is favored by 60%.

This example illustrates two general features of this kind of test. First, having found that 330 voters favored A (that is, 30 less than expected), we calculated the probability that the number favoring A would be 330 *or less*. At first thought, you might have considered the probability that the number favoring A is precisely $\nu = 330$. This probability is extremely small (0.15%, in fact) and even the most probable result ($\nu = 360$) has a low probability (3.3%). To get a proper measure of how unexpected the result $\nu = 330$ is, we have to include $\nu = 330$ *and* any result that is even further below the mean.

Our result $\nu = 330$ was 30 less than the expected result, 360. The probability of a result 30 or more below the mean is sometimes called a "one-tailed probability," because it is the area under one tail of the distribution curve, as in Figure 10.6(a). In some tests, the relevant probability is the "two-tailed probability" of getting a result that differs from the expected mean by 30 or more *in either direc-*

(a)   (b)

Figure 10.6. (a) The "one-tailed" probability for getting a result 30 or more below the mean. (b) The "two-tailed" probability for getting a result that differs from the mean by 30 or more in either direction. (Not to scale.)

---

[7] Strictly speaking, we should have computed the probability for $\nu \leq 330.5$ because the Gauss distribution treats ν as a continuous variable. This number is 2.46σ below the mean, so the correct probability is actually 0.7%, but this small a difference does not affect our conclusion.

tion, that is, the probability of getting $\nu \leqslant 330$ or $\nu \geqslant 390$, as in Figure 10.6(b). Whether you use the one-tailed or two-tailed probability in a statistical test depends on what you consider the interesting alternative to the original hypothesis. Here, we were concerned to show that candidate A was favored by *less* than the claimed 60%, so the one-tailed probability was appropriate. If we were concerned to show that the number favoring A was *different* from 60% (in either direction), the two-tailed probability would be appropriate. In practice, the choice of which probability to use is usually fairly clear. In any case, the experimenter always needs to state clearly the probability and significance level chosen and the calculated value of the probability. With this information readers can judge the significance of the results for themselves.

Quick Check 10.2. If I toss a coin 12 times and get 11 heads, do I have significant evidence that the coin is loaded in favor of heads? (Hint: Assuming the coin is true, the probability of getting heads in a single throw is $p = \frac{1}{2}$. Making this assumption, find the probability that I would have obtained 11 or more heads in 12 tosses.

Principal Definitions and Equations of Chapter 10

THE BINOMIAL DISTRIBUTION

We consider an experiment with various possible outcomes and designate the particular outcome (or outcomes) in which we are interested as a "success." If the probability of success in any one trial is p, then the probability of ν successes in n trials is given by the binomial distribution:

$$Prob(\nu \text{ successes in } n \text{ trials}) = B_{n,p}(\nu)$$

$$= \binom{n}{\nu} p^\nu (1 - p)^{n - \nu}, \qquad \text{[See (10.6)]}$$

where $\binom{n}{\nu}$ denotes the binomial coefficient,

$$\binom{n}{\nu} = \frac{n!}{\nu! \, (n - \nu)!}.$$

If we repeat the whole set of n trials many times, the expected mean number of successes is

$$\bar{\nu} = np \qquad \text{[See (10.9)]}$$

and the standard deviation of ν is

$$\sigma_\nu = \sqrt{np(1 - p)}. \qquad\qquad \text{[See (10.10)]}$$

THE GAUSSIAN APPROXIMATION TO THE BINOMIAL DISTRIBUTION

When n is large, the binomial distribution $B_{n,p}(\nu)$ is well approximated by the Gauss function with the same mean and standard deviation; that is,

$$B_{n,p}(\nu) \approx G_{X,\sigma}(\nu), \qquad\qquad \text{[See (10.12)]}$$

where $X = np$ and $\sigma = \sqrt{np(1 - p)}$.

Problems for Chapter 10

For Section 10.2: Probabilities in Dice Throwing

10.1. ★ Consider the experiment of Section 10.2 in which three dice are thrown. Derive the probabilities for throwing no aces and one ace. Verify all four of the probabilities shown in Figure 10.1.

10.2. ★ Compute the probabilities *Prob*(ν aces in two throws) for $\nu = 0$, 1, and 2 in a throw of two dice. Plot them in a histogram.

10.3. ★★ Compute the probabilities *Prob*(ν aces in four throws) for $\nu = 0$, 1, ..., 4 in a throw of four dice. Plot them in a histogram. You will need to think carefully about the number of different ways in which you can get two aces in a throw of four dice.

For Section 10.3: Definition of the Binomial Distribution

10.4. ★ **(a)** Compute 5!, 6!, and 25!/23! (Please think for a moment before you do the last one.) **(b)** Explain why we traditionally define 0! = 1. [Hint: According to the usual definition, for $n = 1, 2, \ldots$, the factorial function satisfies $n! = (n + 1)!/(n + 1)$.] **(c)** Prove that the binomial coefficient defined by Equation (10.3) is equal to

$$\binom{n}{\nu} = \frac{n!}{\nu!(n - \nu)!}.$$

10.5. ★ Compute the binomial coefficients $\binom{3}{\nu}$ for $\nu = 0$, 1, 2, and 3. Hence, write the binomial expansion (10.5) of $(p + q)^3$.

10.6. ★ Compute the binomial coefficients $\binom{4}{\nu}$ for $\nu = 0$, 1, 2, 3, and 4. Hence, write the binomial expansion (10.5) of $(p + q)^4$.

10.7. ★ Compute and plot a histogram of the binomial distribution function $B_{n,p}(\nu)$ for $n = 4$, $p = 1/2$, and all possible ν.

10.8. ★ Compute and plot a histogram of the binomial distribution function $B_{n,p}(\nu)$ for $n = 4$, $p = 1/5$, and all possible ν.

10.9. ★   I draw six times from a deck of 52 playing cards, replacing each card before making the next draw. Find the probabilities, *Prob*(ν hearts), that I would draw exactly ν hearts in six draws, for $\nu = 0, 1, \ldots, 6$.

10.10. ★   A hospital admits four patients suffering from a disease for which the mortality rate is 80%. Find the probabilities of the following outcomes: **(a)** None of the patients survives. **(b)** Exactly one survives. **(c)** Two or more survive.

10.11. ★★   In the game of Yahtzee, a player throws five dice simultaneously. **(a)** Find the probabilities of throwing ν aces, for $\nu = 0, 1, \ldots, 5$. **(b)** What is the probability of throwing three *or more* aces? **(c)** What is the probability of throwing five of a kind?

10.12. ★★   Prove that the binomial coefficient $\binom{n}{\nu}$ is the number of different orders in which ν successes could be obtained in n trials.

For Section 10.4: Properties of the Binomial Distribution

10.13. ★   Prove that for $p = 1/2$, the binomial distribution is symmetric about $\nu = n/2$; that is,

$$B_{n,1/2}(\nu) = B_{n,1/2}(n - \nu).$$

10.14. ★★   Prove that the mean number of successes,

$$\bar{\nu} = \sum_{\nu=0}^{n} \nu B_{n,p}(\nu),$$

for the binomial distribution is just np. [There are many ways to do this, of which one of the best is this: Write the binomial expansion (10.5) for $(p + q)^n$. Because this expansion is true for any p and q, you can differentiate it with respect to p. If you now set $p + q = 1$ and multiply the result by p, you will have the desired result.]

10.15. ★★   **(a)** The standard deviation for any limiting distribution $f(\nu)$ is defined by

$$\sigma_\nu^2 = \overline{(\nu - \bar{\nu})^2}.$$

Prove that this definition is the same as $\overline{\nu^2} - (\bar{\nu})^2$. **(b)** Use this result to prove that, for the binomial distribution,

$$\sigma_\nu^2 = np(1 - p).$$

(Use the same trick as in Problem 10.14, but differentiate twice.)

10.16. ★★   The Gaussian approximation (10.12) to the binomial distribution is excellent for n large and surprisingly good for n small (especially if p is close to $\frac{1}{2}$). To illustrate this claim, calculate $B_{4,1/2}(\nu)$ (for $\nu = 0, 1, \ldots, 4$) both exactly and using the Gaussian approximation. Compare your results.

10.17. ★★   Use the Gaussian approximation (10.12) to find the probability of getting 15 heads if you throw a coin 25 times. Calculate the same probability exactly and compare answers.

10.18. ★★   Use the Gaussian approximation to find the probability of getting 18 or more heads in 25 tosses of a coin. (In using the Gauss probabilities, you should find the probability for $\nu \geqslant 17.5$.) Compare your approximation with the exact answer, which is 2.16%.

For Section 10.6: Applications; Testing of Hypotheses

10.19. ★   In the test of a ski wax described in Section 10.6, suppose the waxed skis had won 9 of the 10 races. Assuming the wax makes no difference, calculate the probability of 9 or more wins. Do 9 wins give significant (5% level) evidence that the wax is effective? Is the evidence highly significant (1% level)?

10.20. ★★   To test a new fertilizer, a gardener selects 14 pairs of similar plants and treats one plant from each pair with the fertilizer. After two months, 12 of the treated plants are healthier than their untreated partners (and the remaining 2 are less healthy). If, in fact, the fertilizer made no difference, what would be the probability that pure chance led to 12 or more successes? Do the 12 successes give significant (5% level) evidence that the fertilizer helps? Is the evidence highly significant (1% level)?

10.21. ★★   Of a certain kind of seed, 25% normally germinates. To test a new germination stimulant, 100 of these seeds are planted and treated with the stimulant. If 32 of them germinate, can you conclude (at the 5% level of significance) that the stimulant helps?

10.22. ★★   In a certain school, 420 of the 600 students pass a standardized mathematics test, for which the national passing rate is 60%. If the students at the school had no special aptitude for the test, how many would you expect to pass, and what is the probability that 420 or more would pass? Can the school claim that its students are significantly well prepared for the test?

Chapter 11

The Poisson Distribution

This chapter presents a third example of a limiting distribution, the Poisson distribution, which describes the results of experiments in which we count events that occur at random but at a definite average rate. Examples of this kind of counting experiment crop up in almost every area of science; for instance, a sociologist might count the number of babies born in a hospital in a three-day period. An important example in physics is the counting of the decays of a radioactive sample; for instance, a nuclear physicist might decide to count the number of alpha particles given off by a sample of radon gas in a ten-second interval.

This kind of counting experiment was discussed in Section 3.2, where I stated but did not prove the "square-root rule": If you count the occurrences of an event of this type in a chosen time interval T and obtain ν counts, then the best estimate for the true average number in time T is, of course, ν, and the uncertainty in this estimate is $\sqrt{\nu}$.

In Sections 11.1 and 11.2, I introduce the Poisson distribution and explore some of its properties. In particular, I prove in Section 11.2 that the standard deviation of the Poisson distribution is the square root of the expected number of events. This result justifies the square-root rule of Section 3.2. Sections 11.3 and 11.4 describe some applications of the Poisson distribution.

11.1 Definition of the Poisson Distribution

As an example of the Poisson distribution, suppose we are given a sample of radioactive material and use a suitable detector to find the number ν of decay particles ejected in a two-minute interval. If the counter is reliable, our value of ν will have no uncertainty. Nevertheless, if we repeat the experiment, we will almost certainly get a different value for ν. This variation in the number ν does not reflect uncertainties in our counting; rather, it reflects the intrinsically random character of the radioactive decay process.

Each radioactive nucleus has a definite probability for decaying in any two-minute interval. If we knew this probability and the number of nuclei in our sample, we could calculate the *expected average number* of decays in two minutes. Nevertheless, each nucleus decays at a random time, and in any given two-minute interval, the number of decays may be different from the expected average number.

Obviously the question we should ask is this: If we repeat our experiment many times (replenishing our sample if it becomes significantly depleted), what distribution should we expect for the number of decays ν observed in two-minute intervals? If you have studied Chapter 10, you will recognize that the required distribution is the binomial distribution. If there are n nuclei and the probability that any one nucleus decays is p, then the probability of ν decays is just the probability of ν "successes" in n "trials," or $B_{n,p}(\nu)$. In the kind of experiment we are now discussing, however, there is an important simplification. The number of "trials" (that is, nuclei) is enormous ($n \sim 10^{20}$, perhaps), and the probability of "success" (decay) for any one nucleus is tiny (often as small as $p \sim 10^{-20}$). Under these conditions (n large and p small), the binomial distribution can be shown to be indistinguishable from a simpler function called the Poisson distribution. Specifically, it can be shown that

$$Prob(\nu \text{ counts in any definite interval}) = P_\mu(\nu), \tag{11.1}$$

where the *Poisson distribution*, $P_\mu(\nu)$, is given by

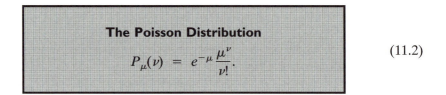

The Poisson Distribution

$$P_\mu(\nu) = e^{-\mu}\frac{\mu^\nu}{\nu!}. \tag{11.2}$$

In this definition, μ is a positive parameter ($\mu > 0$) that, as I will show directly, is just the expected mean number of counts in the time interval concerned, and $\nu!$ denotes the factorial function (with $0! = 1$).

SIGNIFICANCE OF μ AS THE EXPECTED MEAN COUNT

I will not derive the Poisson distribution (11.2) here but simply assert that it *is* the appropriate distribution for the kind of counting experiment in which we are interested.[1] To establish the significance of the parameter μ in (11.2), we have only to calculate the average number of counts, $\bar{\nu}$, expected if we repeat our counting experiment many times. This average is found by summing over all possible values of ν, each multiplied by its probability:

$$\bar{\nu} = \sum_{\nu=0}^{\infty} \nu P_\mu(\nu) = \sum_{\nu=0}^{\infty} \nu e^{-\mu}\frac{\mu^\nu}{\nu!}. \tag{11.3}$$

The first term of this sum can be dropped (because it is zero), and $\nu/\nu!$ can be replaced by $1/(\nu - 1)!$. If we remove a common factor of $\mu e^{-\mu}$, we get

$$\bar{\nu} = \mu e^{-\mu} \sum_{\nu=1}^{\infty} \frac{\mu^{\nu - 1}}{(\nu - 1)!}. \tag{11.4}$$

---

[1] For derivations, see, for example, H. D. Young, *Statistical Treatment of Experimental Data* (McGraw-Hill, 1962), Section 8, or S. L. Meyer, *Data Analysis for Scientists and Engineers* (John Wiley, 1975), p. 207.

The infinite sum that remains is

$$1 + \mu + \frac{\mu^2}{2!} + \frac{\mu^3}{3!} + \cdots = e^{\mu}, \tag{11.5}$$

which is just the exponential function e^{μ} (as indicated). Thus, the exponential $e^{-\mu}$ in (11.4) is exactly canceled by the sum, and we are left with the simple conclusion that

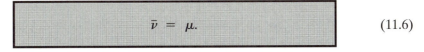

$$\bar{\nu} = \mu. \tag{11.6}$$

That is, the parameter μ that characterizes the Poisson distribution $P_{\mu}(\nu)$ is just the *average number of counts expected if we repeat the counting experiment many times.*

Sometimes, we may know in advance the average rate R at which the events we are counting should occur. In this case, the expected average number of events in a time T is just

$$\mu = \text{rate} \times \text{time} = RT.$$

Conversely, if the rate R is unknown, then by counting the number of events in a time T, we can get an estimate for μ and hence for the rate R as $R_{\text{best}} = \mu_{\text{best}}/T$.

Example: Counting Radioactive Decays

Careful measurements have established that a sample of radioactive thorium emits alpha particles at a rate of 1.5 per minute. If I count the number of alpha particles emitted in two minutes, what is the expected average result? What is the probability that I would actually get this number? What is the probability for observing ν particles for $\nu = 0, 1, 2, 3, 4$, and for $\nu \geqslant 5$?

The expected average count is just the average rate of emissions ($R = 1.5$ per minute) multiplied by the time during which I make my observations ($T = 2$ minutes):

$$(\text{expected average number}) = \mu = 3.$$

This result does not mean, of course, that I should expect to observe exactly three particles in any single trial. On the contrary, the probabilities for observing any number (ν) of particles are given by the Poisson distribution

$$Prob(\nu \text{ particles}) = P_3(\nu) = e^{-3}\frac{3^{\nu}}{\nu!}.$$

In particular, the probability that I would observe exactly three particles is

$$Prob(3 \text{ particles}) = P_3(3) = e^{-3}\frac{3^3}{3!} = 0.22 = 22\%.$$

Notice that although the expected average result is $\nu = 3$, we should expect to get this number only about once in every five trials.

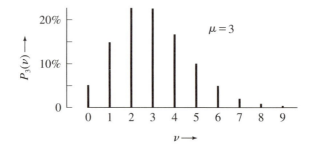

Figure 11.1. The Poisson distribution $P_3(v)$ gives the probabilities of observing v events in a counting experiment for which the expected average count is 3.

The probabilities for any number v can be calculated in the same way and are (as you might want to check):

Number v:	0	1	2	3	4
Probability:	5%	15%	22%	22%	17%

These probabilities (up to $v = 9$) are plotted in Figure 11.1. The simplest way to find the probability for getting 5 or more counts is to add the probabilities for $v = 0, \ldots, 4$ and then subtract the sum from 100% to give[2]

$$Prob(v \geqslant 5) = 100\% - (5 + 15 + 22 + 22 + 17)\%$$
$$= 19\%.$$

Quick Check 11.1. On average, each of the 18 hens in my henhouse lays 1 egg per day. If I check the hens once an hour and remove any eggs that have been laid, what is the average number, μ, of eggs that I find on my hourly visits? Use the Poisson distribution $P_\mu(v)$ to calculate the probabilities that I would find v eggs for $v = 0, 1, 2, 3,$ and $v = 4$ or more. What is the most probable number? What is the probability that I would find exactly μ eggs? Verify the probabilities shown in Figure 11.2.

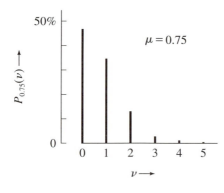

Figure 11.2. The Poisson distribution $P_{0.75}(v)$ gives the probabilities of observing v events in a counting experiment for which the expected average count is 0.75.

---

[2] The correct answer is actually 18.48%, as you can check by keeping a couple of extra decimal places in all the probabilities.

11.2   Properties of the Poisson Distribution

THE STANDARD DEVIATION

The Poisson distribution $P_\mu(\nu)$ gives the probability of getting the result ν in an experiment in which we count events that occur at random but at a definite average rate. We have seen that the parameter μ is precisely the expected average count $\bar{\nu}$. The natural next question is to ask for the standard deviation of the counts ν when we repeat the experiment many times. The standard deviation of any distribution (after a large number of trials) is just the root-mean-square deviation from the mean. That is,

$$\sigma_\nu{}^2 = \overline{(\nu - \bar{\nu})^2},$$

or, using the result of Problems 10.15(a) or 4.5(a),

$$\sigma_\nu{}^2 = \overline{\nu^2} - (\bar{\nu})^2. \tag{11.7}$$

For the Poisson distribution, we have already found that $\bar{\nu} = \mu$ and a similar calculation (Problem 11.9) gives $\overline{\nu^2} = \mu^2 + \mu$. Therefore, Equation (11.7) implies that $\sigma_\nu{}^2 = \mu$ or

$$\sigma_\nu = \sqrt{\mu}. \tag{11.8}$$

That is, the Poisson distribution with mean count μ has standard deviation $\sqrt{\mu}$.

The result (11.8) justifies the square-root rule of Section 3.2. If we carry out a counting experiment once and get the answer ν, we can easily see (using the principle of maximum likelihood, as in Problem 11.11) that the best estimate for the expected mean count is $\mu_{best} = \nu$. From (11.8), it immediately follows that the best estimate for the standard deviation is just $\sqrt{\nu}$. In other words, if we make one measurement of the number of events in a time interval T and get the answer ν, our answer for the expected mean count in time T is

$$\nu \pm \sqrt{\nu}. \tag{11.9}$$

This answer is precisely the square-root rule quoted without proof in Equation (3.2).

Example: More Radioactive Decays

A student monitors the thorium sample of the previous example for 30 minutes and observes 49 alpha particles. What is her answer for the number of particles emitted in 30 minutes? What is her answer for the rate of emission, R, in particles per minute?

According to (11.9), her answer for the number of particles emitted in 30 minutes is

$$(\text{number emitted in 30 minutes}) = 49 \pm \sqrt{49} = 49 \pm 7.$$

To find the rate in particles per minute, she must divide by 30 minutes. Assuming this 30 minutes has no uncertainty, we find

$$R = \frac{49 \pm 7}{30} = 1.6 \pm 0.2 \text{ particles/min.} \tag{11.10}$$

Notice that the square-root rule gives the uncertainty in the actual counted number ($\sigma_\nu = \sqrt{\nu} = 7$, in this case). A common mistake is to calculate the rate of decay $R = \nu/T$ and then to take the uncertainty in R to be $\sqrt{R}$. A glance at (11.10) should convince you that this procedure is simply not correct. The square-root rule applies only to the actual counted number ν, and the uncertainty in $R = \nu/T$ must be found from that in ν using error propagation as in (11.10).

Quick Check 11.2. The farmer of Quick Check 11.1 observes that in a certain ten-hour period his hens lay 9 eggs. Based on this one observation, what would you quote for the number of eggs expected in ten hours? What would you give for the rate R of egg production, in eggs per hour? (Give the uncertainties in both your answers.)

GAUSSIAN APPROXIMATION TO THE POISSON DISTRIBUTION

In Chapter 10, we compared the Gauss distribution with the binomial distribution. We saw that in most ways, the two distributions are very different; nevertheless, under the right conditions, the Gauss distribution gives an excellent, extremely useful, approximation to the binomial distribution. As we will now see, almost exactly the same can be said about the Gauss and *Poisson* distributions.

The Gauss distribution $G_{X,\sigma}(x)$ gives the probabilities of the various values of a *continuous* variable x; by contrast, the Poisson distribution $P_\mu(\nu)$, like the binomial $B_{n,p}(\nu)$, gives the probabilities for a *discrete* variable $\nu = 0, 1, 2, 3, \ldots$. Another important difference is that the Gauss distribution $G_{X,\sigma}(x)$ is specified by *two* parameters, the mean X and the standard deviation σ, whereas the Poisson distribution $P_\mu(\nu)$ is specified by a single parameter, the mean μ, because the width of the Poisson distribution is automatically determined by the mean (namely, $\sigma_\nu = \sqrt{\mu}$). Finally, the Gauss distribution is always bell shaped and symmetric about its mean value, whereas the Poisson distribution has neither of these properties in general. This last point is especially clear in Figure 11.2, which shows the Poisson distribution for $\mu = 0.75$; this curve is certainly not bell shaped, nor is it even approximately symmetric about its mean, 0.75.

Figure 11.1 showed the Poisson distribution for $\mu = 3$. Although this curve is obviously not exactly bell shaped, it is undeniably more nearly so than the curve for $\mu = 0.75$ in Figure 11.2. Figure 11.3 shows the Poisson distribution for $\mu = 9$; this curve is quite nearly bell shaped and close to symmetric about its mean ($\mu = 9$). In fact, it can be proved that as $\mu \to \infty$, the Poisson distribution becomes progres-

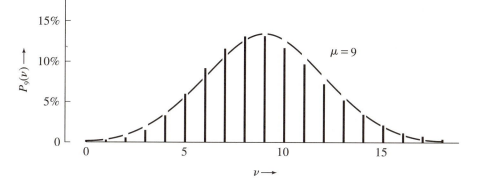

Figure 11.3. The Poisson distribution for $\mu = 9$. The dashed curve is the Gauss distribution with the same mean and standard deviation ($X = 9$ and $\sigma = 3$). As $\mu \to \infty$, the two distributions become indistinguishable; even when $\mu = 9$, they are very close.

sively more bell shaped and approaches the Gauss distribution with the same mean and standard deviation.[3] That is,

$$P_\mu(v) \approx G_{X,\sigma}(v), \qquad \text{(when } \mu \text{ is large)} \tag{11.11}$$

where

$$X = \mu \quad \text{and} \quad \sigma = \sqrt{\mu}.$$

In Figure 11.3, the dashed curve is the Gauss function with $X = 9$ and $\sigma = 3$. You can see clearly how, even when μ is only 9, the Poisson distribution is remarkably close to the appropriate Gauss function; the slight discrepancy reflects the remaining asymmetry in the Poisson function.

The approximation (11.11) is called the *Gaussian approximation to the Poisson distribution.* It is analogous to the corresponding approximation for the binomial distribution (discussed in Section 10.4) and is useful under the same conditions, namely, when the parameters involved are large.

Example: Gaussian Approximation to a Poisson Distribution

To illustrate the Gaussian approximation to the Poisson distribution, consider the Poisson distribution with $\mu = 64$. The probability of 72 counts, for example, is

$$Prob(72 \text{ counts}) = P_{64}(72) = e^{-64} \frac{(64)^{72}}{72!}, \tag{11.12}$$

which a tedious calculation gives as

$$Prob(72 \text{ counts}) = 2.9\%.$$

---

[3] For proof, see S. L. Meyer, *Data Analysis for Scientists and Engineers* (John Wiley, 1975), p. 227.

According to (11.11), however, the probability (11.12) is well approximated by the Gauss function

$$Prob(72 \text{ counts}) \approx G_{64,8}(72),$$

which is easily evaluated to give

$$Prob(72 \text{ counts}) \approx 3.0\%.$$

If we wanted to calculate directly the probability of 72 *or more* counts in the same experiment, an extremely tedious calculation would give

$$Prob(\nu \geq 72) = P_{64}(72) + P_{64}(73) + \ldots$$
$$= 17.3\%.$$

If we use the approximation (11.11), then we have only to calculate the Gaussian probability for getting $\nu \geq 71.5$ (because the Gauss distribution treats ν as a continuous variable). Because 71.5 is 7.5, or 0.94σ, above the mean, the required probability can be found quickly from the table in Appendix B as

$$Prob(\nu \geq 72) \approx Prob_G(\nu \geq 71.5) = Prob_G(\nu \geq X + 0.94\sigma)$$
$$= 17.4\%,$$

by almost any standard an excellent approximation.

11.3 Applications

As I have emphasized, the Poisson distribution describes the distribution of results in a counting experiment in which events are counted that occur at random but at a definite average rate. In an introductory physics laboratory, the two most common examples are counting the disintegrations of radioactive nuclei and counting the arrival of cosmic ray particles.

Another very important example is an experiment to study an expected limiting distribution, such as the Gauss or binomial distributions, or the Poisson distribution itself. A limiting distribution tells us how many events of a particular type are expected when an experiment is repeated several times. (For example, the Gaussian $G_{X,\sigma}(x)$ tells us how many measurements of x are expected to fall in any interval from $x = a$ to $x = b$.) In practice, the observed number is seldom exactly the expected number. Instead, it fluctuates in accordance with the Poisson distribution. In particular, if the expected number of events of some type is n, the observed number can be expected to differ from n by a number of order $\sqrt{n}$. We will make use of this point in Chapter 12.

In many situations, it is reasonable to expect numbers to be distributed approximately according to the Poisson distribution. The number of eggs laid in an hour on a poultry farm and the number of births in a day at a hospital would both be expected to follow the Poisson distribution at least approximately (though they would probably show some seasonal variations as well). To test this assumption, you would need to record the number concerned many times over. After plotting the resulting distribution, you could compare it with the Poisson distribution to see how close the fit is. For a more quantitative test, you would use the chi-squared test described in Chapter 12.

Example: Cosmic Ray Counting

As another example of the Poisson distribution, let us consider an experiment with cosmic rays. These "rays" originate as charged particles, such as protons and alpha particles, that enter the Earth's atmosphere from space. Many of these primary particles collide with atoms in the atmosphere and create further, secondary particles, such as mesons and positrons. Some of the particles (both primary and secondary) travel all the way to ground level and can be detected (with a Geiger counter, for example) in the laboratory. In the following problem, I exploit the fact that the number of cosmic rays hitting any given area in a given time should follow the Poisson distribution.

Student A asserts that he has measured the number of cosmic rays hitting a Geiger counter in one minute. He claims to have made the measurement repeatedly and carefully and to have found that, on average, 9 particles hit the counter per minute, with "negligible" uncertainty. To check this claim, Student B counts how many particles arrive in one minute and gets the answer 12. Does this answer cast serious doubt on A's claim that the expected rate is 9? To make a more careful check, Student C counts the number of particles that arrive in ten minutes. From A's claim, she expects to get 90 but actually gets 120. Does this value cast significant doubt on A's claim?

Let us consider B's result first. If A is right, the expected mean count is 9. Because the distribution of counts should be the Poisson distribution, the standard deviation is $\sqrt{9} = 3$. Student B's result of 12 is, therefore, only one standard deviation away from the mean of 9. This amount is certainly not far enough away to contradict A's claim. More specifically, knowing that the probability of any answer ν is supposed to be $P_9(\nu)$, we can calculate the total probability for getting an answer that differs from 9 by 3 or more. This probability turns out to be 40% (see Problem 11.7). Obviously B's result is not at all surprising, and A has no reason to worry.

Student C's result is quite a different matter. If A is right, C should expect to get 90 counts in ten minutes. Since the distribution should be Poisson, the standard deviation should be $\sqrt{90} = 9.5$. Thus, C's result of 120 is more than *three standard deviations* away from A's prediction of 90. With these large numbers, the Poisson distribution is indistinguishable from the Gauss function, and we can immediately find from the table in Appendix A that the probability of a count more than three standard deviations from the mean is 0.3%. That is, if A is right, it is extremely improbable that C would have observed 120 counts. Turning this statement around, we can say something almost certainly has gone wrong. Perhaps A was just not as careful as he claimed. Perhaps the counter was malfunctioning for A or C, introducing systematic errors into one of the results. Or perhaps A made his measurements at a time when the flux of cosmic rays was truly less than normal.

11.4   Subtracting a Background

To conclude this chapter, I discuss a problem that complicates many counting experiments. Often, the events we want to study are accompanied by other "background" events that cannot be distinguished from the events of interest. For example, in studying the disintegrations of a radioactive source, we usually cannot prevent the detector from registering particles from other radioactive materials in the vicinity or from cosmic rays. This means that the number we count includes the events of interest *plus* these background events, and we must somehow subtract out the unwanted background events. In principle at least, the remedy is straightforward: Having found the total counting rate (due to source and background), we must remove the source and find the rate of events due to the background alone; the rate of events from the source is then just the difference of these two measured rates.

In practice, it is surprisingly easy to make a mistake in this procedure, especially in the error analysis. It is usually convenient to measure the total and background counts using different time intervals. Suppose we count a total of ν_{tot} events (source *plus* background) in a time T_{tot}, and then ν_{bgd} background events in a time T_{bgd}. Obviously, we do not simply subtract ν_{bgd} from ν_{tot} because they refer to different time intervals. Instead, we must first calculate the *rates*

$$R_{tot} = \frac{\nu_{tot}}{T_{tot}} \quad \text{and} \quad R_{bgd} = \frac{\nu_{bgd}}{T_{bgd}} \tag{11.13}$$

and then calculate the rate from the source as the difference

$$R_{sce} = R_{tot} - R_{bgd}. \tag{11.14}$$

In estimating the uncertainties in the quantities involved, you must remember that the square-root rule gives the uncertainties in the *counted numbers* ν_{tot} and ν_{bgd}; the uncertainties in the corresponding rates must be found by error propagation, as in the following example.

Example: Radioactive Decays with a Background

A student decides to monitor the activity of a radioactive source by placing it in a liquid scintillation detector. In the course of 10 minutes, the detector registers 2,540 total counts. To allow for the possibility of unwanted background counts, she removes the source and notes that in 3 minutes, the detector registers a further 95 counts. To find the activity of the source, she calculates the two rates of counting, R_{tot} and R_{bgd} (in counts per minute) and their difference $R_{sce} = R_{tot} - R_{bgd}$. What are her answers with their uncertainties? (Assume the two times have negligible uncertainty.)

According to the square-root rule, the two counted numbers with their uncertainties are

$$\nu_{tot} = 2,540 \pm \sqrt{2,540} = 2,540 \pm 50$$

and

$$\nu_{bgd} = 95 \pm \sqrt{95} = 95 \pm 10.$$

Dividing these numbers by their corresponding times, we find the rates

$$R_{tot} = \frac{\nu_{tot}}{T_{tot}} = \frac{2{,}540 \pm 50}{10} = 254 \pm 5 \text{ counts/min}$$

and

$$R_{bgd} = \frac{\nu_{bgd}}{T_{bgd}} = \frac{95 \pm 10}{3} = 32 \pm 3 \text{ counts/min.}$$

Finally, the rate due to the source alone is

$$R_{sce} = R_{tot} - R_{bgd} = (254 \pm 5) - (32 \pm 3) = 222 \pm 6 \text{ counts/min.}$$

Notice that, in the last step, the errors are combined in quadrature, because they are certainly independent and random.

Principal Definitions and Equations of Chapter 11

THE POISSON DISTRIBUTION

The Poisson distribution describes experiments in which you count events that occur at random but at a definite average rate. If you count for a chosen time interval T, the probability of observing ν events is given by the Poisson function

$$Prob(\nu \text{ counts in time } T) = P_\mu(\nu) = e^{-\mu}\frac{\mu^\nu}{\nu!}, \qquad \text{[See (11.2)]}$$

where the parameter μ is the expected average number of events in time T; that is,

$$\bar{\nu} = \mu \text{ (after many trials)}$$
$$= RT, \qquad \text{[See (11.6)]}$$

where R is the mean rate at which the events occur.

The standard deviation of the observed number ν is

$$\sigma_\nu = \sqrt{\mu}. \qquad \text{[See (11.8)]}$$

THE GAUSSIAN APPROXIMATION TO THE POISSON DISTRIBUTION

When μ is large, the Poisson distribution $P_\mu(\nu)$ is well approximated by the Gauss function with the same mean and standard deviation; that is,

$$P_\mu(\nu) \approx G_{X,\sigma}(\nu),$$

where $X = \mu$ and $\sigma = \sqrt{\mu}$. 　　　　　　　　　　　　　　　　[See (11.11)]

SUBTRACTING A BACKGROUND

The events produced by a source subject to an unavoidable background can be counted in a three-step procedure:

(1) Count the total number ν_{tot} (source plus background) in a time T_{tot}, and calculate the total rate $R_{tot} = \nu_{tot}/T_{tot}$.

(2) Remove the source, and measure the number of background events in a time T_{bgd}; then calculate the background rate $R_{bgd} = \nu_{bgd}/T_{bgd}$.

(3) Calculate the rate of the events from the source as the difference $R_{sce} = R_{tot} - R_{bgd}$.

Finally, the uncertainties in the numbers ν_{tot} and ν_{bgd} are given by the square-root rule, and, from these values, the uncertainties in the three rates can be found using error propagation.

Problems for Chapter 11

For Section 11.1: Definition of the Poisson Distribution

11.1. ★   Compute the Poisson distribution $P_\mu(\nu)$ for $\mu = 0.5$ and $\nu = 0, 1, \ldots, 6$. Plot a bar histogram of $P_{0.5}(\nu)$ against ν.

11.2. ★   **(a)** Compute the Poisson distribution $P_\mu(\nu)$ for $\mu = 1$ and $\nu = 0, 1, \ldots, 6$, and plot your results as a bar histogram. **(b)** Repeat part (a) but for $\mu = 2$.

11.3. ★★   A radioactive sample contains 5.0×10^{19} atoms, each of which has a probability $p = 3.0 \times 10^{-20}$ of decaying in any given five-second interval. **(a)** What is the expected average number, μ, of decays from the sample in five seconds? **(b)** Compute the probability $P_\mu(\nu)$ of observing ν decays in any five-second interval, for $\nu = 0, 1, 2, 3$. **(c)** What is the probability of observing 4 or more decays in any five-second interval?

11.4. ★★   In the course of four weeks, a farmer finds that between 10:00 and 10:30 A.M., his hens lay an average of 2.5 eggs. Assuming the number of eggs laid follows a Poisson distribution with $\mu = 2.5$, on approximately how many days do you suppose he found no eggs laid between 10:00 and 10:30 A.M.? On how many days do you suppose there were 2 or less? 3 or more?

11.5. ★★   A certain radioactive sample is expected to undergo three decays per minute. A student observes the number ν of decays in 100 separate one-minute intervals, with the results shown in Table 11.1. **(a)** Make a histogram of these re-

Table 11.1. Occurrences of numbers of decays in one-minute intervals; for Problem 11.5.

No. of decays ν	0	1	2	3	4	5	6	7	8	9
Times observed	5	19	23	21	14	12	3	2	1	0

sults, plotting f_ν (the fraction of times the result ν was found) against ν. **(b)** On the same plot, show the expected distribution $P_3(\nu)$. Do the data seem to fit the expected distribution? (For a quantitative measure of the fit, you could use the chi-squared test, discussed in Chapter 12.)

11.6. ★★★ **(a)** The Poisson distribution, like all distributions, must satisfy a "normalization condition,"

$$\sum_{\nu=0}^{\infty} P_\mu(\nu) = 1. \tag{11.15}$$

This condition asserts that the total probability of observing *all* possible values of ν must be one. Prove it. [Remember the infinite series (11.5) for e^μ.] **(b)** Differentiate (11.15) with respect to μ, and then multiply the result by μ to give an alternative proof that, after infinitely many trials, $\bar\nu = \mu$ as in Equation (11.6).

For Section 11.2: Properties of the Poisson Distribution

11.7. ★ **(a)** What is the standard deviation σ_ν (after a large number of trials) of the observed counts ν in a counting experiment in which the expected average count is $\mu = 9$? **(b)** Compute the probabilities $P_9(\nu)$ of obtaining ν counts for $\nu = 7$, $8, \ldots,$ 11. **(c)** Hence, find the probability of getting a count ν that differs from the expected mean by one or more standard deviations. **(d)** Would a count of 12 cause you to doubt that the expected mean really is 9?

11.8. ★ A nuclear physicist monitors the disintegrations of a radioactive sample with a Geiger counter. She counts the disintegrations in 15 separate five-second intervals and gets the following numbers:

7, 11, 10, 7, 5, 7, 6, 12, 12, 7, 18, 12, 13, 12, 6.

(a) Her best estimate μ_{best} for the true mean count μ is the average of her 15 counts. (For a proof of this claim, see Problem 11.15.) What is her value for μ_{best}? **(b)** The standard deviation σ_ν of her 15 counts should be close to $\sqrt{\mu}$. What is her standard deviation and how does it compare with $\sqrt{\mu_{best}}$?

11.9. ★★ **(a)** Prove that the average value of ν^2 for the Poisson distribution $P_\mu(\nu)$ is $\overline{\nu^2} = \mu^2 + \mu$. [The easiest way to do this is probably to differentiate the identity (11.15) twice with respect to μ.] **(b)** Hence, prove that the standard deviation of ν is $\sigma_\nu = \sqrt{\mu}$. [Use the identity (11.7).]

11.10. ★★ The average rate of disintegrations from a certain radioactive sample is known to be roughly 20 per minute. If you wanted to measure this rate within 4%, for approximately how long would you plan to count?

11.11. ★★ Consider a counting experiment governed by the Poisson distribution $P_\mu(\nu)$, where the mean count μ is unknown, and suppose that I make a single count and get the value ν. Write down the probability $Prob(\nu)$ for getting this value ν. According to the principle of maximum likelihood, the best estimate for the unknown μ is that value of μ for which $Prob(\nu)$ is largest. Prove that the best estimate μ_{best} is precisely the observed count ν as you would expect. (In this calculation, the unknown μ is the variable and ν is the fixed value I obtained in my one experiment.)

11.12. ★★   The expected mean count in a certain counting experiment is $\mu = 16$. **(a)** Use the Gaussian approximation (11.11) to estimate the probability of getting a count of 10 in any one trial. Compare your answer with the exact value $P_{16}(10)$. **(b)** Use the Gaussian approximation to estimate the probability of getting a count of 10 *or less*. (Remember to compute the Gaussian probability for $\nu \leqslant 10.5$ to allow for the fact that the Gauss distribution treats ν as a continuous variable.) Calculate the exact answer and compare.

Note how, even with μ as small as 16, the Gaussian approximation gives quite good answers and—at least in part (b)—is significantly less trouble to compute than an exact calculation using the Poisson distribution.

11.13. ★★   The expected mean count in a certain counting experiment is $\mu = 400$. Use the Gaussian approximation to find the probability of getting a count anywhere in the range $380 \leqslant \nu \leqslant 420$. Compare your answer with the exact answer, which is 69.47%. Discuss the feasibility of finding this exact probability using your calculator.

11.14. ★★★   In the experiment of Problem 11.5, the expected mean count was known in advance to be $\mu = 3$. Therefore, the mean of the data should be close to 3 and the standard deviation should be close to $\sqrt{3}$. Find the mean and standard deviation of the data and compare them with their expected values.

11.15. ★★★   Consider a counting experiment governed by the Poisson distribution $P_\mu(\nu)$, where the mean count μ is unknown, and suppose that you make N separate trials and get the values

$$\nu_1, \nu_2, \ldots, \nu_N.$$

Write the probability $Prob(\nu_1, \ldots, \nu_N)$ for getting this particular set of values. According to the principle of maximum likelihood, the best estimate for the unknown μ is that value of μ for which $Prob(\nu_1, \ldots, \nu_N)$ is largest. Prove that the best estimate μ_{best} is the average of the observed numbers

$$\mu_{best} = \frac{1}{N} \sum_{i=1}^{N} \nu_i$$

as you would expect. (In this calculation, the unknown μ is the variable, and $\nu_1, \ldots, \nu_N$ are fixed at the values you obtained in your N trials.)

For Section 11.3: Applications

11.16. ★   Using a Geiger counter, a student records 25 cosmic-ray particles in 15 seconds. **(a)** What is her best estimate for the true mean number of particles in 15 seconds, with its uncertainty? **(b)** What should she report for the *rate* (in particles per minute) with its uncertainty?

11.17. ★   A student counts 400 disintegrations from a radioactive sample in 40 seconds. **(a)** What is his best estimate for the true mean number of disintegrations in 40 seconds, with its uncertainty? **(b)** What should he report for the *rate* of disintegrations (per second) with its uncertainty?

11.18. ★★   Sean and Maria monitor the same radioactive sample. Sean counts for one minute and gets 121 counts. Maria counts for six minutes and gets 576 counts. Each then calculates the *rate* in counts per minute. Do their results agree satisfactorily?

For Section 11.4: Subtracting a Background

11.19. ★★   A heating duct at a nuclear plant is suspected of being contaminated with radioactive dust. To check this suspicion, a health physicist places a detector inside the duct and records 7,075 particles of radiation in 120 minutes. To check for background radiation, he moves his detector some distance from the duct and shields it from any radiation coming from the duct; in this new position, he records 3,015 particles in 60 minutes. What is his final answer for the radiation rate (in particles per minute) due to the contents of the duct alone? Does he have significant evidence for radioactive material in the duct?

11.20. ★★   To measure the activity of a rock thought to be radioactive, a physicist puts the rock beside a detector and counts 225 particles in 10 minutes. To check for background, she removes the rock and then records 90 particles in 6 minutes. She converts both these answers into rates, in particles per hour, and takes their difference to give the activity of the rock alone. What is her final answer, in particles per hour, and what is its uncertainty? Does she have significant evidence that the rock is radioactive?

11.21. ★★★   A physicist needs an accurate measurement of the activity of a radioactive source. To plan the best use of his time, he first makes quick approximate measurements of the total counting rate (source plus background) and the background rate. These approximations give him rough values for the true rates, r_{tot} and r_{bgd}. (Notice that I have used lower-case r for the true rates; this will let you use upper-case R for the values he measures in the main experiment.) To get more accurate values, he plans to count the total number ν_{tot} (source plus background) in a long time T_{tot} and the background number ν_{bgd} in a long time T_{bgd}. From these measurements, he will calculate the rates R_{tot} and R_{bgd} and finally the rate R_{sce} due to the source alone.

The detector he is using is available only for a time T, so that the two times T_{tot} and T_{bgd} are constrained not to exceed

$$T_{tot} + T_{bgd} = T. \tag{11.16}$$

Therefore, he must choose the two times T_{tot} and T_{bgd} to minimize the uncertainty in his final answer for R_{sce}.

(a) Show that his final uncertainty will be minimum if he chooses the two times such that

$$\frac{T_{tot}}{T_{bgd}} = \sqrt{\frac{r_{tot}}{r_{bgd}}}. \tag{11.17}$$

[Hint: Write the expected numbers ν_{tot} and ν_{bgd} in terms of the true rates, r_{tot} and r_{bgd}, and the corresponding times. From these calculate all the uncertainties involved. To minimize the uncertainty in R_{sce}, eliminate T_{bgd} using (11.16), and then differentiate with respect to T_{tot} and set the derivative equal to zero.] Because his preliminary measurements of r_{tot} and r_{bgd} tell him the approximate value of the ratio on the right side, (11.17) tells him how to apportion the available time between T_{tot} and T_{bgd}.

(b) If the available time t is two hours and his preliminary measurements have shown that $r_{\text{tot}}/r_{\text{bgd}} \approx 9$, how should he choose T_{tot} and T_{bgd}?

Chapter 12

The Chi-Squared Test
for a Distribution

By now, you should be reasonably familiar with the notion of limiting distributions. These are the functions that describe the expected distribution of results if an experiment is repeated many times. There are many different limiting distributions, corresponding to the many different kinds of experiments possible. Perhaps the three most important limiting distributions in physical science are the three we have already discussed: the Gauss (or normal) function, the binomial distribution, and the Poisson distribution.

This final chapter focuses on how to decide whether the results of an actual experiment are governed by the expected limiting distribution. Specifically, let us suppose that we perform some experiment for which we believe we know the expected distribution of results. Suppose further that we repeat the experiment several times and record our observations. The question we now address is this: How can we decide whether our observed distribution is consistent with the expected theoretical distribution? We will see that this question can be answered using a simple procedure called the *chi-squared,* or χ^2, *test.* (The Greek letter χ is spelled "chi" and pronounced "kie.")

12.1 Introduction to Chi Squared

Let us begin with a concrete example. Suppose we make 40 measurements $x_1, \ldots,$ x_{40} of the range x of a projectile fired from a certain gun and get the results shown in Table 12.1. Suppose also we have reason to believe these measurements are governed by a Gauss distribution $G_{X,\sigma}(x)$, as is certainly very natural. In this type

Table 12.1. Measured values of x (in cm).

731	772	771	681	722	688	653	757	733	742
739	780	709	676	760	748	672	687	766	645
678	748	689	810	805	778	764	753	709	675
698	770	754	830	725	710	738	638	787	712

of experiment, we usually do not know in advance either the center X or the width σ of the expected distribution. Our first step, therefore, is to use our 40 measurements to compute best estimates for these quantities:

$$(\text{best estimate for } X) = \bar{x} = \frac{\sum x_i}{40} = 730.1 \text{ cm} \tag{12.1}$$

and

$$(\text{best estimate for } \sigma) = \sqrt{\frac{\sum(x_i - \bar{x})^2}{39}} = 46.8 \text{ cm.} \tag{12.2}$$

Now we can ask whether the actual distribution of our results $x_1, \ldots, x_{40}$ is consistent with our hypothesis that our measurements were governed by the Gauss distribution $G_{X,\sigma}(x)$ with X and σ as estimated. To answer this question, we must compute how we would expect our 40 results to be distributed if the hypothesis is true and compare this expected distribution with our actual observed distribution. The first difficulty is that x is a continuous variable, so we cannot speak of the expected number of measurements equal to any one value of x. Rather, we must discuss the expected number in some interval $a < x < b$. That is, we must divide the range of possible values into *bins*. With 40 measurements, we might choose bin boundaries at $X - \sigma$, X, and $X + \sigma$, giving four bins as in Table 12.2.

Table 12.2. A possible choice of bins for the data of Table 12.1. The final column shows the number of observations that fell into each bin.

Bin number k	Values of x in bin		Observations O_k
1	$x < X - \sigma$	(or $x < 683.3$)	8
2	$X - \sigma < x < X$	(or $683.3 < x < 730.1$)	10
3	$X < x < X + \sigma$	(or $730.1 < x < 776.9$)	16
4	$X + \sigma < x$	(or $776.9 < x$)	6

We will discuss later the criteria for choosing bins. In particular, they must be chosen so that all bins contain several measured values x_i. In general, I will denote the number of bins by n; for this example with four bins, $n = 4$.

Having divided the range of possible measured values into bins, we can now formulate our question more precisely. First, we can count the number of measurements that fall into each bin k.[1] We denote this number by O_k (where O stands for "observed number"). For the data of our example, the observed numbers O_1, O_2, O_3, O_4 are shown in the last column of Table 12.2. Next, assuming our measurements are distributed normally (with X and σ as estimated), we can calculate the *expected* number E_k of measurements in each bin k. We must then decide how well the observed numbers O_k compare with the expected numbers E_k.

---

[1] If a measurement falls exactly on the boundary between two bins, we can assign half a measurement to each bin.

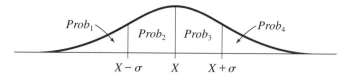

Figure 12.1. The probabilities $Prob_k$ that a measurement falls into each of the bins, $k = 1, 2, 3, 4$, of Table 12.2 are the four areas shown under the Gauss function.

The calculation of the expected numbers E_k is quite straightforward. The *probability* that any one measurement falls in an interval $a < x < b$ is just the area under the Gauss function between $x = a$ and $x = b$. In this example, the probabilities $Prob_1$, $Prob_2$, $Prob_3$, $Prob_4$ for a measurement to fall into each of our four bins are the four areas indicated in Figure 12.1. The two equal areas $Prob_2$ and $Prob_3$ together represent the well-known 68%, so the probability for falling into one of the two central bins is 34%; that is, $Prob_2 = Prob_3 = 0.34$. The outside two areas comprise the remaining 32%; thus $Prob_1 = Prob_4 = 0.16$. To find the expected numbers E_k, we simply multiply these probabilities by the total number of measurements, $N = 40$. Therefore, our expected numbers are as shown in the third column of Table 12.3. That the numbers E_k are not integers serves to remind us that the "expected number" is not what we actually expect in any one experiment; it is rather the expected average number after we repeat our whole series of measurements many times.

Our problem now is to decide how well the expected numbers E_k do represent the corresponding observed numbers O_k (in the last column of Table 12.3). We

Table 12.3. The expected numbers E_k and the observed numbers O_k for the 40 measurements of Table 12.1, with bins chosen as in Table 12.2.

Bin number k	Probability $Prob_k$	Expected number $E_k = NProb_k$	Observed number O_k
1	16%	6.4	8
2	34%	13.6	10
3	34%	13.6	16
4	16%	6.4	6

would obviously not expect *perfect* agreement between E_k and O_k after any finite number of measurements. On the other hand, if our hypothesis that our measurements are normally distributed is correct, we would expect that, in some sense, the deviations

$$O_k - E_k \qquad (12.3)$$

would be *small*. Conversely, if the deviations $O_k - E_k$ prove to be *large,* we would suspect our hypothesis is incorrect.

To make precise the statements that the deviation $O_k - E_k$ is "small" or "large," we must decide how large we would *expect* $O_k - E_k$ to be if the measurements really are normally distributed. Fortunately, this decision is easily made. If we imagine repeating our whole series of 40 measurements many times, then the number O_k of measurements in any one bin k can be regarded as the result of a counting experiment of the type described in Chapter 11. Our many different answers for O_k should have an average value of E_k and would be expected to fluctuate around E_k with a standard deviation of order $\sqrt{E_k}$. Thus, the two numbers to be compared are the deviation $O_k - E_k$ and the expected size of its fluctuations $\sqrt{E_k}$.

These considerations lead us to consider the ratio

$$\frac{O_k - E_k}{\sqrt{E_k}}. \tag{12.4}$$

For some bins k, this ratio will be positive, and for some negative; for a few k, it may be appreciably larger than one, but for most it should be of order one, or smaller. To test our hypothesis (that the measurements are normally distributed), it is natural to square the number (12.4) for each k and then sum over all bins $k = 1, \ldots, n$ (here $n = 4$). This procedure defines a number called *chi squared,*

$$\chi^2 = \sum_{k=1}^{n} \frac{(O_k - E_k)^2}{E_k}. \tag{12.5}$$

This number χ^2 is clearly a reasonable indicator of the agreement between the observed and expected distributions. If $\chi^2 = 0$, the agreement is perfect; that is, $O_k = E_k$ for all bins k, a situation most unlikely to occur. In general, the individual terms in the sum (12.5) are expected to be of order one, and there are n terms in the sum. Thus, if

$$\chi^2 \lesssim n$$

(χ^2 of order n or less), the observed and expected distributions agree about as well as could be expected. In other words, if $\chi^2 \lesssim n$, we have no reason to doubt that our measurements were distributed as expected. On the other hand, if

$$\chi^2 \gg n$$

(χ^2 significantly greater than the number of bins), the observed and expected numbers differ significantly, and we have good reason to suspect that our measurements were not governed by the expected distribution.

In our example, the numbers observed and expected in the four bins and their differences are shown in Table 12.4, and a simple calculation using them gives

$$\begin{aligned}
\chi^2 &= \sum_{k=1}^{4} \frac{(O_k - E_k)^2}{E_k} \\
&= \frac{(1.6)^2}{6.4} + \frac{(-3.6)^2}{13.6} + \frac{(2.4)^2}{13.6} + \frac{(-0.4)^2}{6.4} \\
&= 1.80.
\end{aligned} \tag{12.6}$$

Table 12.4. The data of Table 12.1, shown here with the differences $O_k - E_k$.

Bin number k	Observed number O_k	Expected number $E_k = NProb_k$	Difference $O_k - E_k$
1	8	6.4	1.6
2	10	13.6	−3.6
3	16	13.6	2.4
4	6	6.4	−0.4

Because the value of 1.80 for χ^2 is less than the number of terms in the sum (namely, 4), we have no reason to doubt our hypothesis that our measurements were distributed normally.

Quick Check 12.1.   Each of the 100 students in a class measures the time for a ball to fall from a third-story window. They calculate their mean $\bar{t}$ and standard deviation σ_t and then group their measurements into four bins, chosen as in the example just discussed. Their results are as follows:

less than $(\bar{t} - \sigma_t)$: 19
between $(\bar{t} - \sigma_t)$ and $\bar{t}$: 30
between $\bar{t}$ and $(\bar{t} + \sigma_t)$: 37
more than $(\bar{t} + \sigma_t)$: 14.

Assuming their measurements are normally distributed, what are the expected numbers of measurements in each of the four bins? What is χ^2, and is there reason to doubt that the measurements *are* distributed normally?

12.2   General Definition of Chi Squared

The discussion so far has focused on one particular example, 40 measurements of a continuous variable x, which denoted the range of a projectile fired from a certain gun. We defined the number χ^2 and saw that it is at least a rough measure of the agreement between our observed distribution of measurements and the Gauss distribution we expected our measurements to follow. We can now define and use χ^2 in the same way for many different experiments.

Let us consider any experiment in which we measure a number x and for which we have reason to expect a certain distribution of results. We imagine repeating the measurement many times (N) and, having divided the range of possible results x into n bins, $k = 1, \ldots, n$, we count the number O_k of observations that actually fall into each bin k. Assuming the measurements really are governed by the expected distribution, we next calculate the expected number E_k of measurements in the kth bin. Finally, we calculate χ^2 exactly as in (12.5),

$$\chi^2 = \sum_{k=1}^{n} \frac{(O_k - E_k)^2}{E_k}.$$

(12.7)

The approximate significance of χ^2 is always the same as in our previous example. That is, if $\chi^2 \lesssim n$, the agreement between our observed and expected distributions is acceptable; if $\chi^2 \gg n$, there is significant disagreement.

The procedure for choosing the bins in terms of which χ^2 is computed depends somewhat on the nature of the particular experiment. Specifically, it depends on whether the measured quantity x is continuous or discrete. I will discuss these two situations in turn.

MEASUREMENTS OF A CONTINUOUS VARIABLE

The example discussed in Section 12.1 involved a continuous variable x, and little more needs to be said. The only limiting distribution we have discussed for a continuous variable is the Gauss distribution, but there are, of course, many different distributions that can occur. For example, in many atomic and nuclear experiments, the expected distribution of the measured variable x (actually an energy) is the Lorentzian distribution

$$f(x) \propto \frac{1}{(x - X)^2 + \gamma^2},$$

where X and γ are certain constants. Another example of a continuous distribution, mentioned in Problem 5.6, is the exponential distribution $\frac{1}{\tau} e^{-t/\tau}$, which gives the probability that a radioactive atom (whose expected mean life is τ) will live for a time t.

Whatever the expected distribution $f(x)$, the total area under the graph of $f(x)$ against x is one, and the probability of a measurement between $x = a$ and $x = b$ is just the area between a and b,

$$Prob(a < x < b) = \int_a^b f(x)\, dx.$$

Thus, if the kth bin runs from $x = a_k$ to $x = a_{k+1}$, the expected number of measurements in the kth bin (after N measurements in all) is

$$E_k = N \times Prob(a_k < x < a_{k+1})$$
$$= N \int_{a_k}^{a_{k+1}} f(x)\, dx.$$

(12.8)

When we discuss the quantitative use of the chi-squared test in Section 12.4, we will see that the expected numbers E_k should not be too small. Although there is no definite lower limit, E_k should probably be approximately five or more,

$$E_k \gtrsim 5.$$

(12.9)

We must therefore choose bins in such a way that E_k as given by (12.8) satisfies this condition. We will also see that the number of bins must not be too small. For instance, in the example of Section 12.1, where the expected distribution was a Gauss distribution whose center X and width σ were not known in advance, the chi-squared test cannot work (as we will see) with less than four bins; that is, in this example we needed to have

$$n \geqslant 4. \qquad (12.10)$$

Combining (12.9) and (12.10), we see that we cannot usefully apply the chi-squared test to this kind of experiment if our total number of observations is less than about 20.

MEASUREMENT OF A DISCRETE VARIABLE

Suppose we measure a discrete variable, such as the now-familiar number of aces when we throw several dice. In practice, the most common discrete variable is an integer (such as the number of aces), and we will denote the discrete variable by v instead of x (which we use for a continuous variable). If we throw five dice, the possible values of v are $v = 0, 1, \ldots, 5$, and we do not actually need to group the possible results into bins. We can simply count how many times we got each of the six possible results. In other words, we can choose six bins, each of which contains just one result.

Nonetheless, it is often desirable to group several different results into one bin. For instance, if we threw our five dice 200 times, then (according to the probabilities found in Problem 10.11) the expected distribution of results is as shown in the first two columns of Table 12.5. We see that here the expected numbers of throws giving four and five aces are 0.6 and 0.03, respectively, both much less than the five or so occurrences required in each bin if we want to use the chi-squared test. This difficulty is easily remedied by grouping the results $v = 3, 4$, and 5 into a single bin. This grouping leaves us with four bins, $k = 1, 2, 3, 4$, which are shown with their corresponding expected numbers E_k, in the last two columns of Table 12.5.

Table 12.5. Expected occurrence of v aces ($v = 0, 1, \ldots, 5$) after throwing five dice 200 times.

Result	Expected occurrences	Bin number k	Expected number E_k
No aces	80.4	1	80.4
One	80.4	2	80.4
Two	32.2	3	32.2
Three	6.4 ⎫		
Four	0.6 ⎬	4	7.0
Five	0.03 ⎭		

Having chosen bins as just described, we could count the observed occurrences O_k in each bin. We could then compute χ^2 and see whether the observed and expected distributions seem to agree. In this experiment, we know that the expected distribution is certainly the binomial distribution $B_{5,1/6}(v)$ *provided* the dice are true (so that p really is $\frac{1}{6}$). Thus, our test of the distribution is, in this case, a test of whether the dice are true or loaded.

In any experiment involving a discrete variable, the bins can be chosen to contain just one result each, provided the expected number of occurrences for each bin is at least the needed five or so. Otherwise, several different results should be grouped together into a single larger bin that does include enough expected occurrences.

OTHER FORMS OF CHI SQUARED

The notation χ^2 has been used earlier in the book, in Equations (7.6) and (8.5); it could also have been used for the sum of squares in (5.41). In all these cases, χ^2 is a sum of squares with the general form

$$\chi^2 = \sum_1^n \left(\frac{\text{observed value} - \text{expected value}}{\text{standard deviation}}\right)^2. \qquad (12.11)$$

In all cases, χ^2 is an indicator of the agreement between the observed and expected values of some variable. If the agreement is good, χ^2 will be of order n; if it is poor, χ^2 will be much greater than n.

Unfortunately, we can use χ^2 to test this agreement only if we know the expected values and the standard deviation, and can therefore calculate (12.11). Perhaps the most common situation in which these values are known accurately enough is the kind of test discussed in this chapter, namely, a test of a distribution, in which E_k is given by the distribution, and the standard deviation is $\sqrt{E_k}$. Nevertheless, the chi-squared test is of very wide application. Consider, for example, the problem discussed in Chapter 8, the measurement of two variables x and y, where y is expected to be some definite function of x,

$$y = f(x)$$

(such as $y = A + Bx$). Suppose we have N measured pairs (x_i, y_i), where the x_i have negligible uncertainty and the y_i have known uncertainties σ_i. Here, the expected value of y_i is $f(x_i)$, and we could test how well y fits the function $f(x)$ by calculating

$$\chi^2 = \sum_1^N \left(\frac{y_i - f(x_i)}{\sigma_i}\right)^2.$$

All our previous remarks about the expected value of χ^2 would apply to this number, and the quantitative tests described in the following sections could be used. This important application will not be pursued here, because only rarely in the introductory physics laboratory would the uncertainties σ_i be known reliably enough (but see Problem 12.14).

12.3 Degrees of Freedom and Reduced Chi Squared

I have argued that we can test agreement between an observed and an expected distribution by computing χ^2 and comparing it with the number of bins used in collecting the data. A slightly better procedure, however, is to compare χ^2, not with the number of bins n, but instead with the *number of degrees of freedom,* denoted d. The notion of degrees of freedom was mentioned briefly in Section 8.3, and we must now discuss it in more detail.

In general, the number of degrees of freedom d in a statistical calculation is defined as the number of observed data *minus* the number of parameters computed from the data and used in the calculation. For the problems considered in this chapter, the observed data are the numbers of observations O_k in the n bins, $k = 1, \ldots, n$. Thus, the number of observed data is just n, the number of bins. Therefore, in the problems considered here,

$$d = n - c,$$

where n is the number of bins and c is the number of parameters that had to be calculated from the data to compute the expected numbers E_k. The number c is often called the number of *constraints,* as I will explain shortly.

The number of constraints c varies according to the problem under consideration. Consider first the dice-throwing experiment of Section 12.2. If we throw five dice and are testing the hypothesis that the dice are true, the expected distribution of numbers of aces is the binomial distribution $B_{5,1/6}(\nu)$, where $\nu = 0, \ldots, 5$ is the number of aces in any one throw. Both parameters in this function—the number of dice, five, and the probability of an ace, $\frac{1}{6}$—are known in advance and do not have to be calculated from the data. When we calculate the expected number of occurrences of any particular ν, we must multiply the binomial probability by the total number of throws N (in our example, $N = 200$). This parameter *does* depend on the data. Specifically, N is just the sum of the numbers O_k,

$$N = \sum_{k=1}^{n} O_k. \tag{12.12}$$

Thus, in calculating the expected results of our dice experiment, we have to calculate just one parameter (N) from the data. The number of constraints is, therefore,

$$c = 1,$$

and the number of degrees of freedom is

$$d = n - 1.$$

In Table 12.5, the results of the dice experiment were grouped into four bins (that is, $n = 4$), so that experiment had 3 degrees of freedom.

The equation (12.12) illustrates well the curious terminology of constraints and degrees of freedom. Once the number N has been determined, we can regard (12.12) as an equation that "constrains" the values of $O_1, \ldots, O_n$. More specifically, we can say that, because of the constraint (12.12), only $n - 1$ of the numbers $O_1, \ldots, O_n$ are independent. For instance, the first $n - 1$ numbers $O_1, \ldots, O_{n-1}$ could take any value (within certain ranges), but the last number O_n would be completely determined by Equation (12.12). In this sense, only $n - 1$ of the data are *free* to take on independent values, so we say there are only $n - 1$ independent degrees of freedom.

In the first example in this chapter, the range x of a projectile was measured 40 times ($N = 40$). The results were collected into four bins ($n = 4$) and compared with what we would expect for a Gauss distribution $G_{X,\sigma}(x)$. Here, there were *three* constraints and hence only one degree of freedom,

$$d = n - c = 4 - 3 = 1.$$

The first constraint is the same as (12.12): The total number of observations N is the sum of the observations O_k in all the bins. But here there were two more constraints, because (as is usual in this kind of experiment) we did not know in advance the parameters X and σ of the expected Gauss distribution $G_{X,\sigma}(x)$. Thus, before we could calculate the expected numbers E_k, we had to estimate X and σ using the data. Therefore, there were three constraints in all, so in this example

$$d = n - 3. \tag{12.13}$$

Incidentally, this result explains why we had to use at least four bins in this experiment. We will see that the number of degrees of freedom must always be one or more, so, from (12.13), we clearly had to choose $n \geqslant 4$.

The examples considered here will always have at least one constraint (namely, the constraint $N = \Sigma O_k$, involving the total number of measurements), and there may be one or two more. Thus, the number of degrees of freedom, d, will range from $n - 1$ to $n - 3$ (in our examples). When n is large, the difference between n and d is fairly unimportant, but when n is small (as it often is, unfortunately), there is obviously a significant difference.

Armed with the notion of degrees of freedom, we can now begin to make our chi-squared test more precise. It can be shown (though I will not do so) that the *expected* value of χ^2 is precisely d, the number of degrees of freedom,

$$(\text{expected average value of } \chi^2) = d. \tag{12.14}$$

This important equation does not mean that we really expect to find $\chi^2 = d$ after any one series of measurements. It means instead that if we could repeat our whole series of measurements infinitely many times and compute χ^2 each time, the average of these values of χ^2 would be d. Nonetheless, even after just *one* set of measurements, a comparison of χ^2 with d is an indicator of the agreement. In particular, if our expected distribution was the *correct* distribution, χ^2 would be very unlikely to be a lot larger than d. Turning this statement around, if we find $\chi^2 \gg d$, we can assert that our expected distribution was most unlikely to be correct.

We have *not* proved the result (12.14), but we can see that some aspects of the result are reasonable. For example, because $d = n - c$, we can rewrite (12.14) as

$$(\text{expected average value of } \chi^2) = n - c. \tag{12.15}$$

That is, for any given n, the expected value of χ^2 will be smaller when c is larger (that is, if we calculate more parameters from the data). This result is just what we should expect. In the example of Section 12.1, we used the data to calculate the center X and width σ of the expected distribution $G_{X,\sigma}(x)$. Naturally, because X and σ were chosen to fit the data, we would expect to find a somewhat better agreement between the observed and expected distributions; that is, these two extra constraints would be expected to reduce the value of χ^2. This reduction is just what (12.15) implies.

The result (12.14) suggests a slightly more convenient way to think about our chi-squared test. We introduce a *reduced chi squared* (or *chi squared per degree of freedom*), which we denote by $\tilde{\chi}^2$ and define as

$$\tilde{\chi}^2 = \chi^2/d. \tag{12.16}$$

Because the expected value of χ^2 is d, we see that the

$$(\text{expected average value of } \tilde{\chi}^2) = 1. \tag{12.17}$$

Thus, whatever the number of degrees of freedom, our test can be stated as follows: If we obtain a value of $\tilde{\chi}^2$ of order one or less, then we have no reason to doubt our expected distribution; if we obtain a value of $\tilde{\chi}^2$ much larger than one, our expected distribution is unlikely to be correct.

Quick Check 12.2.   For the experiment of Quick Check 12.1, what is the number of degrees of freedom, and what is the value of the reduced chi squared, $\tilde{\chi}^2$?

12.4   Probabilities for Chi Squared

Our test for agreement between observed data and their expected distribution is still fairly crude. We now need a *quantitative* measure of agreement. In particular, we need some guidance on where to draw the boundary between agreement and disagreement. For example, in the experiment of Section 12.1, we made 40 measurements of a certain range x whose distribution should, we believed, be Gaussian. We collected our data into four bins, and found that $\chi^2 = 1.80$. With three constraints, there was only one degree of freedom ($d = 1$), so the reduced chi squared, $\tilde{\chi}^2 = \chi^2/d$, is also 1.80,

$$\tilde{\chi}^2 = 1.80.$$

The question is now: Is a value of $\tilde{\chi}^2 = 1.80$ sufficiently larger than one to rule out our expected Gauss distribution or not?

To answer this question, we begin by supposing that our measurements *were* governed by the expected distribution (a Gaussian, in this example). With this assumption, we can calculate the *probability* of obtaining a value of $\tilde{\chi}^2$ as large as, or larger than, our value of 1.80. Here, this probability turns out to be

$$Prob(\tilde{\chi}^2 \geqslant 1.80) \approx 18\%,$$

as we will soon see. That is, if our results were governed by the expected distribution, there would be an 18% probability of obtaining a value of $\tilde{\chi}^2$ greater than or equal to our actual value 1.80. In other words, in this experiment a value of $\tilde{\chi}^2$ as large as 1.80 is not at all unreasonable, so we would have no reason (based on this evidence) to reject our expected distribution.

Our general procedure should now be reasonably clear. After completing any series of measurements, we calculate the reduced chi squared, which we now denote by $\tilde{\chi}_o^2$ (where the subscript o stands for "observed," because $\tilde{\chi}_o^2$ is the value actually observed). Next, assuming our measurements do follow the expected distribution, we compute the probability

$$Prob(\tilde{\chi}^2 \geq \tilde{\chi}_o^2) \qquad (12.18)$$

of finding a value of $\tilde{\chi}^2$ greater than or equal to the observed value $\tilde{\chi}_o^2$. If this probability is high, our value $\tilde{\chi}_o^2$ is perfectly acceptable, and we have no reason to reject our expected distribution. If this probability is unreasonably low, a value of $\tilde{\chi}^2$ as large as our observed $\tilde{\chi}_o^2$ is very unlikely (if our measurements were distributed as expected), and our expected distribution is correspondingly unlikely to be correct.

As always with statistical tests, we have to decide on the boundary between what is reasonably probable and what is not. Two common choices are those already mentioned in connection with correlations. With the boundary at 5%, we would say that our observed value $\tilde{\chi}_o^2$ indicates a significant disagreement if

$$Prob(\tilde{\chi}^2 \geq \tilde{\chi}_o^2) < 5\%,$$

and we would reject our expected distribution at the 5% significance level. If we set the boundary at 1%, then we could say that the disagreement is highly significant if $Prob(\tilde{\chi}^2 \geq \tilde{\chi}_o^2) < 1\%$ and reject the expected distribution at the 1% significance level.

Whatever level you choose as your boundary for rejection, the level chosen should be stated. Perhaps even more important, you should state the probability $Prob(\tilde{\chi}^2 \geq \tilde{\chi}_o^2)$, so that your readers can judge its reasonableness for themselves.

The calculation of the probabilities $Prob(\tilde{\chi}^2 \geq \tilde{\chi}_o^2)$ is too complicated to describe in this book. The results can be tabulated easily, however, as in Table 12.6 or in the more complete table in Appendix D. The probability of getting any particular values of $\tilde{\chi}^2$ depends on the number of degrees of freedom. Thus, we will write the probability of interest as $Prob_d(\tilde{\chi} \geq \tilde{\chi}_o^2)$ to emphasize its dependence on d.

The usual calculation of the probabilities $Prob_d(\tilde{\chi}^2 \geq \tilde{\chi}_o^2)$ treats the observed numbers O_k as continuous variables distributed around their expected values E_k according to a Gauss distribution. In the problems considered here, O_k is a discrete variable distributed according to the Poisson distribution.[2] Provided all numbers involved are reasonably large, the discrete character of the O_k is unimportant, and the Poisson distribution is well approximated by the Gauss function. Under these conditions, the tabulated probabilities $Prob_d(\tilde{\chi}^2 \geq \tilde{\chi}_o^2)$ can be used safely. For this reason, we have said the bins must be chosen so that the expected count E_k in each bin is reasonably large (at least five or so). For the same reason, the number of bins should not be too small.

---

[2]I have argued that finding the number O_k amounts to a counting experiment and hence that O_k should follow a Poisson distribution. If the bin k is too large, then this argument is not strictly correct, because the probability of a measurement in the bin is not much less than one (which is one of the conditions for the Poisson distribution, as mentioned in Section 11.1), so we must have a reasonable number of bins.

Table 12.6. The percentage probability $Prob_d(\tilde{\chi}^2 \geqslant \tilde{\chi}_o^2)$ of obtaining a value of $\tilde{\chi}^2$ greater than or equal to any particular value $\tilde{\chi}_o^2$, assuming the measurements concerned are governed by the expected distribution. Blanks indicate probabilities less than 0.05%. For a more complete table, see Appendix D.

	$\tilde{\chi}_o^2$												
d	0	0.25	0.5	0.75	1.0	1.25	1.5	1.75	2	3	4	5	6
1	100	62	48	39	32	26	22	19	16	8	5	3	1
2	100	78	61	47	37	29	22	17	14	5	2	0.7	0.2
3	100	86	68	52	39	29	21	15	11	3	0.7	0.2	—
5	100	94	78	59	42	28	19	12	8	1	0.1	—	—
10	100	99	89	68	44	25	13	6	3	0.1	—	—	—
15	100	100	94	73	45	23	10	4	1	—	—	—	—

With these warnings, we now give the calculated probabilities $Prob_d(\tilde{\chi}^2 \geqslant \tilde{\chi}_o^2)$ for a few representative values of d and $\tilde{\chi}_o^2$ in Table 12.6. The numbers in the left column give six choices of d, the number of degrees of freedom (d = 1, 2, 3, 5, 10, 15). Those in the other column heads give possible values of the observed $\tilde{\chi}_o^2$. Each cell in the table shows the percentage probability $Prob_d(\tilde{\chi}^2 \geqslant \tilde{\chi}_o^2)$ as a function of d and $\tilde{\chi}_o^2$. For example, with 10 degrees of freedom ($d = 10$), we see that the probability of obtaining $\tilde{\chi}^2 \geqslant 2$ is 3%,

$$Prob_{10}(\tilde{\chi}^2 \geqslant 2) \;=\; 3\%.$$

Thus, if we obtained a reduced chi squared of 2 in an experiment with 10 degrees of freedom, we could conclude that our observations differed significantly from the expected distribution and reject the expected distribution at the 5% significance level (though not at the 1% level).

The probabilities in the second column of Table 12.6 are all 100%, because $\tilde{\chi}^2$ is always certain to be greater than or equal to 0. As $\tilde{\chi}_o^2$ increases, the probability of getting $\tilde{\chi}^2 \geqslant \tilde{\chi}_o^2$ diminishes, but it does so at a rate that depends on d. Thus, for 2 degrees of freedom ($d = 2$), $Prob_d(\tilde{\chi}^2 \geqslant 1)$ is 37%, whereas for $d = 15$, $Prob_d(\tilde{\chi}^2 \geqslant 1)$ is 45%. Note that $Prob_d(\tilde{\chi}^2 \geqslant 1)$ is always appreciable (at least 32%, in fact), so a value for $\tilde{\chi}_o^2$ of 1 or less is perfectly reasonable and never requires rejection of the expected distribution.

The minimum value of $\tilde{\chi}_o^2$ that does require questioning the expected distribution depends on d. For 1 degree of freedom, we see that $\tilde{\chi}_o^2$ can be as large as 4 before the disagreement becomes significant (5% level). With 2 degrees of freedom, the corresponding boundary is $\tilde{\chi}_o^2 = 3$; for $d = 5$, it is closer to 2 ($\tilde{\chi}_o^2 = 2.2$, in fact), and so on.

Armed with the probabilities in Table 12.6 (and Appendix D), we can now assign a quantitative significance to the value of $\tilde{\chi}_o^2$ obtained in any particular experiment. Section 12.5 gives some examples.

Quick Check 12.3. Each student in a large class times a glider on an air track as it coasts the length of the track. They calculate their mean time and standard deviation and then divide their data into six bins. Assuming their measurements ought to be normally distributed, they calculate the numbers of measurements expected in each bin and the reduced chi squared, for which they get 4.0. If their measurements really were normally distributed, what would have been the probability of getting a value of $\tilde{\chi}^2$ this large? Is there reason to think the measurements were *not* normally distributed?

12.5   Examples

We have already analyzed rather completely the example of Section 12.1. In this section, we consider three more examples to illustrate the application of the chi-squared test.

Example: Another Example of the Gauss Distribution

The example of Section 12.1 involved a measurement for which the results were expected to be distributed normally. The normal, or Gauss, distribution is so common that we consider briefly another example. Suppose an anthropologist is interested in the heights of the natives on a certain island. He suspects that the heights of the adult males should be normally distributed and measures the heights of a sample of 200 men. Using these measurements, he calculates the mean and standard deviation and uses these numbers as best estimates for the center X and width parameter σ of the expected normal distribution $G_{X,\sigma}(x)$. He now chooses eight bins, as shown in the first two columns of Table 12.7, and groups his observations, with the results shown in the third column.

Table 12.7. Measurements of the heights of 200 adult males.

Bin number k	Heights in bin	Observed number O_k	Expected number E_k
1	less than $X - 1.5\sigma$	14	13.4
2	between $X - 1.5\sigma$ and $X - \sigma$	29	18.3
3	between $X - \sigma$ and $X - 0.5\sigma$	30	30.0
4	between $X - 0.5\sigma$ and X	27	38.3
5	between X and $X + 0.5\sigma$	28	38.3
6	between $X + 0.5\sigma$ and $X + \sigma$	31	30.0
7	between $X + \sigma$ and $X + 1.5\sigma$	28	18.3
8	more than $X + 1.5\sigma$	13	13.4

Our anthropologist now wants to check whether these results are consistent with the expected normal distribution $G_{X,\sigma}(x)$. To this end, he first calculates the probability $Prob_k$ that any one man has height in any particular bin k (assuming a normal distribution). This probability is the integral of $G_{X,\sigma}(x)$ between the bin boundaries and is easily found from the table of integrals in Appendix B. The expected number E_k in each bin is then $Prob_k$ times the total number of men sampled (200). These numbers are shown in the final column of Table 12.7.

To calculate the expected numbers E_k, the anthropologist had to use three parameters calculated from his data (the total number in the sample and his estimates for X and σ). Thus, although there are eight bins, he had three constraints; so the number of degrees of freedom is $d = 8 - 3 = 5$. A simple calculation using the data of Table 12.7 gives for his reduced chi squared

$$\tilde{\chi}^2 = \frac{1}{d} \sum_{i=1}^{8} \frac{(O_k - E_k)^2}{E_k} = 3.5.$$

Because this value is appreciably larger than one, we immediately suspect that the islanders' heights do not follow the normal distribution. More specifically, we see from Table 12.6 that, if the islanders' heights were distributed as expected, then the probability $Prob_5(\tilde{\chi}^2 \geqslant 3.5)$ of obtaining $\tilde{\chi}^2 \geqslant 3.5$ is approximately 0.5%. By any standards, this value is very improbable, and we conclude that the islanders' heights are very unlikely to be normally distributed. In particular, at the 1% (or highly significant) level, we can reject the hypothesis of a normal distribution of heights.

Example: More Dice

In Section 12.2, we discussed an experiment in which five dice were thrown many times and the number of aces in each throw recorded. Suppose we make 200 throws and divide the results into bins as discussed before. Assuming the dice are true, we can calculate the expected numbers E_k as before. These numbers are shown in the third column of Table 12.8.

Table 12.8. Distribution of numbers of aces in 200 throws of 5 dice.

Bin number k	Results in bin	Expected number E_k	Observed number O_k
1	no aces	80.4	60
2	one ace	80.4	88
3	two aces	32.2	39
4	3, 4, or 5 aces	7.0	13

In an actual test, five dice were thrown 200 times and the numbers in the last column of Table 12.8 were observed. To test the agreement between the observed and expected distributions, we simply note that there are three degrees of freedom (four bins minus one constraint) and calculate

$$\tilde{\chi}^2 = \frac{1}{3} \sum_{k=1}^{4} \frac{(O_k - E_k)^2}{E_k} = 4.16.$$

Referring back to Table 12.6, we see that with three degrees of freedom, the probability of obtaining $\tilde{\chi}^2 \geqslant 4.16$ is approximately 0.7%, *if* the dice are true. We conclude that the dice are almost certainly not true. Comparison of the numbers E_k and O_k in Table 12.8 suggests that at least one die is loaded in favor of the ace.

Example: An Example of the Poisson Distribution

As a final example of the use of the chi-squared test, let us consider an experiment in which the expected distribution is the Poisson distribution. Suppose we arrange a Geiger counter to count the arrival of cosmic-ray particles in a certain region. Suppose further that we count the number of particles arriving in 100 separate one-minute intervals, and our results are as shown in the first two columns of Table 12.9.

Table 12.9. Numbers of cosmic-ray particles observed in 100 separate one-minute intervals.

Counts ν in one minute	Occurrences	Bin number k	Observations O_k in bin k	Expected number E_k
None	7	1	7	7.5
One	17	2	17	19.4
Two	29	3	29	25.2
Three	20	4	20	21.7
Four	16	5	16	14.1
Five	8			
Six	1	6	11	12.1
Seven	2			
Eight or more	0			
Total	100			

Inspection of the numbers in column two immediately suggests that we group all counts $\nu \geq 5$ into a single bin. This choice of six bins ($k = 1, \ldots, 6$) is shown in the third column and the corresponding numbers O_k in column four.

The hypothesis we want to test is that the number ν is governed by a Poisson distribution $P_\mu(\nu)$. Because the expected mean count μ is unknown, we must first calculate the average of our 100 counts. This value is easily found to be $\bar{\nu} = 2.59$, which gives us our best estimate for μ. Using this value $\mu = 2.59$, we can calculate the probability $P_\mu(\nu)$ of any particular count ν and hence the expected numbers E_k as shown in the final column.

In calculating the numbers E_k, we used two parameters based on the data, the total number of observations (100), and our estimate of μ ($\mu = 2.59$). (Note that because the Poisson distribution is completely determined by μ, we did not have to estimate the standard deviation σ. Indeed, because $\sigma = \sqrt{\mu}$, our estimate for μ automatically gives us an estimate for σ.) There are, therefore, two constraints, which reduces our six bins to four degrees of freedom, $d = 4$.

A simple calculation using the numbers in the last two columns of Table 12.9 now gives for the reduced chi squared

$$\tilde{\chi}^2 = \frac{1}{d} \sum_{k=1}^{6} \frac{(O_k - E_k)^2}{E_k} = 0.35.$$

Because this value is less than one, we can conclude immediately that the agreement between our observations and the expected Poisson distribution is satisfactory. More specifically, we see from the table in Appendix D that a value of $\tilde{\chi}^2$ as large as 0.35 is very probable; in fact

$$Prob_4(\tilde{\chi}^2 \geqslant 0.35) \approx 85\%.$$

Thus, our experiment gives us absolutely no reason to doubt the expected Poisson distribution.

The value of $\tilde{\chi}^2 = 0.35$ found in this experiment is actually appreciably less than one, indicating that our observations fit the Poisson distribution very well. This small value does *not*, however, give stronger evidence that our measurements are governed by the expected distribution than would a value $\tilde{\chi}^2 \approx 1$. If the results really are governed by the expected distribution, and if we were to repeat our series of measurements many times, we would expect many different values of $\tilde{\chi}^2$, fluctuating about the average value one. Thus, if the measurements are governed by the expected distribution, a value of $\tilde{\chi}^2 = 0.35$ is just the result of a large chance fluctuation away from the expected mean value. In no way does it give extra weight to our conclusion that our measurements do seem to follow the expected distribution.

If you have followed these three examples, you should have no difficulty applying the chi-squared test to any problems likely to be found in an elementary physics laboratory. Several further examples are included in the problems below. You should certainly test your understanding by trying some of them.

Principal Definitions and Equations of Chapter 12

DEFINITION OF CHI SQUARED

If we make n measurements for which we know, or can calculate, the expected values and the standard deviations, then we define χ^2 as

$$\chi^2 = \sum_1^n \left(\frac{\text{observed value} - \text{expected value}}{\text{standard deviation}} \right)^2.$$
[See 12.11)]

In the experiments considered in this chapter, the n measurements were the numbers, $O_1, \ldots, O_n$, of times that the value of some quantity x was observed in each of n bins. In this case, the expected number E_k is determined by the assumed distribution of x, and the standard deviation is just $\sqrt{E_k}$; therefore,

$$\chi^2 = \sum_{k=1}^n \frac{(O_k - E_k)^2}{E_k}.$$
[See (12.7)]

If the assumed distribution of x is correct, then χ^2 should be of order n. If $\chi^2 \gg n$, the assumed distribution is probably incorrect.

DEGREES OF FREEDOM AND REDUCED CHI SQUARED

If we were to repeat the whole experiment many times, the mean value of χ^2 should be equal to d, the number of *degrees of freedom,* defined as

$$d = n - c,$$

where c is the number of *constraints,* the number of parameters that had to be calculated from the data to compute χ^2.

The *reduced* χ^2 is defined as

$$\tilde{\chi}^2 = \chi^2/d. \qquad \text{[See (12.16)]}$$

If the assumed distribution is correct, $\tilde{\chi}^2$ should be of order 1; if $\tilde{\chi}^2 \gg 1$, the data do not fit the assumed distribution satisfactorily.

PROBABILITIES FOR CHI SQUARED

Suppose you obtain the value $\tilde{\chi}_o^2$ for the reduced chi squared in an experiment. If $\tilde{\chi}_o^2$ is appreciably greater than one, you have reason to doubt the distribution on which your expected values E_k were based. From the table in Appendix D, you can find the probability,

$$Prob_d(\tilde{\chi}^2 \geq \tilde{\chi}_o^2),$$

of getting a value $\tilde{\chi}^2$ as large as $\tilde{\chi}_o^2$, assuming the expected distribution is correct. If this probability is small, you have reason to reject the expected distribution; if it is less than 5%, you would reject the assumed distribution at the 5%, or significant, level; if the probability is less than 1%, you would reject the distribution at the 1%, or highly significant, level.

Problems for Chapter 12

For Section 12.1: Introduction to Chi Squared

12.1. ★ Each member of a class of 50 students is given a piece of the same metal (or what is said to be the same metal) and told to find its density ρ. From the 50 results, the mean $\bar{\rho}$ and standard deviation σ_ρ are calculated, and the class decides to test whether the results are normally distributed. To this end, the measurements are grouped into four bins with boundaries at $\bar{\rho} - \sigma_\rho$, $\bar{\rho}$, and $\bar{\rho} + \sigma_\rho$, and the results are shown in Table 12.10.

Table 12.10. Observed densities of 50 pieces of metal arranged in four bins; for Problem 12.1.

Bin number k	Range of bin	Observed number O_k
1	less than $\bar{\rho} - \sigma_\rho$	12
2	between $\bar{\rho} - \sigma_\rho$ and $\bar{\rho}$	13
3	between $\bar{\rho}$ and $\bar{\rho} + \sigma_\rho$	11
4	more than $\bar{\rho} + \sigma_\rho$	14

Assuming the measurements were normally distributed, with center $\bar{p}$ and width σ_p, calculate the expected number of measurements E_k in each bin. Hence, calculate χ^2. Do the measurements seem to be normally distributed?

12.2. ★★ Problem 4.13 reported 30 measurements of a time t, with mean $\bar{t} = 8.15$ sec and standard deviation $\sigma_t = 0.04$ sec. Group the values of t into four bins with boundaries at $\bar{t} - \sigma_t$, $\bar{t}$, and $\bar{t} + \sigma_t$, and count the observed number O_k in each bin $k = 1, 2, 3, 4$. Assuming the measurements were normally distributed with center at $\bar{t}$ and width σ_t, find the expected number E_k in each bin. Calculate χ^2. Is there any reason to doubt the measurements are normally distributed?

For Section 12.2: General Definition of Chi Squared

12.3. ★ A gambler decides to test a die by throwing it 240 times. Each throw has six possible outcomes, $k = 1, 2, \ldots, 6$, where k is the face showing, and the distribution of his throws is as shown in Table 12.11. If you treat each possible

Table 12.11. Number of occurrences of each face showing on a die thrown 240 times; for Problem 12.3.

Face showing k:	1	2	3	4	5	6
Occurrences O_k:	20	46	35	45	42	52

result k as a separate bin, what is the expected number E_k in each bin, assuming that the die is true? Compute χ^2. Does the die seem likely to be loaded?

12.4. ★★ I throw three dice together a total of 400 times, record the number of sixes in each throw, and obtain the results shown in Table 12.12. Assuming the dice

Table 12.12. Number of occurrences for each result for three dice thrown together 400 times; for Problem 12.4.

Result	Bin number k	Occurrences O_k
No sixes	1	217
One six	2	148
Two or three sixes	3	35

are true, use the binomial distribution to find the expected number E_k for each of the three bins and then calculate χ^2. Do I have reason to suspect the dice are loaded?

12.5. ★★ As a radioactive specimen decays, its activity decreases exponentially as the number of radioactive atoms diminishes. Some radioactive species have mean lives in the millions, and even billions, of years, and for such species the exponential decay of activity is not readily apparent. On the other hand, many species have mean lives of minutes or hours, and for such species the exponential decay is easily observed. The following problem illustrates this latter case.

A student wants to verify the exponential decay law for a material with a known mean life of τ. Starting at time $t = 0$, she counts the decays from a sample in successive intervals of length T; that is, she establishes bins with $0 < t < T$, and $T < t < 2T$, and so on, and she counts the number of decays in each bin. **(a)** According to the exponential decay law, the number of radioactive atoms that remain after time t is $N(t) = N_0 e^{-t/\tau}$, where N_0 is the number of atoms at $t = 0$. Deduce that the number of decays expected in the kth time bin $[(k - 1) T < t < kT]$ is

$$E_k = N_0(e^{T/\tau} - 1)e^{-kT/\tau}. \tag{12.19}$$

(b) For convenience, the student chooses her time interval T equal to the known value of τ, and she gets the results shown in Table 12.13. What are the expected

Table 12.13. Observed number of decays in successive time intervals of a radioactive sample. Note that all decays that occurred after $t = 4T$ have been included in a single bin; for Problem 12.5.

Bin number k	Time interval	Observed decays O_k
1	$0 < t < T$	528
2	$T < t < 2T$	180
3	$2T < t < 3T$	71
4	$3T < t < 4T$	20
5	$4T < t$	16

numbers E_k, according to (12.19), and what is her value for χ^2? Are her results consistent with the exponential decay law? (Note that the initial number N_0 is easily found from the data, because it must be equal to the total number of decays after $t = 0$.)

For Section 12.3: Degrees of Freedom and Reduced Chi Squared

12.6. ★ **(a)** For the experiment of Problem 12.1, find the number of constraints c and the number of degrees of freedom d. **(b)** Suppose now that the accepted value ρ_{acc} of the density was known and that the students decided to test the hypothesis that the results were governed by a normal distribution centered on ρ_{acc}. For this test, how many constraints would there be, and how many degrees of freedom?

12.7. ★ For each of Problems 12.2 to 12.4, find the number of constraints c and the number of degrees of freedom d.

For Sections 12.4 and 12.5: Probabilities for Chi Squared and Examples

12.8. ★ If we observe the distribution of results in some experiment and know the expected distribution, we can calculate the observed $\tilde{\chi}_o^2$ and find the probability $Prob_d(\tilde{\chi}^2 \geq \tilde{\chi}_o^2)$. If this probability is less than 5%, we can then reject the expected distribution at the 5% level. For example, with two degrees of freedom ($d = 2$), any value of $\tilde{\chi}_o^2$ greater than 3.0 would justify rejection of the expected distribution at the 5% level; that is, for $d = 2$, $\tilde{\chi}_o^2 = 3.0$ is the critical value above which we can reject the distribution at the 5% level. Use the probabilities in Appendix D to make a table of the corresponding critical values (5% level) of $\tilde{\chi}_o^2$, for $d = 1, 2, 3, 4, 5, 10, 15, 20, 30$.

12.9. ★   As in Problem 12.8, make a table of the critical values of $\tilde{\chi}_o^2$ for rejection of the expected distribution, but this time at the 1% level, for $d = 1, 2, 3, 4, 5, 10, 15, 20, 30$.

12.10. ★★   For the data of Problem 12.1, compute $\tilde{\chi}^2$. If the measurements were normally distributed, what was the probability of getting a value of $\tilde{\chi}^2$ this large or larger? At the 5% significance level, can you reject the hypothesis that the measurements were normally distributed? At the 1% level? (See Appendix D for the needed probabilities.)

12.11. ★★   In Problem 12.3, find the value of $\tilde{\chi}^2$. Can we conclude that the die was loaded at the 5% significance level? At the 1% level? (See Appendix D for the necessary probabilities.)

12.12. ★★   In Problem 12.4, find the value of $\tilde{\chi}^2$. If the dice really are true, what is the probability of getting a value of $\tilde{\chi}^2$ this large or larger? Explain whether the evidence suggests the dice are loaded. (See Appendix D for the necessary probabilities.)

12.13. ★★   Calculate χ^2 for the data of Problem 11.5, assuming the observations should follow the Poisson distribution with mean count $\mu = 3$. (Group all values $\nu \geqslant 6$ into a single bin.) How many degrees of freedom are there? (Don't forget that μ was given in advance and didn't have to be calculated from the data.) What is $\tilde{\chi}^2$? Are the data consistent with the expected Poisson distribution?

12.14. ★★   The chi-squared test can be used to test how well a set of measurements (x_i, y_i) of two variables fits an expected relation $y = f(x)$, provided the uncertainties are known reliably. Suppose y and x are expected to satisfy a linear relation

$$y = f(x) = A + Bx. \tag{12.20}$$

(For instance, y might be the length of a metal rod and x its temperature.) Suppose that A and B are predicted theoretically to be $A = 50$ and $B = 6$, and that five measurements have produced the values shown in Table 12.14. The uncertainty in

Table 12.14. Five measurements of two variables expected to fit the relation $y = A + Bx$; for Problem 12.14.

x (no uncertainty):	1	2	3	4	5
y (all ± 4):	60	56	71	66	86

the measurements of x is negligible; all the measurements of y have the same standard deviation, which is known to be $\sigma = 4$.

(a) Make a table of the observed and expected values of y_i, and calculate χ^2 as

$$\chi^2 = \sum_{i=1}^{5} \left(\frac{y_i - f(x_i)}{\sigma} \right)^2.$$

(b) Because no parameters were calculated from the data, there are no constraints and hence five degrees of freedom. Calculate $\tilde{\chi}^2$, and use Appendix D to find the probability of obtaining a value of $\tilde{\chi}^2$ this large, assuming y does satisfy (12.20). At the 5% level, would you reject the expected relation (12.20)? (Note that if the constants A and B were not known in advance, you would have to calculate them from the data by the method of least squares. You would then proceed as before, except that there would now be two constraints and only three degrees of freedom.)

12.15. ★★★ Two dice are thrown together 360 times, and the total score is recorded for each throw. The possible totals are 2, 3, . . . , 12, and their numbers of occurrences are as shown in Table 12.15. (a) Calculate the probabilities for each

Table 12.15. Observed occurrences of total scores for two dice thrown together 360 times; for Problem 12.15.

Total score:	2	3	4	5	6	7	8	9	10	11	12
Occurrences:	6	14	23	35	57	50	44	49	39	27	16

total and hence the expected number of occurrences (assuming the dice are true). (b) Calculate χ^2, d, and $\tilde{\chi}^2 = \chi^2/d$. (c) Assuming the dice are true, what is the probability of getting a value of $\tilde{\chi}^2$ this large or larger? (d) At the 5% level of significance, can you reject the hypothesis that the dice are true? At the 1% level?

12.16. ★★★ A certain long-lived radioactive sample is alleged to produce an average of 2 decays per minute. To check this claim, a student counts the numbers of decays in 40 separate one-minute intervals and obtains the results shown in Table 12.16. (The mean life is so long that any depletion of the sample is negligible over

Table 12.16. Observed number of decays in one-minute intervals of a radioactive sample; for Problem 12.16.

Number of decays ν:	0	1	2	3	4	5 or more
Times observed:	11	12	11	4	2	0

the course of all these measurements.) (a) If the Poisson distribution that governs the decays really does have $\mu = 2$, what numbers E_k should the student expect to have found? (Group all observations with $\nu \geqslant 3$ into a single bin.) Calculate χ^2, d, and $\tilde{\chi}^2 = \chi^2/d$. (Don't forget that μ was not calculated from the data.) At the 5% significance level, would you reject the hypothesis that the sample follows the Poisson distribution with $\mu = 2$? (b) The student notices that the actual mean of his results is $\bar{\nu} = 1.35$, and he therefore decides to test whether the data fit a Poisson distribution with $\mu = 1.35$. In this case, what are d and $\tilde{\chi}^2$? Are the data consistent with this new hypothesis?

12.17. ★★★   Chapter 10 described a test for fit to the binomial distribution. We considered n trials, each with two possible outcomes: "success" (with probability p) and "failure" (with probability $1 - p$). We then tested whether the observed number of successes, ν, was compatible with some assumed value of p. Provided the numbers involved are reasonably large, we can also treat this same problem with the chi-squared test, with just two bins—$k = 1$ for successes and $k = 2$ for failures — and one degree of freedom. In the following problem, you will use both methods and compare results. When the numbers are large, you will find the agreement is excellent; when they are small, it is less so but still good enough that chi squared is a useful indicator.

(a) A soup manufacturer believes he can introduce a different dumpling into his chicken dumpling soup without noticeably affecting the flavor. To test this hypothesis, he makes 16 cans labeled "Style X" that contain the old dumpling and 16 cans labeled "Style Y" that contain the new dumpling. He sends one of each type to 16 tasters and asks them which they prefer. If his hypothesis is correct, we should expect eight tasters to prefer X and eight to prefer Y. In the actual test, the number who favored X was $\nu = 11$. Calculate χ^2 and the probability of getting a value this large or larger. Does the test indicate a significant difference between the two kinds of dumpling? Now, calculate the corresponding probability exactly, using the binomial distribution, and compare your results. (Note that the chi-squared test includes deviations away from the expected numbers in both directions. Therefore, for this comparison you should calculate the two-tailed probability for values of ν that deviate from 8 by 3 or more in either direction; that is, $\nu = 11, 12, \ldots, 16$ *and* $\nu = 5, 4, \ldots, 1$.)

(b) Repeat part (a) for the next test, in which the manufacturer makes 400 cans of each style and the number of tasters who prefer X is 225. (In calculating the binomial probabilities, use the Gaussian approximation.)

(c) In part (a), the numbers were small enough that the chi-squared test was fairly crude. (It gave a probability of 13%, compared with the correct value of 21%.) With one degree of freedom, you can improve on the chi-squared test by using an *adjusted* chi-squared, defined as

$$\text{(adjusted } \chi^2) \; = \; \sum_{k=1}^{2} \frac{\left(|O_k - E_k| - \frac{1}{2}\right)^2}{E_k}.$$

Calculate the adjusted χ^2 for the data of part (a) and show that the use of this value, instead of the ordinary χ^2, in the table of Appendix D gives a more accurate value of the probability.[3]

---

[3] We have not justified the use of the adjusted chi squared here, but this example does illustrate its superiority. For more details, see H. L. Alder and E. B. Roessler, *Introduction to Probability and Statistics*, 6th ed. (W. H. Freeman, 1977), p. 263.

Appendixes

Appendix A

Normal Error Integral, I

If the measurement of a continuous variable x is subject to many small errors, all of them random, the expected distribution of results is given by the normal, or Gauss, distribution,

$$G_{X,\sigma}(x) = \frac{1}{\sigma\sqrt{2\pi}} e^{-(x-X)^2/2\sigma^2},$$

where X is the true value of x, and σ is the standard deviation.

The integral of the normal distribution function, $\int_a^b G_{X,\sigma}(x)dx$, is called the *normal error integral,* and is the probability that a measurement falls between $x = a$ and $x = b$,

$$Prob(a \leqslant x \leqslant b) = \int_a^b G_{X,\sigma}(x)\,dx.$$

Table A shows this integral for $a = X - t\sigma$ and $b = X + t\sigma$. This gives the probability of a measurement within t standard deviations on either side of X,

$$
\begin{aligned}
Prob(\text{within } t\sigma) &= Prob(X - t\sigma \leqslant x \leqslant X + t\sigma) \\
&= \int_{X-t\sigma}^{X+t\sigma} G_{X,\sigma}(x)\,dx = \frac{1}{\sqrt{2\pi}} \int_{-t}^{t} e^{-z^2/2}\,dz.
\end{aligned}
$$

This function is sometimes denoted erf(t), but this notation is also used for a slightly different function.

The probability of a measurement *outside* the same interval can be found by subtraction;

$$Prob(\text{outside } t\sigma) = 100\% - Prob(\text{within } t\sigma).$$

For further discussions, see Section 5.4 and Appendix B.

Table A. The percentage probability,
$Prob(\text{within } t\sigma) = \int_{X-t\sigma}^{X+t\sigma} G_{X,\sigma}(x)\,dx$,
as a function of t.

t	0.00	0.01	0.02	0.03	0.04	0.05	0.06	0.07	0.08	0.09
0.0	0.00	0.80	1.60	2.39	3.19	3.99	4.78	5.58	6.38	7.17
0.1	7.97	8.76	9.55	10.34	11.13	11.92	12.71	13.50	14.28	15.07
0.2	15.85	16.63	17.41	18.19	18.97	19.74	20.51	21.28	22.05	22.82
0.3	23.58	24.34	25.10	25.86	26.61	27.37	28.12	28.86	29.61	30.35
0.4	31.08	31.82	32.55	33.28	34.01	34.73	35.45	36.16	36.88	37.59
0.5	38.29	38.99	39.69	40.39	41.08	41.77	42.45	43.13	43.81	44.48
0.6	45.15	45.81	46.47	47.13	47.78	48.43	49.07	49.71	50.35	50.98
0.7	51.61	52.23	52.85	53.46	54.07	54.67	55.27	55.87	56.46	57.05
0.8	57.63	58.21	58.78	59.35	59.91	60.47	61.02	61.57	62.11	62.65
0.9	63.19	63.72	64.24	64.76	65.28	65.79	66.29	66.80	67.29	67.78
1.0	68.27	68.75	69.23	69.70	70.17	70.63	71.09	71.54	71.99	72.43
1.1	72.87	73.30	73.73	74.15	74.57	74.99	75.40	75.80	76.20	76.60
1.2	76.99	77.37	77.75	78.13	78.50	78.87	79.23	79.59	79.95	80.29
1.3	80.64	80.98	81.32	81.65	81.98	82.30	82.62	82.93	83.24	83.55
1.4	83.85	84.15	84.44	84.73	85.01	85.29	85.57	85.84	86.11	86.38
1.5	86.64	86.90	87.15	87.40	87.64	87.89	88.12	88.36	88.59	88.82
1.6	89.04	89.26	89.48	89.69	89.90	90.11	90.31	90.51	90.70	90.90
1.7	91.09	91.27	91.46	91.64	91.81	91.99	92.16	92.33	92.49	92.65
1.8	92.81	92.97	93.12	93.28	93.42	93.57	93.71	93.85	93.99	94.12
1.9	94.26	94.39	94.51	94.64	94.76	94.88	95.00	95.12	95.23	95.34
2.0	95.45	95.56	95.66	95.76	95.86	95.96	96.06	96.15	96.25	96.34
2.1	96.43	96.51	96.60	96.68	96.76	96.84	96.92	97.00	97.07	97.15
2.2	97.22	97.29	97.36	97.43	97.49	97.56	97.62	97.68	97.74	97.80
2.3	97.86	97.91	97.97	98.02	98.07	98.12	98.17	98.22	98.27	98.32
2.4	98.36	98.40	98.45	98.49	98.53	98.57	98.61	98.65	98.69	98.72
2.5	98.76	98.79	98.83	98.86	98.89	98.92	98.95	98.98	99.01	99.04
2.6	99.07	99.09	99.12	99.15	99.17	99.20	99.22	99.24	99.26	99.29
2.7	99.31	99.33	99.35	99.37	99.39	99.40	99.42	99.44	99.46	99.47
2.8	99.49	99.50	99.52	99.53	99.55	99.56	99.58	99.59	99.60	99.61
2.9	99.63	99.64	99.65	99.66	99.67	99.68	99.69	99.70	99.71	99.72
3.0	99.73									
3.5	99.95									
4.0	99.994									
4.5	99.9993									
5.0	99.99994									

Appendix B

Normal Error Integral, II

In certain calculations, a convenient form of the normal error integral is

$$Q(t) = \int_X^{X+t\sigma} G_{X,\sigma}(x)\, dx$$

$$= \frac{1}{\sqrt{2\pi}} \int_0^t e^{-z^2/2}\, dz.$$

(This integral is, of course, just half the integral tabulated in Appendix A.) The probability $Prob(a \leqslant x \leqslant b)$ of a measurement in any interval $a \leqslant x \leqslant b$ can be found from $Q(t)$ by a single subtraction or addition. For example,

$$Prob(X + \sigma \leqslant x \leqslant X + 2\sigma) = Q(2) - Q(1).$$

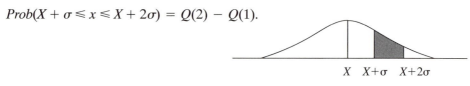

Similarly,

$$Prob(X - 2\sigma \leqslant x \leqslant X + \sigma) = Q(2) + Q(1).$$

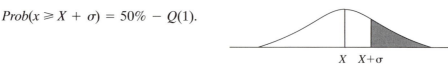

The probability of a measurement greater than any $X + t\sigma$ is just $0.5 - Q(t)$. For example,

$$Prob(x \geqslant X + \sigma) = 50\% - Q(1).$$

Table B. The percentage probability,
$Q(t) = \int_X^{X+t\sigma} G_{X,\sigma}(x)\,dx,$
as a function of t.

t	0.00	0.01	0.02	0.03	0.04	0.05	0.06	0.07	0.08	0.09
0.0	0.00	0.40	0.80	1.20	1.60	1.99	2.39	2.79	3.19	3.59
0.1	3.98	4.38	4.78	5.17	5.57	5.96	6.36	6.75	7.14	7.53
0.2	7.93	8.32	8.71	9.10	9.48	9.87	10.26	10.64	11.03	11.41
0.3	11.79	12.17	12.55	12.93	13.31	13.68	14.06	14.43	14.80	15.17
0.4	15.54	15.91	16.28	16.64	17.00	17.36	17.72	18.08	18.44	18.79
0.5	19.15	19.50	19.85	20.19	20.54	20.88	21.23	21.57	21.90	22.24
0.6	22.57	22.91	23.24	23.57	23.89	24.22	24.54	24.86	25.17	25.49
0.7	25.80	26.11	26.42	26.73	27.04	27.34	27.64	27.94	28.23	28.52
0.8	28.81	29.10	29.39	29.67	29.95	30.23	30.51	30.78	31.06	31.33
0.9	31.59	31.86	32.12	32.38	32.64	32.89	33.15	33.40	33.65	33.89
1.0	34.13	34.38	34.61	34.85	35.08	35.31	35.54	35.77	35.99	36.21
1.1	36.43	36.65	36.86	37.08	37.29	37.49	37.70	37.90	38.10	38.30
1.2	38.49	38.69	38.88	39.07	39.25	39.44	39.62	39.80	39.97	40.15
1.3	40.32	40.49	40.66	40.82	40.99	41.15	41.31	41.47	41.62	41.77
1.4	41.92	42.07	42.22	42.36	42.51	42.65	42.79	42.92	43.06	43.19
1.5	43.32	43.45	43.57	43.70	43.82	43.94	44.06	44.18	44.29	44.41
1.6	44.52	44.63	44.74	44.84	44.95	45.05	45.15	45.25	45.35	45.45
1.7	45.54	45.64	45.73	45.82	45.91	45.99	46.08	46.16	46.25	46.33
1.8	46.41	46.49	46.56	46.64	46.71	46.78	46.86	46.93	46.99	47.06
1.9	47.13	47.19	47.26	47.32	47.38	47.44	47.50	47.56	47.61	47.67
2.0	47.72	47.78	47.83	47.88	47.93	47.98	48.03	48.08	48.12	48.17
2.1	48.21	48.26	48.30	48.34	48.38	48.42	48.46	48.50	48.54	48.57
2.2	48.61	48.64	48.68	48.71	48.75	48.78	48.81	48.84	48.87	48.90
2.3	48.93	48.96	49.98	49.01	49.04	49.06	49.09	49.11	49.13	49.16
2.4	49.18	49.20	49.22	49.25	49.27	49.29	49.31	49.32	49.34	49.36
2.5	49.38	49.40	49.41	49.43	49.45	49.46	49.48	49.49	49.51	49.52
2.6	49.53	49.55	49.56	49.57	49.59	49.60	49.61	49.62	49.63	49.64
2.7	49.65	49.66	49.67	49.68	49.69	49.70	49.71	49.72	49.73	49.74
2.8	49.74	49.75	49.76	49.77	49.77	49.78	49.79	49.79	49.80	49.81
2.9	49.81	49.82	49.82	49.83	49.84	49.84	49.85	49.85	49.86	49.86
3.0	49.87									
3.5	49.98									
4.0	49.997									
4.5	49.9997									
5.0	49.99997									

Appendix C

Probabilities for Correlation Coefficients

The extent to which N points $(x_1, y_1), \ldots, (x_N, y_N)$ fit a straight line is indicated by the linear correlation coefficient

$$r = \frac{\Sigma(x_i - \bar{x})(y_i - \bar{y})}{\sqrt{\Sigma(x_i - \bar{x})^2 \Sigma(y_i - \bar{y})^2}},$$

which always lies in the interval $-1 \leq r \leq 1$. Values of r close to ± 1 indicate a good linear correlation; values close to 0 indicate little or no correlation.

A more quantitative measure of the fit can be found by using Table C. For any given observed value r_o, $Prob_N(|r| \geq |r_o|)$ is the probability that N measurements of two uncorrelated variables would give a coefficient r as large as r_o. Thus, if we obtain a coefficient r_o for which $Prob_N(|r| \geq |r_o|)$ is small, it is correspondingly unlikely that our variables are uncorrelated; that is, a correlation is indicated. In particular, if $Prob_N(|r| \geq |r_o|) \leq 5\%$, the correlation is called *significant*; if it is less than 1%, the correlation is called *highly significant*.

For example, the probability that 20 measurements ($N = 20$) of two uncorrelated variables would yield $|r| \geq 0.5$ is given in the table as 2.5%. Thus, if 20 measurements gave $r = 0.5$, we would have *significant* evidence of a linear correlation between the two variables. For further discussion, see Sections 9.3 to 9.5.

The values in Table C were calculated from the integral

$$Prob_N(|r| \geq |r_o|) = \frac{2\Gamma[(N-1)/2]}{\sqrt{\pi}\Gamma[(N-2)/2]} \int_{|r_o|}^{1} (1 - r^2)^{(N-4)/2}\, dr.$$

See, for example, E. M. Pugh and G. H. Winslow, *The Analysis of Physical Measurements* (Addison-Wesley, 1966), Section 12-8.

Table C. The percentage probability $Prob_N(|r| \geq r_0)$ that N measurements of two uncorrelated variables give a correlation coefficient with $|r| \geq r_0$, as a function of N and r_0. (Blanks indicate probabilities less than 0.05%.)

N	0	0.1	0.2	0.3	0.4	0.5	0.6	0.7	0.8	0.9	1
3	100	94	87	81	74	67	59	51	41	29	0
4	100	90	80	70	60	50	40	30	20	10	0
5	100	87	75	62	50	39	28	19	10	3.7	0
6	100	85	70	56	43	31	21	12	5.6	1.4	0
7	100	83	67	51	37	25	15	8.0	3.1	0.6	0
8	100	81	63	47	33	21	12	5.3	1.7	0.2	0
9	100	80	61	43	29	17	8.8	3.6	1.0	0.1	0
10	100	78	58	40	25	14	6.7	2.4	0.5		0
11	100	77	56	37	22	12	5.1	1.6	0.3		0
12	100	76	53	34	20	9.8	3.9	1.1	0.2		0
13	100	75	51	32	18	8.2	3.0	0.8	0.1		0
14	100	73	49	30	16	6.9	2.3	0.5	0.1		0
15	100	72	47	28	14	5.8	1.8	0.4			0
16	100	71	46	26	12	4.9	1.4	0.3			0
17	100	70	44	24	11	4.1	1.1	0.2			0
18	100	69	43	23	10	3.5	0.8	0.1			0
19	100	68	41	21	9.0	2.9	0.7	0.1			0
20	100	67	40	20	8.1	2.5	0.5	0.1			0
25	100	63	34	15	4.8	1.1	0.2				0
30	100	60	29	11	2.9	0.5					0
35	100	57	25	8.0	1.7	0.2					0
40	100	54	22	6.0	1.1	0.1					0
45	100	51	19	4.5	0.6						0

	0	0.05	0.1	0.15	0.2	0.25	0.3	0.35	0.4	0.45
50	100	73	49	30	16	8.0	3.4	1.3	0.4	0.1
60	100	70	45	25	13	5.4	2.0	0.6	0.2	
70	100	68	41	22	9.7	3.7	1.2	0.3	0.1	
80	100	66	38	18	7.5	2.5	0.7	0.1		
90	100	64	35	16	5.9	1.7	0.4	0.1		
100	100	62	32	14	4.6	1.2	0.2			

Appendix D

Probabilities for Chi Squared

If a series of measurements is grouped into bins $k = 1, \ldots, n$, we denote by O_k the number of measurements observed in the bin k. The number *expected* (on the basis of some assumed or expected distribution) in the bin k is denoted by E_k. The extent to which the observations fit the assumed distribution is indicated by the reduced chi squared, $\tilde{\chi}^2$, defined as

$$\tilde{\chi}^2 = \frac{1}{d} \sum_{k=1}^{n} \frac{(O_k - E_k)^2}{E_k},$$

where d is the number of degrees of freedom, $d = n - c$, and c is the number of constraints (see Section 12.3). The expected average value of $\tilde{\chi}^2$ is 1. If $\tilde{\chi}^2 \gg 1$, the observed results do not fit the assumed distribution; if $\tilde{\chi}^2 \lesssim 1$, the agreement is satisfactory.

This test is made quantitative with the probabilities shown in Table D. Let $\tilde{\chi}_o^2$ denote the value of $\tilde{\chi}^2$ actually obtained in an experiment with d degrees of freedom. The number $Prob_d(\tilde{\chi}^2 \geq \tilde{\chi}_o^2)$ is the probability of obtaining a value of $\tilde{\chi}^2$ as large as the observed $\tilde{\chi}_o^2$, if the measurements really did follow the assumed distribution. Thus, if $Prob_d(\tilde{\chi}^2 \geq \tilde{\chi}_o^2)$ is large, the observed and expected distributions are consistent; if it is small, they probably disagree. In particular, if $Prob_d(\tilde{\chi}^2 \geq \tilde{\chi}_o^2)$ is less than 5%, we say the disagreement is *significant* and reject the assumed distribution at the 5% level. If it is less than 1%, the disagreement is called *highly significant,* and we reject the assumed distribution at the 1% level.

For example, suppose we obtain a reduced chi squared of 2.6 (that is, $\tilde{\chi}_o^2 = 2.6$) in an experiment with six degrees of freedom ($d = 6$). According to Table D, the probability of getting $\tilde{\chi}^2 \geq 2.6$ is 1.6%, if the measurements were governed by the assumed distribution. Thus, at the 5% level (but not quite at the 1% level), we would reject the assumed distribution. For further discussion, see Chapter 12.

Table D. The percentage probability $Prob_d(\tilde{\chi}^2 \geq \tilde{\chi}_o^2)$ of obtaining a value of $\tilde{\chi}^2 \geq \tilde{\chi}_o^2$ in an experiment with d degrees of freedom, as a function of d and $\tilde{\chi}_o^2$. (Blanks indicate probabilities less than 0.05%.)

d	0	0.5	1.0	1.5	2.0	2.5	3.0	3.5	4.0	4.5	5.0	5.5	6.0	8.0	10.0
1	100	48	32	22	16	11	8.3	6.1	4.6	3.4	2.5	1.9	1.4	0.5	0.2
2	100	61	37	22	14	8.2	5.0	3.0	1.8	1.1	0.7	0.4	0.2		
3	100	68	39	21	11	5.8	2.9	1.5	0.7	0.4	0.2	0.1			
4	100	74	41	20	9.2	4.0	1.7	0.7	0.3	0.1	0.1				
5	100	78	42	19	7.5	2.9	1.0	0.4	0.1						

	0	0.2	0.4	0.6	0.8	1.0	1.2	1.4	1.6	1.8	2.0	2.2	2.4	2.6	2.8	3.0
1	100	65	53	44	37	32	27	24	21	18	16	14	12	11	9.4	8.3
2	100	82	67	55	45	37	30	25	20	17	14	11	9.1	7.4	6.1	5.0
3	100	90	75	61	49	39	31	24	19	14	11	8.6	6.6	5.0	3.8	2.9
4	100	94	81	66	52	41	31	23	17	13	9.2	6.6	4.8	3.4	2.4	1.7
5	100	96	85	70	55	42	31	22	16	11	7.5	5.1	3.5	2.3	1.6	1.0
6	100	98	88	73	57	42	30	21	14	9.5	6.2	4.0	2.5	1.6	1.0	0.6
7	100	99	90	76	59	43	30	20	13	8.2	5.1	3.1	1.9	1.1	0.7	0.4
8	100	99	92	78	60	43	29	19	12	7.2	4.2	2.4	1.4	0.8	0.4	0.2
9	100	99	94	80	62	44	29	18	11	6.3	3.5	1.9	1.0	0.5	0.3	0.1
10	100	100	95	82	63	44	29	17	10	5.5	2.9	1.5	0.8	0.4	0.2	0.1
11	100	100	96	83	64	44	28	16	9.1	4.8	2.4	1.2	0.6	0.3	0.1	0.1
12	100	100	96	84	65	45	28	16	8.4	4.2	2.0	0.9	0.4	0.2	0.1	
13	100	100	97	86	66	45	27	15	7.7	3.7	1.7	0.7	0.3	0.1	0.1	
14	100	100	98	87	67	45	27	14	7.1	3.3	1.4	0.6	0.2	0.1		
15	100	100	98	88	68	45	26	14	6.5	2.9	1.2	0.5	0.2	0.1		
16	100	100	98	89	69	45	26	13	6.0	2.5	1.0	0.4	0.1			
17	100	100	99	90	70	45	25	12	5.5	2.2	0.8	0.3	0.1			
18	100	100	99	90	70	46	25	12	5.1	2.0	0.7	0.2	0.1			
19	100	100	99	91	71	46	25	11	4.7	1.7	0.6	0.2	0.1			
20	100	100	99	92	72	46	24	11	4.3	1.5	0.5	0.1				
22	100	100	99	93	73	46	23	10	3.7	1.2	0.4	0.1				
24	100	100	100	94	74	46	23	9.2	3.2	0.9	0.3	0.1				
26	100	100	100	95	75	46	22	8.5	2.7	0.7	0.2					
28	100	100	100	95	76	46	21	7.8	2.3	0.6	0.1					
30	100	100	100	96	77	47	21	7.2	2.0	0.5	0.1					

The values in Table D were calculated from the integral

$$Prob_d(\tilde{\chi}^2 \geq \tilde{\chi}_o^2) = \frac{2}{2^{d/2}\Gamma(d/2)} \int_{\chi_o}^{\infty} x^{d-1} e^{-x^2/2}\, dx.$$

See, for example, E. M. Pugh and G. H. Winslow, *The Analysis of Physical Measurements* (Addison-Wesley, 1966), Section 12-5.

Appendix E

Two Proofs Concerning Sample Standard Deviations

In Chapter 5, I quoted without proof two important results concerning measurements of a quantity x normally distributed with width σ: (1) The best estimate of σ based on N measurements of x is the sample standard deviation, σ_x, of the N measurements, as defined by (5.45) with the factor of $(N - 1)$ in the denominator. (2) The fractional uncertainity in σ_x as an estimate of σ is $1/\sqrt{2(N - 1)}$, as in (5.46). The proofs of these two results are surprisingly awkward and were omitted from Chapter 5. For those who like to see proofs, I give them here.

Consider an experiment that consists of N measurements (all using the same method) of a quantity x, with the results

$$x_1, x_2, \ldots, x_N.$$

The sample standard deviation of these N measurements is defined by

$$\sigma_x^2 = \frac{\Sigma(x_i - \bar{x})^2}{N - 1}$$

$$= \frac{SS}{N - 1}, \tag{E1}$$

where I have introduced the notation SS for the *Sum of Squares,*

$$
\begin{aligned}
SS &= \text{Sum of Squares} \\
&= \sum_1^N (x_i - \bar{x})^2 \\
&= \sum_1^N (x_i)^2 - \frac{1}{N}\left(\sum_1^N x_i\right)^2.
\end{aligned}
\tag{E2}
$$

In writing the last line of (E2), I have used the identity (4.28) from Problem 4.5. Equation (E1) actually defines σ_x *squared*, which is called the *sample variance* of the N measurements. To avoid numerous square root signs, I will work mostly with the variance rather than the standard deviation itself.

Our proofs will be simplified by noting that neither the true width σ nor the sample standard deviation σ_x is changed if we subtract any fixed constant from our measured quantity x. In particular, we can subtract from x its true value X, which gives a quantity normally distributed about a true value of zero. In other words, *we can (and will) assume that the true value of our measured quantity x is $X = 0$.*

To discuss our best estimate for σ and its uncertainty, we must imagine repeating our whole experiment an enormous ("infinite") number of times $\alpha = 1, 2, 3, \ldots$. In this way we generate an immense array of measurements, which we can display as in Table E1.

Table E1. Results of a large number of experiments, $\alpha = 1, 2, 3, \ldots$, each of which consists of N measurements of a quantity x. The ith measurement in the αth experiment is denoted by $x_{\alpha i}$ and is shown in the ith column of the αth row.

Experiment number	1st measurement	2nd measurement		ith measurement		Nth measurement
1	x_{11}	x_{12}	....	x_{1i}	....	x_{1N}
2	x_{21}	x_{22}	....	x_{2i}	....	x_{2N}
....	....	....	....	....	....	....
α	$x_{\alpha 1}$	$x_{\alpha 2}$	....	$x_{\alpha i}$	....	$x_{\alpha N}$
....	....	....	....	....	....	....
....	....	....	....	....	....	....

For each of the infinitely many experiments, we can calculate the mean and variance. For example, for the αth experiment (or αth "sampling"), we find the sample mean $\bar{x}_\alpha$ by adding all the numbers in the row of the αth experiment and dividing by N. To find the corresponding sample variance, we compute the sum of squares

$$SS_\alpha = \sum_{i=1}^{N} (x_{\alpha i} - \bar{x}_\alpha)^2$$

and divide by $(N - 1)$. Here again, the sum runs over all entries in the appropriate row of data.

We use the usual symbol of a bar, as in $\bar{x}_\alpha$, to indicate an average over all measurements in one sample (that is, in one row of our data). We also need to consider the average of all numbers in a given column of our data (that is, an average over all $\alpha = 1, 2, 3, \ldots$), and we denote this kind of average by two brackets $\langle . . \rangle$. For example, we denote by $\langle x_{\alpha i} \rangle$, or just $\langle x \rangle_i$, the average of all the measurements in the ith column. Because this value is the average of infinitely many measurements of x, it equals the true value X, which we have arranged to be zero. Thus,

$$\langle x \rangle_i = X = 0.$$

Similarly, if we average the *squares* of all the values in any column, we will get the true variance

$$\langle x^2 \rangle_i = \sigma^2. \qquad (E3)$$

[Remember that $X = 0$, so that $\langle x^2 \rangle_i$ is the same as $\langle (x - X)^2 \rangle_i$, which is just σ^2.] Armed with these ideas, we are ready for our two proofs.

I   Best Estimate for the Width σ

We want to show that, based on N measurements of x, the best estimate for the true width σ is the sample standard deviation of the N measurements. We do this by proving the following proposition: If we calculate the sum of squares SS_α for each sample α and then average the sums over all $\alpha = 1, 2, 3, \ldots$, the result is $(N-1)$ times the true variance σ^2:

$$\langle SS \rangle = (N-1)\sigma^2. \tag{E4}$$

Dividing (E4) by $(N-1)$, we see the true variance is $\sigma^2 = \langle SS \rangle/(N-1)$. Here, $\langle SS \rangle$ is the average of the sums of squares from infinitely many sets of N measurements. This, in turn, means that the best estimate for σ^2 based on a *single* set of N measurements is just $SS/(N-1)$, where SS is the sum of squares for that single set of measurements. This is just the variance given in Equation (E1). Thus, if we can prove (E4), we will have established the desired result.

To prove Equation (E4), we start with the sum of squares SS_α for the αth sample. Using (E2), we can write this sum as

$$SS_\alpha = \sum_i (x_{\alpha i})^2 - \frac{1}{N}\sum_i \sum_j x_{\alpha i} x_{\alpha j}. \tag{E5}$$

This equation differs from (E2) only in that I have added the subscripts α and have written out the square in the last term of (E2) as a product of two sums. The double sum in (E5) consists of two parts: First, there are N terms with $i = j$, which can be combined with the single sum in (E5). This leaves $N(N-1)$ terms with $i \neq j$. Thus, we can rewrite (E5) as

$$SS_\alpha = \left(1 - \frac{1}{N}\right)\sum_i (x_{\alpha i})^2 - \frac{1}{N}\sum_i \sum_{j \neq i} x_{\alpha i} x_{\alpha j}. \tag{E6}$$

The expression (E6) is the sum of squares for a single sample α. To find $\langle SS \rangle$, we have only to average (E6) over all values of α. Because the average of any sum equals the sum of the corresponding averages, (E6) implies that

$$\langle SS \rangle = \frac{N-1}{N}\sum_i \langle (x_{\alpha i})^2 \rangle - \frac{1}{N}\sum_i \sum_{j \neq i} \langle x_{\alpha i} x_{\alpha j} \rangle. \tag{E7}$$

The N terms in the first sum are all the same, and each is equal to σ^2 [as in (E3)]. Therefore, the first sum is just $N\sigma^2$. With $i \neq j$, the terms of the double sum in (E7) are all zero, $\langle x_{\alpha i} x_{\alpha j} \rangle = 0$. (Remember that x is normally distributed about $X = 0$. Thus, for each possible value of $x_{\alpha i}$, positive and negative values of $x_{\alpha j}$ are equally likely and cancel each other out.) Thus, the whole second sum is zero, and we are left with

$$\langle SS \rangle = (N-1)\sigma^2,$$

which is precisely the result (E4) we needed to prove.

2   Uncertainty in the Estimate for the Width σ

We have shown that, based on N measurements of x, the best estimate for the true variance σ^2 is the sample variance of the N measurements, as defined in (E1), $SS/(N-1)$. Therefore, the fractional uncertainty in σ^2 is the same as that in the sum of squares SS:

(fractional uncertainty in estimate for σ^2) $=$ (fractional uncertainty in SS).   (E8)

Because σ is the square root of σ^2, the fractional uncertainty in σ is just half of this:

$$\text{(fractional uncertainty in estimate for } \sigma)$$
$$= \frac{1}{2}\text{(fractional uncertainty in } SS).\tag{E9}$$

To find the uncertainty in SS, we use the result (11.7) that the uncertainty in any quantity q is $\sqrt{\langle q^2 \rangle - \langle q \rangle^2}$, where, as usual, the brackets $\langle .\,. \rangle$ denote the average of the quantity concerned after an infinite number of measurements. Therefore, the uncertainty in the sum of squares SS is

$$\text{(uncertainty in } SS) = \sqrt{\langle SS^2 \rangle - \langle SS \rangle^2}.\tag{E10}$$

We already know [Equation (E4)] that the second average in this square root is $\langle SS \rangle = (N-1)\sigma^2$, so all that remains is to find the first term $\langle SS^2 \rangle$.

For any one set of N measurements, the sum of squares SS is given by (E6). Squaring this expression, we find that (I omit all subscripts α to reduce the clutter)

$$SS^2 = \left(\frac{N-1}{N}\right)^2 \sum_i (x_i)^2 \sum_j (x_j)^2$$
$$- 2\frac{N-1}{N^2}\sum_i (x_i)^2 \sum_j \sum_{k \neq j} x_j x_k\tag{E11}$$
$$+ \frac{1}{N^2}\sum_j \sum_{k \neq j} x_j x_k \sum_m \sum_{n \neq m} x_m x_n$$
$$= A + B + C.$$

To find the average value of (E11), we need to evaluate each of the three averages $\langle A \rangle$, $\langle B \rangle$, and $\langle C \rangle$. The double sum in A contains N terms with $i = j$, each of which averages to $\langle x^4 \rangle$, and $N(N-1)$ terms with $i \neq j$, each of which averages to $\langle x^2 \rangle^2$. Because x is distributed normally about $X = 0$, $\langle x^2 \rangle = \sigma^2$ (as we already knew), and a straightforward integration shows that $\langle x^4 \rangle = 3\sigma^4$. Thus, the double sum in A averages to $3N\sigma^4 + N(N-1)\sigma^4 = N(N+2)\sigma^4$, and

$$\langle A \rangle = \frac{(N-1)^2(N+2)}{N}\sigma^4.$$

Every term in the sum of B can easily be seen to contain an odd power of x_k (either x_k or x_k^3). Because x is normally distributed about 0, the average of any odd power is zero, so

$$\langle B \rangle = 0.$$

The quadruple sum in the term C contains $N(N - 1)$ terms in which $j = m$ and $k = n$, each of which averages to $\langle x^2 \rangle^2 = \sigma^4$. There is the same number of terms in which $j = n$ and $k = m$, each of which also averages to σ^4. All the remaining terms contain an odd power of x and average to zero. Thus, the quadruple sum in C averages to $2N(N - 1)\sigma^4$, and

$$\langle C \rangle = \frac{2(N - 1)}{N} \sigma^4.$$

Adding together the last three equations and inserting the result into (E11), we conclude that

$$\langle SS^2 \rangle = \frac{(N - 1)^2(N + 2) + 2(N - 1)}{N} \sigma^4 = (N^2 - 1)\sigma^4.$$

Inserting this result into (E10) [and replacing $\langle SS \rangle$ by $(N - 1)\sigma^2$], we find that

$$(\text{uncertainty in } SS) = \sqrt{(N^2 - 1) - (N - 1)^2}\, \sigma^2 = \sqrt{2(N - 1)}\, \sigma^2.$$

If we divide through by $\langle SS \rangle = (N - 1)\sigma^2$, this result implies that the fractional uncertainty in SS is $\sqrt{2/(N - 1)}$. Finally, we know [from (E9)] that the fractional uncertainty in our value of σ is half of that in SS. So,

$$(\text{fractional uncertainty in estimate for } \sigma) = \frac{1}{\sqrt{2(N - 1)}},$$

which is the required result.

Bibliography

I have found the following books useful. They are arranged approximately according to their mathematical level and the completeness of their coverage.

A beautifully clear introduction to statistical methods that contrives to use no calculus is O. L. Lacy, *Statistical Methods in Experimentation* (Macmillan, 1953).

A more advanced book on statistics that is also very clear and uses no calculus is H. L. Alder and E. B. Roessler, *Introduction to Probability and Statistics,* 6th ed. (W. H. Freeman, 1977).

Four books at approximately the level of this book that cover many of the same topics are:

D. C. Baird, *Experimentation; An Introduction to Measurement Theory and Experiment Design,* 2nd ed. (Prentice Hall, 1988);

N. C. Barford, *Experimental Measurements; Precision, Error, and Truth* (Addison-Wesley, 1967);

G. L. Squires, *Practical Physics* (Cambridge, 1985).

H. D. Young, *Statistical Treatment of Experimental Data* (McGraw-Hill, 1962).

Numerous further topics and derivations can be found in the following more advanced books:

P. R. Bevington and K. D. Robinson, *Data Reduction and Error Analysis for the Physical Sciences* (McGraw-Hill, 1992);

S. L. Meyer, *Data Analysis for Scientists and Engineers* (John Wiley, 1975);

E. M. Pugh and G. H. Winslow, *The Analysis of Physical Measurements* (Addison-Wesley, 1966).

ANSWERS
to
Quick Checks
and
Odd-Numbered Problems

Chapter 2

QUICK CHECKS

QC2.1. (a) 110 ± 2 mm. **(b)** between 3.02 and 3.08 amps.

QC2.2. (a) 8.12 ± 0.03 m/s. **(b)** $31{,}234 \pm 2$ m or $(3.1234 \pm 0.0002) \times 10^4$ m.
(c) $(5.68 \pm 0.03) \times 10^{-7}$ kg.

QC2.3. 43 ± 5 grams.

QC2.4. (a) $0.04 = 4\%$. **(b)** $0.1 = 10\%$. **(c)** 4.58 ± 0.09 J.

QC2.5. Percent uncertainties are 1% and 3%; area $= 30$ cm^2 $\pm 4\% = 30$ ± 1 cm^2.

PROBLEMS

2.1. 210 ± 5 cm; 36.0 ± 0.5 mm; 5.3 ± 0.1 V; 2.4 ± 0.1 s.

2.3. (a) 5.03 ± 0.04 m. **(b)** There is a strong case for retaining an extra digit and quoting 1.5 ± 1 s. **(c)** $(-3.2 \pm 0.3) \times 10^{-19}$ C. **(d)** $(5.6 \pm 0.7) \times 10^{-7}$ m. **(e)** $(3.27 \pm 0.04) \times 10^3$ grams $\cdot$ cm/s.

2.5. Discrepancy $= 2$ mm, which is *not* significant.

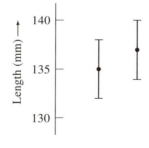

Figure A 2.5.

2.7. (a) Probably the only reasonable conclusion at this stage is 1.9 ± 0.1 gram/cm^3. **(b)** The discrepancy is 0.05 gram/cm^3; because this value is less than the uncertainty, it is *not* significant.

2.9. Length, $l = 93.5 \pm 0.1$ cm.

2.11. The column headed $L - L'$ should read: 0.3 ± 0.9, -0.6 ± 1.5, -2.2 ± 2 (which could be rounded to -2 ± 2), 1 ± 4, 1 ± 4, -4 ± 4. The difference $L - L'$ should theoretically be zero. In all cases but one, the measured value of $L - L'$ is smaller than its uncertainty. In the one exceptional case (-2.2 ± 2), it is only slightly larger. Therefore, the observed values are consistent with the expected value zero.

2.13. (Best estimate for $M + m$) = $M_{best} + m_{best}$. The largest probable value for $M + m$ is $(M_{best} + m_{best}) + (\delta M + \delta m)$, and the smallest probable value is the same expression except for a minus sign. Therefore, the answer for $M + m$ is $(M_{best} + m_{best}) \pm (\delta M + \delta m)$.

2.15. The straight line shown in Figure A2.15 passes through the origin and through, or close to, all error bars. Therefore, the data are consistent with P being proportional to T.

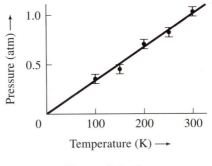

Figure A 2.15.

2.17. The graph of P against I in Figure A2.17 looks as if it would fit a parabola, but it is hard to be sure. The plot of P against I^2 clearly fits a straight line through the origin and is, therefore, consistent with P being proportional to I^2.

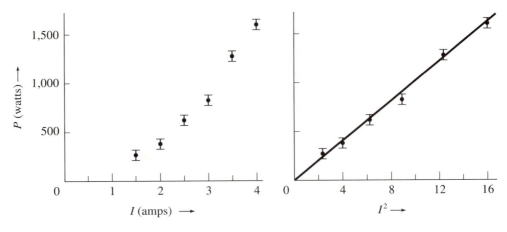

Figure A 2.17.

2.19. (a) In Figure A2.19(a), which includes the origin, it is impossible to say whether T varies with A. Figure A2.19(b) has a much enlarged vertical scale and clearly shows that T does vary with A. Obviously, the best choice of scale

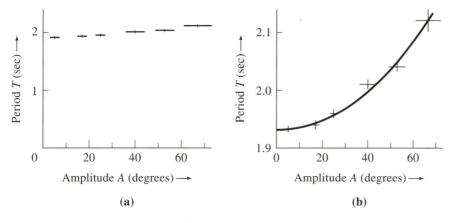

(a)   **(b)**

Figure A 2.19.

for the purpose at hand must be considered carefully. **(b)** If either picture were redrawn with error bars of 0.3s (up and down), there would be no evidence for variation of T with A.

2.21. (a) 40%. **(b)** 4%. **(c)** 0.6%. **(d)** 6%.

2.23. For the meter stick, $\delta l = 0.5$ mm and $\delta l/l = 2.5\%$, which is not precise enough, for the microscope, $\delta l = 0.05$ mm and $\delta l/l = 0.25\%$, which is precise enough.

2.25. (a) The answers for $v_f - v_i$ are 4.0 ± 0.3 cm/s and 0.6 ± 0.4 cm/s. **(b)** The percent uncertainties are 8% and 70%.

2.27. (a) 6.1 ± 0.1, which has two significant figures. **(b)** 1.12 ± 0.02. This value has three figures with at least *some* significance. **(c)** 9.1 ± 0.2, which has two significant figures.

2.29. (a) $q = 2{,}700$ N·s, with an uncertainty of 10% or 270 (or approximately 300) N·s. **(b)** $q = 12$ ft·lb, with an uncertainty of 10% or 1 ft·lb.

2.31. (a) $q_{best} = 10 \times 20 = 200$; (highest probable value of q) $= 11 \times 21 = 231$; (lowest probable value of q) $= 9 \times 19 = 171$. The rule (2.28) gives $q = 200 \pm 30$, which agrees well. **(b)** (Highest probable value of q) $= 18 \times 35 = 630$; (lowest probable value of q) $= 2 \times 5 = 10$. The rule (2.28) gives $q = 200 \pm 300$ (that is, $q_{max} = 500$ and $q_{min} = -100$). The reason this result is so badly incorrect is that the rule (2.28) applies only when all fractional uncertainties are small compared with 1. This condition (which *is* usually met in practice) is not met here.

Chapter 3

QUICK CHECKS

QC3.1. (a) 33 ± 6. **(b)** 910 ± 30. **(c)** The percent uncertainty is 18% in part (a) but only 3% in part (b). His patience in counting for a longer time has been rewarded by a smaller fractional uncertainty.

QC3.2. Percent uncertainties in x, y, and z are 2.5%, 2%, and 2.5%. Hence, $q = 10.0 \pm 7\% = 10.0 \pm 0.7$.

QC3.3. 15.7 ± 0.3 cm.

QC3.4. The percent uncertainty in l is 1%; $V = 8.0$ cm^3 $\pm 3\% = 8.0 \pm 0.2$ cm^3.

QC3.5. $V = 195$ ml; $\delta V = 7$ ml if the uncertainties are independent and random, and $\delta V = 10$ ml otherwise.

QC3.6. $q = 230 \pm 3$; $r = 500 \pm 6.5\% = 500 \pm 30$.

QC3.7. $q = 20 \pm 2$.

QC3.8. $q = 10.0 \pm 0.3$.

QC3.9. $q = 20 \pm 6$.

PROBLEMS

3.1. (a) 28 ± 5. **(b)** 310 ± 20. **(c)** The fractional uncertainty is 18% for A, 6% for B. B's absolute uncertainty is larger, but her patience has been rewarded with a smaller fractional uncertainty.

3.3. The observation is that the number of cases in four years is $20 \pm \sqrt{20} = 20 \pm 4.5$. This value is consistent with the expected number 16, and there is no significant evidence that the rate is abnormally high.

3.5. (and **3.19**)

	Problem 3.5 Straight sum	Problem 3.19 Quadratic sum
(a)	3 ± 7	3 ± 5
(b)	40 ± 18	40 ± 13
(c)	0.5 ± 0.1	0.50 ± 0.07
(d)	300 ± 21	300 ± 12

3.7. (a) 35 ± 4 cm $= 35$ cm $\pm 10\%$. **(b)** 11 ± 4 cm $= 11 \pm 40\%$. **(c)** 36 ± 9 cm $= 36$ cm $\pm 25\%$. **(d)** 110 ± 40 grams·cm/s $= 110$ grams·cm/s $\pm 30\%$.

3.9. $c = 18.8 \pm 0.3$ cm; $r = 3.00 \pm 0.05$ cm.

3.11. (a) 0.48 ± 0.02 s (or 4%). **(b)** 0.470 ± 0.005 s (or 1%). **(c)** No. In the first place, the pendulum will eventually stop, unless it is driven. Even if it is driven, other effects will eventually become important and thwart our quest for greater accuracy. For example, if we time for several hours, the reliability of the stopwatch may become a limiting factor, and the period τ may *vary* because of changing temperature, humidity, and other variables.

3.13. Volume $= 34 \pm 4 \text{ m}^3$.

3.15. **(a)** Student A: 32 ± 6. **(b)** Student B: 790 ± 30. **(c)** A's rate is 16 ± 3 counts/min, and B's is 13.1 ± 0.5 counts/min. These two measurements are compatible, but B has been rewarded with a smaller uncertainty.

3.17. Those cases in which the second uncertainty could be ignored are indicated with an asterisk.

		Uncertainty	
Case	Answer	Quadratic sum	Straight sum
a	9.3	0.7*	0.7*
b	7.9	0.7*	0.8
c	9.7	0.7*	0.9
d	7.5	0.8	1

3.19. See answer to Problem 3.5.

3.21. **(a)** $v = 0.847 \pm 0.003$ m/s. **(b)** $p = 0.602 \pm 0.003$ kg·m/s.

3.23. $L = 0.74 \pm 0.03$ kg·m^2/s.

3.25. The rule (3.18) holds only if the uncertainties in the various factors are independent. When we write x^2 as $x \times x$, it is certainly *not* true that the two factors (x and x) are independent, and the rule (3.18) does not hold.

3.27. Refractive index, $n = 1.52 \pm 0.03$.

3.29. **(a)** $\delta h/h = 0.1\%$. Because h is proportional to a fractional power of λ, the percent uncertainty in h is *smaller* than that in λ. **(b)** The measure value of 6.644×10^{-34} J·s has an uncertainty of 0.1%, or 0.007×10^{-34} J·s. The discrepancy is 0.018×10^{-34} and is more than twice the uncertainty. Therefore, the result is somewhat unsatisfactory.

3.31. **(a)** $\sin(\theta) = 0.82 \pm 0.02$. [Don't forget that $\delta\theta$ must be expressed in radians when using $\delta \sin(\theta) = |\cos(\theta)| \delta\theta$.] **(b)** $f_{best} = e^{a_{best}}$, $\delta f = f_{best} \delta a$, $e^a = 20 \pm 2$. **(c)** $f_{best} = \ln(a_{best})$, $\delta f = \delta a/a_{best}$, $\ln(a) = 1.10 \pm 0.03$.

3.33. **(a)** $q = -1.38$ and $\delta q = 0.0791 \approx 0.08$. **(b)** $\delta q = 0.0787 \approx 0.08$. The two methods agree to two significant figures.

3.35. **(a)** 30 ± 1. **(b)** 120 ± 30. **(c)** 2.0 ± 0.2.

3.37. **(a)** 0.08 ± 0.01 m/s^2. **(b)** This measurement does not agree at all well with the predicted value 0.13 ± 0.01 m/s^2.

3.39. **(a)** $E = 0.247 \pm 0.004$ J. **(b)** $E = 0.251 \pm 0.003$ J. **(c)** The margins of error for the two measurements overlap, and they are, therefore, consistent with conservation of energy.

3.41.

i (deg)	n	$\delta n/n$	δn
10	1.42	17%	0.2
20	1.52	9%	0.14
30	1.46	6%	0.09
50	1.58	3%	0.05
70	1.53	2%	0.03

All but one of the measurements straddle the given value, 1.50. The only exception is for $i = 50$ deg ($n = 1.58 \pm 0.05$), and even here the given value, 1.50, is only slightly outside the margins of error. Therefore, the measurements *are* consistent with the given value.

As the angle i increases, the fractional uncertainty in n *decreases*; this decrease occurs mainly because the absolute uncertainties are constant, so the fractional uncertainties decrease as the angles increase.

3.43. (a) 1 and 1. **(b)** y and x. **(c)** $2xy^3$ and $3x^2y^2$.

3.45. (a) If $q = xy$, the two partial derivatives are y and x. The rule (3.47) gives

$$\delta q = \sqrt{y^2\delta x^2 + x^2\delta y^2}.$$

Dividing the equation by $|q|$ produces exactly the rule (3.18) for the fractional uncertainty in xy. In a similar way, (3.48) gives

$$\delta q \leqslant |y|\,\delta x + |x|\,\delta y,$$

and, when divided by $|q|$, gives the rule (3.19) for the fractional uncertainty in xy. **(b)** If $q = x^n y^m$, then, after division by $|q|$, the rules (3.47) and (3.48) become

$$\frac{\delta q}{|q|} = \sqrt{\left(n\frac{\delta x}{x}\right)^2 + \left(m\frac{\delta y}{y}\right)^2} \quad \text{and} \quad \frac{\delta q}{|q|} \leqslant |n|\frac{\delta x}{|x|} + |m|\frac{\delta y}{|y|}.$$

These equations agree exactly with the answers from our old rules. **(c)** If q depends only on x, then both (3.47) and (3.48) reduce to the old rule (3.23) for a function of one variable.

3.47. Uncertainty, $\delta a = \dfrac{2g}{(M + m)^2}\sqrt{m^2(\delta M)^2 + M^2(\delta m)^2}$; $a = 3.3 \pm 0.1$ m/s^2.

3.49. Either method gives the uncertainty as $\delta f = \dfrac{\sqrt{q^4(\delta p)^2 + p^4(\delta q)^2}}{(p + q)^2}$.

Chapter 4

QUICK CHECKS

QC4.1. Mean = 23 s; SD = 1.8 s.

QC4.2. 16.0 ± 0.7.

PROBLEMS

4.1. Mean = 12; (sample SD) = 1; (population SD) = 0.8.

4.3. Mean = 85.7; SD (sample) = 2.2.

4.5. (a)

$$
\begin{aligned}
\Sigma(x_i - \bar{x})^2 &= \Sigma(x_i^2 - 2\bar{x}x_i + \bar{x}^2) \\
&= \Sigma x_i^2 - 2\bar{x}\Sigma x_i + N\bar{x}^2 \\
&= \Sigma x_i^2 - 2\bar{x}N\bar{x} + N\bar{x}^2 \\
&= \Sigma x_i^2 - N\bar{x}^2 = \Sigma x_i^2 - \frac{1}{N}(\Sigma x_i)^2.
\end{aligned}
$$

(b) With the data of Problem 4.1, either side of (4.28) gives 2.

4.7. (a) Mean = 15.4; SD = 3.5. **(b)** Expected SD = $\sqrt{15.4}$ = 3.9, compared with observed SD = 3.5. **(c)** By the rule (3.26) for the uncertainty in a power, the fractional uncertainty in $\sqrt{\nu}$ is

$$
\frac{\delta\sqrt{\nu}}{\sqrt{\nu}} = \frac{1}{2}\frac{\delta\nu}{\nu} = \frac{1}{2}\frac{\sqrt{\nu}}{\nu} = \frac{1}{2\sqrt{\nu}}.
$$

Therefore, $\delta\sqrt{\nu} = \frac{1}{2}$. This result means the expected value of the SD is 3.9 ± 0.5, which is consistent with the observed 3.5.

4.9. Mean = 1.7 s; SD = 0.1 s; probability = 32%.

4.11. Uncertainty in any single measurement = SD = 0.2 μC.

4.13. (a) Mean = 8.149 sec; SD = 0.039 sec. **(b)** Outside one SD, we expect to find 32% of the measurements, or 9.6, and we got 8. **(c)** Outside two SD, we expect to find 5% of the measurements, or 1.5, and we got 2.

4.15. The best estimate is the mean = 12.0, and uncertainty in this estimate is the SDOM = 0.6.

4.17 (a) (Final answer for time) = mean ± SDOM = 8.149 ± 0.007 s. **(b)** The data have three significant figures, whereas the final answer has four; this result is what we should expect with a large number of measurements because the SDOM is then much smaller than the SD.

4.19. (a) (Mean of 20 counts) = 10.70; (uncertainty in this answer) = SDOM = 0.76. **(b)** (Total count in 40 s) = 214 ± $\sqrt{214}$ = 214 ± 14.6; dividing by 20, we get (mean count in 2 s) = 10.70 ± 0.73. **(c)** Suppose we make N counts, $\nu_1, \ldots, \nu_N$, of the particles in time T. Then, using either argument gives approximately

$$
\text{(best estimate for count in time } T) = \frac{1}{N}\Sigma\nu_i \pm \frac{1}{N}\sqrt{\Sigma\nu_i}.
$$

4.21. (Mean of $A_1, \ldots, A_N$) = 1,221.2; SD = 0.88; SDOM = 0.29. Thus, the final answer this way is 1,221.2 ± 0.3 mm³, compared with 1,221.2 ± 0.4 mm³ the other way.

4.23. By the rule (3.26) for the uncertainty in a power,

$$
\frac{\delta e}{e} = \frac{3}{2}\frac{\delta\eta}{\eta} = 0.6\%.
$$

4.25. (a) 336 ± 15 m/s; the 1% systematic uncertainty in f is negligible beside the 4.5% uncertainty in λ. **(b)** 336 ± 11 m/s; here, the systematic uncertainty dominates.

4.27. (a) $N = 3.18 \pm 0.09$ (or 3%); fractional uncertainty in d is 3%. **(b)** 3%.

Chapter 5

QUICK CHECKS

QC5.1.

Grade:	F	D	C	B	A
s_k:	0	1	2	3	4
n_k:	2	0	7	4	7

$\bar{s} = 2.7$.

QC5.2. See Figure AQC5.2.

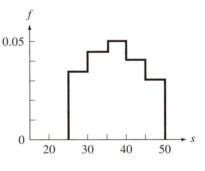

Figure AQC 5.2.

QC5.3. See Figure AQC5.3.

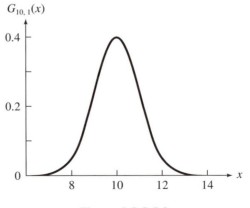

Figure AQC 5.3

QC5.4. *Prob*(between 7 and 13) $= 87\%$; *Prob*(outside 7 to 13) $= 13\%$.

QC5.5. (Answer for σ_q) $= 6 \pm 3\ \mu C$; $\sigma_q - \delta\sigma_q = 3\ \mu C$; $\sigma_q + \delta\sigma_q = 9\ \mu C$.

QC5.6. *Prob*(outside 2.4σ) $= 1.64\%$; therefore, a discrepancy of 2.4σ is significant at the 5% and 2% levels but not at the 1% level.

PROBLEMS

5.1. See Figure A5.1. $\bar{s} = 2.7$.

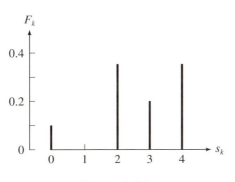

Figure A 5.1.

5.3. (and **5.15.**) See Figure A5.3. The broken curve in part (c) is the Gauss function for Problem 5.15.

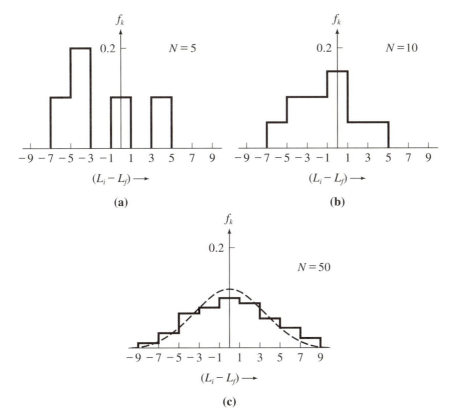

Figure A 5.3.

5.5. (a) *Prob*(outside $-a$ to a) $= 0$. **(b)** *Prob*$(x > 0) = 1/2$. **(c)** $C = 1/a$.
(d) See Figure A5.5.

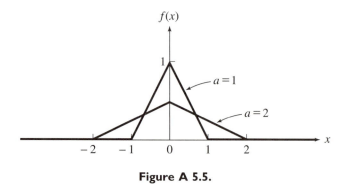

Figure A 5.5.

5.7. *Prob*$(t > \tau) = 1/e = 37\%$; *Prob*$(t > 2\tau) = 1/e^2 = 14\%$.

5.9. (a) $C = 1/(2a)$. **(b)** See Figure A5.9. All values between $-a$ and a are equally likely; no measurements fall outside this range. **(c)** Mean $= 0$; SD $= a/\sqrt{3}$.

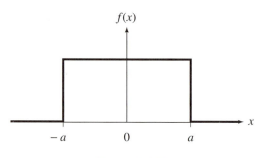

Figure A 5.9.

5.11. See Figure A5.11.

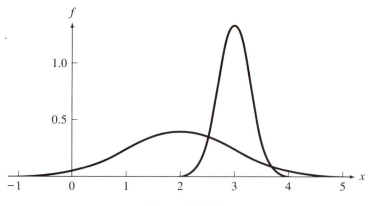

Figure A 5.11.

5.13. Ignoring uninteresting constants,

$$\frac{d^2G}{dx^2} \propto \left[1 - \frac{(x - X)^2}{\sigma^2}\right] e^{-(x-X)^2/2\sigma^2}.$$

This vanishes when $x = X \pm \sigma$.

5.15. See answer to Problem 5.3.

5.17. The 68–95 rule states that, for normally distributed measurements, 68% is the probability of a result within 1σ of X, and 95% is that of a result within 2σ of X. The 68–95–99.7 rule is the same but includes 99.7% as the probability of a result within 3σ of X.

5.19. (a) 68%. **(b)** 38%. **(c)** 95%. **(d)** 48%. **(e)** 14%. **(f)** Between $y = 22.3$ and 23.7.

5.21. (a) 72. **(b)** 5′9″.

5.23. (a) After two integrations by parts, the integral (5.16) reduces to τ^2; therefore $\sigma_t = \tau$. **(b)** The required probability is given by the integral of the distribution function over the range $\bar{t} \pm \sigma_t$, that is, from $t = 0$ to 2τ. This integral is $1 - e^{-2} = 0.86 = 86\%$.

5.25. (a) Best estimate = 77, SD = 2. **(b)** 4%.

5.27. (a) Mean = 53, $\sigma = 1.41 \pm 0.45$; that is, σ could well be anywhere from 0.96 to 1.86. **(b)** If $\sigma = 1.41$, the probability asked for is 7%; if $\sigma = 0.96$, the same probability is only 0.9%; if $\sigma = 1.86$, the probability is 18%. Probabilities of this kind are very sensitive to the value of σ; after a small number of measurements (like 6), our estimate for σ is very uncertain, and calculated probabilities of this kind are, therefore, very unreliable.

5.29. See Figure A5.29.

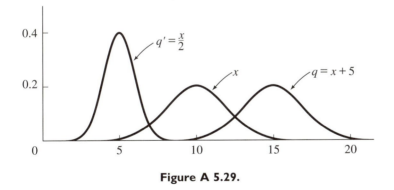

Figure A 5.29.

5.31. (a) SD = 7.04. **(b)** $\bar{t}_1 = 74.25$; $\bar{t}_2 = 67.75$, etc. If $\bar{t}$ denotes the average of any group of four measurements, we should expect $\sigma_{\bar{t}} = \sigma_t/\sqrt{4} = 3.52$; in fact, the SD of the 10 means is 3.56. **(c)** See Figure A5.31.

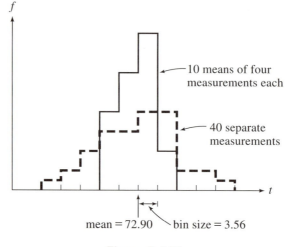

mean = 72.90 ⟍ bin size = 3.56

Figure A 5.31.

5.33. (Best estimate) = 77.0; uncertainty = 0.7; (uncertainty in uncertainty) = 0.2; that is, $t = 77.0 \pm (0.7 \pm 0.2)$.

5.35. The student's answer is 3σ away from the expected answer; the probability for a result this deviant is 0.3%. His discrepancy is highly significant, and there is very likely some undetected systematic error.

5.37. The observed energy change, $E_f - E_i$, is -15 MeV, with a standard deviation of 9.5 MeV. If the measurement was normally distributed around 0, with $\sigma = 9.5$, then the observed value differs from the true value by 15/9.5, or 1.6, standard deviations. Because $Prob$(outside 1.6σ) = 11%, the observed value is perfectly reasonable, and there is no reason to doubt conservation of energy.

Chapter 6

QUICK CHECKS

QC6.1. $Prob = 1.24\%$; the number expected to be this deviant is $1.24\% \times 20 = 0.25$, and according to Chauvenet's criterion, she should reject the suspect value.

PROBLEMS

6.1. (a) $Prob = 4.5\%$; (expected number) = 2.3; no, she should not reject it.
(b) Yes.

6.3. (Expected number) = 0.9; no, she should not reject it.

6.5. (a) Mean = 19.1; SD = 6.3. (b) Reject. (c) New mean = 17.7; SD = 4.3.

6.7. (a) Mean = 25.5; SD = 4.23. (b) Yes, reject; (expected number = 0.46).
(c) $\delta\sigma_x = 1.34$. (d) If $\sigma = \sigma_x - \delta\sigma_x$, the expected number is 0.05, and he would definitely reject the measurement. If $\sigma = \sigma_x + \delta\sigma_x$, the expected number is 1.06, and he would definitely *not* reject the measurement. The range of possible values of σ spans from "definitely reject" to "definitely don't reject." In other words with this few measurements (6), Chauvenet's criterion is ambiguous and should be avoided, if possible.

Chapter 7

QUICK CHECKS

QC7.1. Weights = 4 and 1; weighted average = 10.4 ns.

QC7.2. Uncertainty = 0.4 ns.

PROBLEMS

7.1. Voltage = 1.2 ± 0.1.

7.3. (a) Yes, the two measurements are consistent; their weighted average is 334.4 $\pm$ 0.9 m/s. **(b)** These results are also consistent (more so, in fact); their weighted average is 334.08 $\pm$ 0.98, which should be rounded to 334 $\pm$ 1. The second measurement is not worth including.

7.5. (a) $R_{\text{wav}} = 76 \pm 4$ ohms. **(b)** Approximately 26 measurements.

7.7. If $\sigma_1 = \ldots = \sigma_N = \sigma$, say, then $w_1 = \ldots = w_N = 1/\sigma^2 = w$, say. With all the w's equal, Equation (7.10) for x_{wav} reduces to $\Sigma wx_i/\Sigma w = w\Sigma x_i/Nw$; the factors of w cancel to give the ordinary average. According to (7.12), σ_{wav} = $1/\sqrt{\Sigma w} = 1/\sqrt{Nw}$; because $w = 1/\sigma^2$, this result reduces to $\sigma_{\text{wav}} = \sigma/\sqrt{N}$, which is the ordinary SDOM.

7.9. (a) The details of your spreadsheet will depend on what program you use, but its general appearance should be something like this:

Trial	x_i	σ_i	$w_i = 1/\sigma_i^2$	$w_i x_i$
1	11	1	1.00	11.00
2	12	1	1.00	12.00
3	10	3	0.11	1.11
Σ			2.11	24.11

$\sigma_{\text{wav}} = \Sigma wx/\Sigma w$	$\sigma_{\text{wav}} = 1/\sqrt{\Sigma w}$
11.42	0.69

The cells in the fourth column contain a formula to calculate w_i as $1/\sigma_i^2$. Similarly, the fifth column has a formula to give $w_i x_i$. The cell for x_{wav} (11.42 here) has a formula to calculate x_{wav} as shown, and the cell for σ_{wav} has a formula to calculate σ_{wav}.

(b) Making a similar spreadsheet with 20 rows is easy. If you put in just three data, however, the blank cells for σ_i will be treated as zeros when the program tries to evaluate w_i as $1/\sigma_i^2$, and the program will return an error message. To avoid this difficulty, the formula for w_i needs a conditional command that returns $1/\sigma_i^2$ if the cell for σ_i contains an entry but returns a blank if the cell for σ_i is blank.

Chapter 8

QUICK CHECKS

QC8.1. The best fit is $y = 4 + 3x$, as shown in Figure AQC8.1.

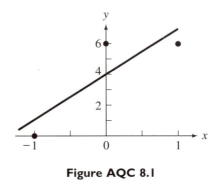

Figure AQC 8.1

PROBLEMS

8.1. The best fit is $y = 7.75 - 1.25x$, as shown by the solid line in Figure A8.1. (The dashed line is for Problem 8.17.)

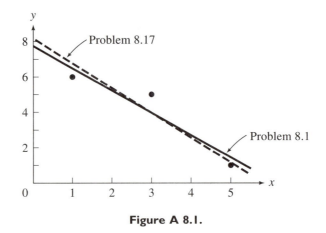

Figure A 8.1.

8.3. If you multiply (8.8) by $\sum x_i^2$ and (8.9) by $\sum x_i$ and then subtract, you will get the given solution for A. Similarly, (8.8) times $\sum x_i$ minus (8.9) times N gives the solution for B.

8.5. The argument parallels closely that leading from (8.2) to (8.12) in Section 8.2; the only important change is that $A = 0$ throughout. Thus, $Prob(y_1, \ldots, y_N) \propto \exp(-\chi^2/2)$ as in (8.4), and χ^2 is given by (8.5), except that $A = 0$. Differentiation with respect to B gives (8.7) (again with $A = 0$), and the solution is $B = (\sum x_i y_i)/(\sum x_i^2)$.

8.7. $l_o = 3.69$ cm; $k = 162$ N/m.

8.9. As in Problem 8.5, the argument closely parallels that leading from (8.2) to (8.12). As in Equation (8.4), $Prob(y_1, \ldots, y_N) \propto \exp(-\chi^2/2)$, but because the measurements have different uncertainties, $\chi^2 = \sum w_i(y_i - A - Bx_i)^2$. (Remember that $w_i = 1/\sigma_i^2$.) The argument then continues as before.

8.11. The details depend on the program and layout you choose, but the general appearance of the worksheet for the data of Problem 8.1 will be as follows: (To save space, my spreadsheet accommodates only six data.)

Trial	x_i	y_i	x_i^2	$x_i y_i$
1	1	6	1	6
2	3	5	9	15
3	5	1	25	5
4			0	0
5			0	0
6			0	0
Σ	9	12	35	26

N	$\Delta = N\Sigma x^2 - (\Sigma x)^2$	$A = (\Sigma x^2 \Sigma y - \Sigma x \Sigma xy)/\Delta$	$B = (N\Sigma xy - \Sigma x \Sigma y)/\Delta$
3	24	7.75	-1.25

Notice that the blank entries in the x_i and y_i columns get treated as zeros; thus (unless you take precautions to prevent this effect), the formulas in columns 4 and 5 return zeros in rows that have no data. Fortunately, these zeros do not affect the answers. The cell for N (3 in this example) contains an instruction to count the number of entries in the x_i column. The cells for Δ, A, and B contain formulas to calculate the appropriate numbers.

8.13. (a) $s_0 = 8.45$ cm; $v = 1.34$ cm/s; $\sigma_s = 0.64$ cm. **(b)** If $\delta s \approx 1$ cm, then δs and σ_s are roughly the same, and the data are consistent with a straight line, as shown in Figure A8.13. **(c)** If $\delta s \approx 0.1$ cm (approximately the size of the dots on the graph), then the data are not consistent with a straight line. (σ_s is approximately 6 times bigger than δs.) From the graph, the glider appears to be slowing down. (Friction?)

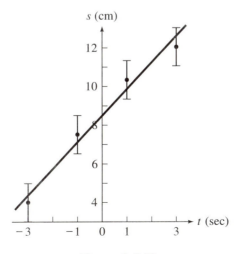

Figure A 8.13.

8.15. $A = -3.90$; $B = 9.03$; $\lambda = 18.1 \pm 0.1$ cm.

8.17. (a) As in Problem 8.1, $A = 7.8 \pm 1.5$; $B = -1.25 \pm 0.4$. **(b)** If we fit to $x = A' + B'y$, then we get $A' = 5.86$ and $B' = -0.714$; rewriting this equation as $y = A + Bx$, we get $A = 8.2$ and $B = -1.4$. These results are consistent with the answers in part (a), as we would expect, based on the discussion of Section 8.4. See Figure A8.1.

8.19. The constant A depends on all of the x_i and y_i, but only the y_i have uncertainties (denoted σ_i). Therefore, by the error propagation formula (3.47),

$$\sigma_A{}^2 = \sum \left(\frac{\partial A}{\partial y_i}\sigma_i\right)^2.$$

The required derivative is

$$\frac{\partial A}{\partial y_i} = \frac{(\sum w_j x_j{}^2)w_i - (\sum w_j x_j)w_i x_i}{\Delta}.$$

If you substitute the second equation into the first and have the courage to wade through some algebra, this process should give the advertised answer for σ_A. [Remember the definition (8.39) of Δ and that $w_i = 1/\sigma_i{}^2$.] A similar argument gives σ_B.

8.21. The probability of obtaining the observed values is given by (8.25), with χ^2 given by (8.26). The maximum probability corresponds to the minimum value of χ^2. The derivatives of χ^2 with respect to A, B, and C are $(-2/\sigma^2)\sum(y_i - A - Bx_i - Cx_i{}^2)$, $(-2/\sigma^2)\sum(y_i - A - Bx_i - Cx_i{}^2)x_i$, and $(-2/\sigma^2)\sum(y_i - A - Bx_i - Cx_i{}^2)x_i{}^2$. Setting these three derivatives equal to zero gives the three equations (8.27).

8.23. The argument closely parallels that of Section 8.2. The probability of getting the observed values $y_1, \ldots, y_N$ is proportional to $\exp(-\chi^2/2)$, where

$$\chi^2 = \sum [y_i - Af(x_i) - Bg(x_i)]^2/\sigma^2.$$

The probability is largest when χ^2 is minimum. To find this minimum, we differentiate χ^2 with respect to A and B and set the derivatives equal to zero. This process gives the required (8.41).

8.25. $\tau = 45.1$ min; $\nu_0 = 494$.

Chapter 9

QUICK CHECKS

QC9.1. $\sigma_q = 3.7$, including covariance; but $\sigma_q = 2.7$, ignoring covariance.

QC9.2. $r = 0.94$.

QC9.3. Because $Prob_{20}(|r| \geqslant 0.6) = 0.5\%$, the correlation is both significant *and* highly significant.

PROBLEMS

9.1. $\sigma_{xy} = 1.75$.

9.3. (a) $\sigma_x^2 = 1.5$; $\sigma_y^2 = 3.5$; $\sigma_{xy} = 1.75$. **(b)** 2.92. **(c)** 2.24. **(d)** 2.92.

9.5. (a) If tape i has shrunk (compared to the average), both answers x_i and y_i will be greater than average; if the tape has stretched, both will be less than average. Either way, the product $(x_i - \bar{x})(y_i - \bar{y})$ will be positive. Because every term in the sum that defines σ_{xy} is positive, the same is true of σ_{xy} itself. **(b)** Because $x_i = \lambda_i X$ for every measurement i, it follows that $\bar{x} = \bar{\lambda} X$ and then that $\sigma_x = \sigma_\lambda X$. In the same way, $\sigma_y = \sigma_\lambda Y$ and, finally, $\sigma_{xy} = \sigma_\lambda^2 XY = \sigma_x \sigma_y$.

9.7. This calculation is simplest if you note that the function $A(t)$ is the same as $\sigma_x^2 + 2t\sigma_{xy} + t^2 \sigma_y^2$.

9.9. $r = 0.9820$ either way.

9.11. (a) $Prob_5(|r| \geqslant 0.7) = 19\%$, so the evidence for a linear relation is not significant. **(b)** $Prob_{20}(|r| \geqslant 0.5) = 2\%$, so the evidence for a linear relation is significant at the 5% level.

9.13. Because $r = 0.73$ and $Prob_{10}(|r| \geqslant 0.73) \approx 1.5\%$, the correlation is significant at the 5% level.

9.15. See Figure A9.15. Because $r = 0.98$ and $Prob_6(|r| \geqslant 0.98) \approx 0.3\%$, the correlation is both significant and highly significant.

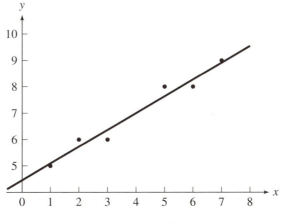

Figure A 9.15.

Chapter 10

QUICK CHECKS

QC10.1. $Prob$(3 hearts) $= 1.56\%$; $Prob$(2 hearts) $= 14.06\%$; $Prob$(2 or 3 hearts) $= 15.62\%$.

QC10.2. $Prob$(11 or 12 heads) $= 0.32\%$, and there is significant (even highly significant) evidence that the coin is loaded toward heads.

PROBLEMS

10.1. *Prob*(0 aces) = 57.87%; *Prob*(1 ace) = 34.72%.

10.3. *Prob*(ν aces in 4 throws) = 48.23, 38.58, 11.57, 1.54, and 0.08% for $\nu = 0$, 1, ..., 4. See Figure A10.3.

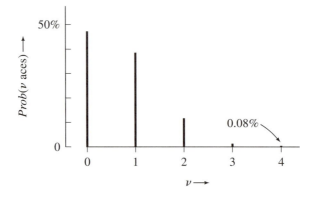

Figure A 10.3.

10.5. The binomial coefficients $\binom{3}{\nu}$ are 1, 3, 3, 1 for $\nu = 0, 1, 2, 3$.

$$(p + q)^3 = \sum_{\nu=0}^{3} \binom{3}{\nu} p^\nu q^{(3-\nu)} = q^3 + 3pq^2 + 3p^2q + p^3.$$

10.7. $B_{4,1/2}(\nu) = 6.25, 25, 37.5, 25, 6.25\%$ for $\nu = 0, 1, \ldots, 4$. See Figure 10.2.

10.9. *Prob*(ν hearts in 6 draws) = 17.8, 35.6, 29.7, 13.2, 3.3, 0.4, 0.02% for $\nu = 0, 1, \ldots, 6$.

10.11. (a) 40, 40, 16, 3.2, 0.3, 0.01%. **(b)** 3.5%. **(c)** 0.08%.

10.13.

$$B_{n,1/2}(n - \nu) = \frac{n!}{(n - \nu)![n - (n - \nu)]!} \left(\frac{1}{2}\right)^n$$

$$= \frac{n!}{\nu!(n - \nu)!} \left(\frac{1}{2}\right)^n = B_{n,1/2}(\nu).$$

10.15. (a)

$$\sigma_\nu^2 = \overline{(\nu - \bar{\nu})^2}$$
$$= \Sigma f(\nu)(\nu - \bar{\nu})^2 = \Sigma f(\nu)(\nu^2 - 2\nu\bar{\nu} + \bar{\nu}^2)$$
$$= \Sigma f(\nu)\nu^2 - 2\bar{\nu}\Sigma f(\nu)\nu + \bar{\nu}^2 \Sigma f(\nu)$$
$$= \overline{\nu^2} - 2\bar{\nu}\bar{\nu} + \bar{\nu}^2 = \overline{\nu^2} - \bar{\nu}^2.$$

(b) We already know that $\bar{\nu} = np$. To find $\overline{\nu^2}$, we observe that for any values of p and q,

$$(p + q)^n = \Sigma \binom{n}{\nu} p^\nu q^{n-\nu}.$$

If you differentiate this equation twice with respect to p, multiply through by p^2, and set $p + q = 1$, you will find that

$$n(n - 1)p^2 = \Sigma \nu(\nu - 1)\binom{n}{\nu}p^{\nu}q^{n-\nu}$$
$$= \Sigma \nu^2 B_{n,p}(\nu) - \Sigma \nu B_{n,p}(\nu) = \overline{\nu^2} - \overline{\nu}.$$

Because $\overline{\nu} = np$, this result tells us that $\overline{\nu^2} = n(n - 1)p^2 + np$. Substitution into the result of part (a) gives σ_{ν}^2 as $np(1 - p)$.

10.17. 9.68% (Gaussian approximation); 9.74% (exact).

10.19. Because *Prob*(9 or 10 wins) $= 1.07\%$, the evidence is significant but just not highly significant.

10.21. *Prob*(ν germinations) $= B_{100,1/4}(\nu) \approx G_{25,4.33}(\nu)$. Therefore,

$$Prob(32 \text{ or more germinations}) \approx Prob_{\text{Gauss}}(\nu \geqslant 31.5)$$
$$= Prob_{\text{Gauss}} \text{ (more than } 1.5\sigma \text{ above mean)} = 6.7\%.$$

The test is just *not* significant.

Chapter 11

QUICK CHECKS

QC11.1. Average number in 1 hour is $\mu = 0.75$. *Prob*(ν eggs in 1 hour) $= 47.2$, 35.4, 13.3, 3.3% for $\nu = 0, \ldots, 3$. *Prob*(4 or more eggs) $= 0.7\%$. The most probable number is 0; the probability of getting exactly 0.75 eggs is 0.

QC11.2. The mean number expected in 10 hours is 9 ± 3. The rate $R = 0.9 \pm 0.3$ eggs/hour.

PROBLEMS

11.1. $P_{0.5}(\nu) = 60.65, 30.33, 7.58, 1.26, 0.16, 0.02, 0.001$ for $\nu = 0, 1, \ldots, 6$, as shown in Figure A11.1.

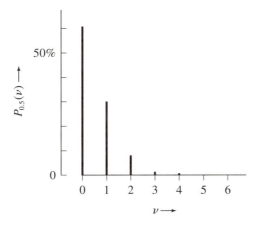

Figure A 11.1.

11.3. (a) $\mu = 1.5$. **(b)** $P_{1.5}(\nu) = 22.3, 33.5, 25.1, 12.6\%$, for $\nu = 0, 1, 2, 3$. **(c)** *Prob*(4 or more decays) $= 6.6\%$.

11.5. The vertical bars in Figure A11.5 show the observed distribution. The expected Poisson distribution, $P_3(\nu)$, has been connected by a continuous curve to guide the eye. The fit seems very good.

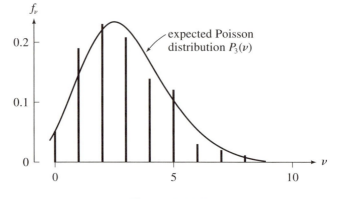

Figure A 11.5.

11.7. (a) 3. **(b)** $P_9(7) = 11.7\%$, etc. **(c)** *Prob*($\nu \leqslant 6$ or $\nu \geqslant 12$) $= 40.3\%$. **(d)** A result as deviant as 12 is not surprising and gives no reason to question that $\mu = 9$.

11.9. (a) If you differentiate (11.15) twice with respect to μ and multiply the result by μ^2, you get the following:

$$0 = \mu^2 - 2\mu\Sigma\nu P_\mu(\nu) + \Sigma(\nu^2 - \nu)P_\mu(\nu)$$
$$= \mu^2 - 2\mu\bar{\nu} + (\overline{\nu^2} - \bar{\nu}) = \overline{\nu^2} - \mu^2 - \mu,$$

from which the desired result follows. **(b)** Substituting into (11.7), you should get

$$\sigma_\nu^2 = (\mu^2 + \mu) - \mu^2 = \mu,$$

as claimed.

11.11. *Prob*(ν) $= e^{-\mu}\mu^\nu/\nu!$. This value is maximum when its derivative with respect to μ is 0. The derivative is $(\nu - \mu)e^{-\mu}\mu^{(\nu-1)}/\nu!$, which vanishes when $\mu = \nu$, as claimed.

11.13. Because the SD is 20, $P_{400}(\nu) \approx G_{400,20}(\nu)$. The simplest argument approximates *Prob*($380 \leqslant \nu \leqslant 420$) by the corresponding Gaussian probability; this is just the probability for a result within one SD of the mean and is well known to be 68.27%. This value compares quite well with the exact 69.47%. We do even better if we calculate the Gaussian probability for $379.5 \leqslant \nu \leqslant 420.5$ (to take account that the Gauss distribution treats ν as a continuous variable); this value is easily found with the help of Appendix A to be close to 69.46%, in remarkable agreement. The exact distribution is $e^{-400}400^\nu/\nu!$; most calculators overflow when calculating any one of these three factors with ν in the range from 380 to 420.

11.15. The probability of getting the observed values is

$$Prob(\nu_1, \ldots, \nu_N) = P_\mu(\nu_1)P_\mu(\nu_2)\ldots P_\mu(\nu_N)$$
$$= e^{-N\mu}\frac{\mu^{(\nu_1+\ldots+\nu_N)}}{\nu_1!\ldots\nu_N!}.$$

To find the value of μ for which this probability is a maximum, differentiate it with respect to μ and note that the derivative is zero at $\mu = \Sigma\nu_i/N$.

11.17. (a) 400 ± 20. **(b)** 10.0 ± 0.5 events/second.

11.19. 8.75 ± 1.1 particles/min. This value is strong evidence that the duct is contaminated.

11.21. (a) The number he will count in time T_{tot} is approximately $\nu_{tot} = r_{tot}T_{tot}$, and its uncertainty will be $\delta\nu_{tot} = \sqrt{\nu_{tot}} = \sqrt{r_{tot}T_{tot}}$. He next calculates R_{tot} as ν_{tot}/T_{tot}, so its uncertainty is $\delta R_{tot} = \delta\nu_{tot}/T_{tot} = \sqrt{r_{tot}}/T_{tot}$. In exactly the same way, he finds R_{bgd}, and its uncertainty is $\delta R_{bgd} = \sqrt{r_{bgd}}/T_{bgd}$. Finally, he calculates the source's rate as $R_{sce} = R_{tot} - R_{bgd}$, with an uncertainty given by

$$(\delta R_{sce})^2 = (\delta R_{tot})^2 + (\delta R_{bgd})^2 = \frac{r_{tot}}{T_{tot}} + \frac{r_{bgd}}{T_{bgd}}.$$

To make the uncertainty as small as possible, he substitutes $T_{bgd} = T - T_{tot}$ and then differentiates the equation with respect to T_{tot}. The resulting derivative is $-r_{tot}/T_{tot}^2 + r_{bdg}/T_{bgd}^2$, and this derivative is 0 when $T_{tot}/T_{bgd} = \sqrt{r_{tot}/r_{bgd}}$, as claimed. **(b)** $T_{tot} = 1.5$ h; $T_{bgd} = 0.5$ h.

Chapter 12

QUICK CHECKS

QC12.1. Expected numbers $= 16, 34, 34, 16$; $\chi^2 = 1.6$. There is no reason to doubt the measurements are normally distributed.

QC12.2. $d = 1$; $\tilde{\chi}^2 = 1.6$.

QC12.3. 0.7%. The evidence is highly significant that the measurements were *not* normally distributed.

PROBLEMS

12.1. Expected numbers $= 7.9, 17.1, 17.1, 7.9$; $\chi^2 = 10.0$. Because $\chi^2 \gg n$, the measurements were unlikely to be normally distributed.

12.3. Expected number $= 40$ in each bin; $\chi^2 = 15.85$. Because $\chi^2 \gg n$, the die is probably loaded.

12.5. (a) The number of decays in the kth bin is the number that decay between times $(k-1)T$ and kT; this should be

$$E_k = N([k-1]T) - N(kT)$$
$$= N_0 e^{-(k-1)T/\tau} - N_0 e^{-kT/\tau} = N_0(e^{T/\tau} - 1)e^{-kT/\tau}.$$

(b) Expected numbers $= 515.2, 189.5, 69.7, 25.6, 15.0$; $\chi^2 = 1.8$. Her results are consistent with the exponential decay law.

12.7. For Problem 12.2, $c = 3$, $d = 1$; for Problem 12.3, $c = 1$, $d = 5$; for Problem 12.4, $c = 1$, $d = 2$.

12.9.

d:	1	2	3	4	5	10	15	20	30
$\tilde{\chi}_{crit}^2$:	6.6	4.6	3.8	3.3	3.0	2.3	2.0	1.9	1.7

12.11. $\tilde{\chi}^2 = 3.17$. The probability of getting a value of $\tilde{\chi}^2$ this large is 0.7%, so we can conclude that the die is loaded at both the 5% and 1% levels.

12.13. $\chi^2 = 2.74$; $d = 6$; $\tilde{\chi}^2 = 0.46$. The data are consistent with the expected Poisson distribution.

12.15. (a) The probabilities for totals of 2, 3, . . . , 12 are $\frac{1}{36}, \frac{2}{36}, \cdots, \frac{6}{36}, \cdots, \frac{1}{36}$. **(b)** $\chi^2 = 19.8$; $d = 10$; $\tilde{\chi}^2 = 1.98$. **(c)** The probability of getting a value of $\tilde{\chi}^2$ this large or larger is 3.1%. **(d)** At the 5% level, we can conclude that the dice are loaded; at the 1% level, we cannot.

12.17. (a) $\chi^2 = 2.25$; $Prob(\chi^2 \geqslant 2.25) = 13\%$, and the difference is not significant. The exact calculation gives $Prob(\nu \geqslant 11) + Prob(\nu \leqslant 5) = 21\%$. The two methods agree only very approximately. **(b)** $\chi^2 = 6.25$; $Prob(\chi^2 \geqslant 6.25) = 1.2\%$, and the difference is significant at the 5% level. The "exact" calculation (using the Gaussian approximation) gives $Prob(\nu \geqslant 224.5) + Prob(\nu \leqslant 175.5) = 1.4\%$. The two methods agree quite well. **(c)** The adjusted χ^2 is 1.56, and $Prob(\chi^2 \geqslant 1.56) = 21.2\%$, in excellent agreement with the exact answer 21.0%.

Index

For definitions of important symbols and summaries of principal formulas, see the insides of the front and back covers.

Principal Formulas in Part II

Weighted Averages (Chapter 7)

If $x_1, \ldots, x_N$ are measurements of the same quantity x, with known uncertainties $\sigma_1, \ldots, \sigma_N$, then the best estimate for x is the weighted average

$$x_{\text{wav}} = \frac{\sum w_i x_i}{\sum w_i},$$

(p. 175)

where the weight $w_i = 1/\sigma_i^2$.

Least-Squares Fit to a Straight Line (Chapter 8)

If $(x_1, y_1), \ldots, (x_N, y_N)$ are measured pairs of data, then the best straight line $y = A + Bx$ to fit these N points has

$$A = [(\sum x_i^2)(\sum y_i) - (\sum x_i)(\sum x_i y_i)]/\Delta,$$
$$B = [N(\sum x_i y_i) - (\sum x_i)(\sum y_i)]/\Delta,$$

where

$$\Delta = N(\sum x_i^2) - (\sum x_i)^2.$$

(p. 184)

Covariance and Correlation (Chapter 9)

The covariance σ_{xy} of N pairs $(x_1, y_1), \ldots, (x_N, y_N)$ is

$$\sigma_{xy} = \frac{1}{N} \sum (x_i - \bar{x})(y_i - \bar{y}).$$

(p. 212)

The coefficient of linear correlation is

$$r = \frac{\sigma_{xy}}{\sigma_x \sigma_y} = \frac{\sum (x_i - \bar{x})(y_i - \bar{y})}{\sqrt{\sum (x_i - \bar{x})^2 \sum (y_i - \bar{y})^2}}$$

(p. 217)

Values of r near 1 or -1 indicate strong linear correlation; values near 0 indicate little or no correlation. (For a table of probabilities for r, see Appendix C.)

Binomial Distribution (Chapter 10)

If the probability of "success" in one trial is p, then the probability of ν successes in n trials is given by the binomial distribution

$$Prob(\nu \text{ successes in } n \text{ trials}) = B_{n,p}(\nu) = \frac{n!}{\nu!(n - \nu)!} p^\nu (1 - p)^{n - \nu}.$$

(p. 230)